T0331219

Four-Dimensional Manifolds and Projective Structure

Four-Dimensional Manifolds and Projective Structure may be considered first as an introduction to differential geometry and, in particular, to 4–dimensional manifolds, and secondly as an introduction to the study of projective structure and projective relatedness in manifolds.

The initial chapters mainly cover the elementary aspects of set theory, linear algebra, topology, Euclidean geometry, manifold theory, and differential geometry, including the idea of a metric and a connection on a manifold and the concept of curvature. After this, the author dives deeper into 4-dimensional manifolds covering each of the positive definite, Lorentz and neutral signature cases and introducing, and making use of, the holonomy group for connections associated with metrics of each of these signatures. A brief interlude on some key aspects of geometrical symmetry is then given and this is followed by a detailed description of projective relatedness, that is, the relationship between two symmetric connections (and their associated metrics) which give rise to the same geodesic paths.

Features:
- Offers a detailed, straightforward discussion of the basic properties of (4-dimensional) manifolds.
- Introduces holonomy theory, and makes use of it, in a novel manner.
- Suitable for postgraduates and researchers, including master's degree and PhD students.

Graham Hall, FRSE, is Professor Emeritus in the Institute of Mathematics of the University of Aberdeen, Scotland, UK. He received his BSc (1968) and PhD (1971) from the University of Newcastle upon Tyne and came to Aberdeen in 1973. He also served as head of department in Aberdeen from 1992 to 1995. His interests lie in classical mathematical relativity theory and differential geometry. He is the author of the text *Symmetries and Curvature Structure in General Relativity* (World Scientific, 2004) and has contributed to, or edited, several other volumes. He has also delivered over 200 invited talks on these topics at many universities and academies in Europe, North and South America, Asia, Africa, and Australasia, and has published over 180 papers in scientific research journals.

Prof Hall is a Fellow of the Royal Society of Edinburgh and of the Royal Astronomical Society and serves on the editorial board of many scientific research journals worldwide.

Four-Dimensional Manifolds and Projective Structure

Graham Hall, FRSE
University of Aberdeen, Scotland, UK

CRC Press
Taylor & Francis Group
Boca Raton London New York

CRC Press is an imprint of the
Taylor & Francis Group, an **informa** business

A CHAPMAN & HALL BOOK

First edition published 2023
by CRC Press
6000 Broken Sound Parkway NW, Suite 300, Boca Raton, FL 33487-2742

and by CRC Press
4 Park Square, Milton Park, Abingdon, Oxon, OX14 4RN

CRC Press is an imprint of Taylor & Francis Group, LLC

ISBN: 978-0-367-90042-7 (hbk)
ISBN: 978-1-032-52235-7 (pbk)
ISBN: 978-1-003-02316-6 (ebk)

DOI: 10.1201/ 9781003023166

Typeset in CMR10
by KnowledgeWorks Global Ltd.

Publisher's note: This book has been prepared from camera-ready copy provided by the authors.

Contents

Preface

This book may be considered first as an introduction to differential geometry and, in particular, to 4-dimensional manifolds, and second as an introduction to the study of projective structure and projective relatedness in manifolds. It arose out of collaborations in the University of Aberdeen, Scotland, UK between the author and his PhD students, postdocs and research visitors and from overseas visits by the author to other workers in the field, and to conferences. Chapters 1, 2 and the first part of chapter 3 deal mainly with the elementary aspects of set theory, linear algebra, topology, Euclidean geometry, manifold theory and differential geometry, including the idea of a metric and a connection on a manifold and the concept of curvature. The second part of chapter 3 specialises in the case of 4-dimensional manifolds and, in particular, in the positive definite case for the metric. Chapter 4 deals with the case of Lorentz signature and chapter 5 with the so-called neutral signature case. These chapters deal with the associated (metric) connection, the elementary properties of the curvature and Weyl conformal tensors and also the sectional curvature function and the close relationships between these geometrical objects. It introduces, and makes use of, the holonomy group of such a manifold for connections associated with metrics of each of these three possible signatures. For this purpose, useful representations of the Lie algebras $o(4)$, $o(1,3)$ and $o(2,2)$ in the language of bivectors (skew-symmetric second order tensors) are constructed. A study of the algebraic properties of certain tensors is also provided. Chapter 6 is a brief interlude on those aspects of geometrical symmetry which are needed to understand chapter 7. Chapter 7, the final chapter, provides a detailed description of projective relatedness, that is, the relationship between two symmetric connections (and between their associated metrics) which give rise to the same geodesic paths. This leads to the introduction and description of the Weyl projective tensor. This topic is of significant current interest and an attempt is made to show that, with the help of holonomy theory and a certain classification of the curvature tensor, a systematic study of this subject may be made, at least in the 4-dimensional case.

The author wishes to put on record his special thanks to three of his former research students Dr David Lonie, Dr Zhixiang Wang, and Dr Bahar Kırık for many illuminating discussions and collaborations on these topics and for several of the ideas contained in this book and, additionally to David Lonie, for his MAPLE calculations and technical help in preparing the manuscript and Bahar Kırık for help with the proof reading. He has also benefitted from useful discussions with many other colleagues especially Dr John Pulham, Dr Michael Crabb, Prof Vladimir Matveev and others too numerous to mention here. Finally he expresses a very special gratitude to Aileen Sylvester for her patience, guidance and understanding.

Chapter 1

Algebra, Topology and Geometry

1.1 Notation

This chapter will be devoted to a very brief summary of some topics in set theory, algebra, topology and geometry which will be needed in what is to follow. Only those topics which, within reasonable bounds of completeness, are strictly needed will be discussed since they may (mostly) be found in more detail in many standard texts. The material here is heavily conditioned by the necessities of manifold theory. The opportunity will also be taken here to introduce the notation required. Recommended texts for this material are [1], [2] for algebra, [3],[4] and [5] for topology and [6], [7] and [8] for geometry.

The notation, followed will be a fairly standard one. Set membership is denoted by $\in$, non-membership by $\notin$, the empty set by $\emptyset$ and the members of a non-empty set will, where appropriate, sometimes be listed inside brackets $\{\}$. The symbol $\Rightarrow$ means "implies" whilst $\Leftrightarrow$ means "implies and is implied by" or "is equivalent to" or "if and only if". The symbol $\forall$ means "for all" and $\exists$ means "there exist(s)". For sets A and B the inclusion of A as a *subset* of B is denoted by $A \subset B$ and this includes the possibility of equality, $A = B$, which is equivalent to $A \subset B$ *and* $B \subset A$. If $A \subset B$ and $A \neq B$, A is *properly contained in* B or a *proper subset of* B. The union and intersection of A and B are denoted, respectively, by $A \cup B$ and $A \cap B$, and, if $A \neq \emptyset \neq B$, their Cartesian product, denoted by $A \times B$, is the set $\{(a, b) : a \in A, b \in B\}$ where (a, b) denotes an ordered pair. (Those occasions where such a non-empty restriction is needed will usually be taken as obvious to avoid repetition.) An obvious extended version of this applies when a finite number > 2 of sets are involved. The Cartesian product of n copies of a set A is denoted A^n. Also the symbol $\mathbb{R}$ will denote the set of all real numbers, $\mathbb{C}$ the set of all complex numbers, $\mathbb{Q}$ the subset of $\mathbb{R}$ consisting of all the rational numbers, $\mathbb{Z}$ the set of all integers and $\mathbb{N}$ is the subset of $\mathbb{Z}$ given by $\mathbb{N} = \{1, 2, ...\}$. Thus one has the sets $\mathbb{R}^n$ and $\mathbb{C}^n$ for $n \in \mathbb{N}$. The *set-theoretic difference* of sets A and B is written $A \setminus B$ and is $\{x \in A : x \notin B\}$. If A and B are sets, a function (or map, or mapping) f from A to B is denoted by $f : A \to B$. Such a map is said to be *onto* (or *surjective*) if given any $b \in B$ $\exists a \in A$ such that $f(a) = b$, *one-to-one* (or *injective*) if $f(a) = f(b) \Rightarrow a = b$ and a *bijection* (or *bijective*) if it is both injective and surjective. If $f : A \to B$ is bijective, it gives rise to a

DOI: 10.1201/ 9781003023166-1

unique map from B to A denoted f^{-1} and called the *inverse* of f and given for $b \in B$ by $f^{-1}(b) = a \Leftrightarrow f(a) = b$. For sets A and B if $f : A \to B$ is a map and $C \subset A$ and $D \subset B$, the symbol $f(C) \equiv \{f(a) : a \in C\}$ is the *image* of C under f whilst $f^{-1}D$ denotes the *inverse image* of D under f and is given by $f^{-1}D \equiv \{a \in A : f(a) \in D\}$ whilst the map $i : C \to A$ given by $i(p) = p$ for each $p \in C$ is called the *(natural) inclusion (map)* of C into A. The restriction of the map $f : A \to B$ to the subset $C \subset A$ is denoted by $f_{|C}$. If A, B and C are sets and $f : A \to B$ and $g : B \to C$ then the composite map $g \circ f$ is the map $A \to C$ given by $g \circ f(a) = g(f(a))$ for $a \in A$ (and in the previous sentence, $f_{|C} = f \circ i : C \to B$). A *binary relation* or *binary operation* on a set X is a map $X \times X \to X$.

If X is a set and $A \subset X$, the *complement of A* (in X) is the set $X \setminus A$. If X is understood, the complement of A is denoted and defined by $\mathcal{C}(A) \equiv \{x \in X, x \notin A\}$. The rules for manipulating subsets A and B of an (understood) set X are those of de Morgan and are (i) $\mathcal{C}(A \cup B) = \mathcal{C}(A) \cap \mathcal{C}(B)$ and (ii) $\mathcal{C}(A \cap B) = \mathcal{C}(A) \cup \mathcal{C}(B)$. If $A, B, ..., C$ are subsets of a set X such that the intersection of any two of them is the empty set then $A \cup B \cdots \cup C$ is said to be a *disjoint union* (of $A, B, ..., C$), and then if $X = A \cup B \cup \cdots \cup C$, one refers to this as a *partition* or a *disjoint decomposition* of X.

Let X be a set. An *equivalence relation* on X is a subset $R \subset X \times X$ such that for $x, y, z \in X$ (i) $(x, x) \in R$, $\forall x \in X$, (ii) $(x, y) \in R \Rightarrow (y, x) \in R$ and (iii) $(x, y), (y, z) \in R \Rightarrow (x, z) \in R$. Then if $(x, y) \in R$ one sometimes writes $x \sim y$. If, for $x \in X$ one defines the subset $A_x \equiv \{y \in X \Leftrightarrow x \sim y\}$ of X the collection of all such subsets, called *equivalence classes under* $\sim$, is such that for $x, x' \in X$ either $A_x = A_{x'}$ or $A_x \cap A_{x'} = \emptyset$ and the union of all such equivalence classes equals X and is a partition of X. The collection of all such equivalence classes is denoted by $X/\sim$ and called the *quotient set* arising from X and $\sim$ and this leads to a natural projection map $\mu : X \to X/\sim$ which maps $x \in X$ to the unique equivalence class containing x.

A set A is called *finite* if there exists a bijective map f from A to the set $\{1, 2, ..., n\}$ for some $n \in \mathbb{N}$, *denumerable* if there exists a bijective map from A to $\mathbb{N}$ and *countable* if it is either finite or denumerable. If A is not finite it is said to be *infinite*.

The subset S^n of $\mathbb{R}^{n+1}$ given by $S^n \equiv \{(x_1, ..., x_{n+1}) : x_1^2 + \cdots + x_{n+1}^2 = 1\}$ is called the *n-sphere*. The symbol δ_i^j or δ_{ij}, for non-negative integers i, j, denotes the *Kronecker delta* and takes the value 1 if $i = j$ and zero otherwise. The symbol $\equiv$ means "is equal to by definition" or "is identically equal to". The end of a proof will be denoted $\square$.

1.2   Groups

A *group* is a pair $(G, .)$ where G is a non-empty set and $.$ a binary operation $G \times G \to G$, $(a, b) \to a.b$, for $a, b \in G$ (called the *group product*) satisfying

(**G1**) the associative law $a.(b.c) = (a.b).c$,

(**G2**) there exists $e \in G$ such that $a.e = e.a = a$, $\forall a \in G$,

(**G3**) for each $a \in G$, $\exists a^{-1} \in G$ such that $a.a^{-1} = a^{-1}.a = e$.

The member $e \in G$ is called the *identity* of G and is unique in satisfying (**G2**) above. In (**G3**) the member $a^{-1} \in G$ is called the *inverse* of a and, given a, is also unique. Unless confusion may arise, the pair $(G, .)$ is usually written as G and $a.b$ is usually written as ab.

If $H \subset G$ is such that H with the inherited operation . from G is a group then H is called a *subgroup* of G. This is written $H < G$ and it is remarked that the identities of G and H coincide and that if $a \in H$, the inverse of a is the same whether taken in G or H. Thus a subset H of G is a subgroup of G if and only if either $a, b \in H \Rightarrow ab \in H$ *and* $a^{-1} \in H$, or, $a, b \in H \Rightarrow ab^{-1} \in H$. The subset $\{e\}$ of G is a subgroup of G called the *trivial* subgroup. A group G satisfying the property that for any $a, b \in G$, $ab = ba$, is called *Abelian* and, generally, if $a.b = b.a$, a and b are said to *commute*.

The set $\mathbb{R}$ is an Abelian group under the usual addition with identity 0 and, for $r \in \mathbb{R}$, inverse $-r$, and $\mathbb{R} \setminus \{0\}$ is an Abelian group under the usual multiplication with identity 1 and inverse r^{-1}. Similar comments apply to $\mathbb{C}$ and $\mathbb{C} \setminus \{0\}$. The set $GL(n, \mathbb{R})$ of $n \times n$ non-singular real matrices is, with the usual identity matrix, matrix multiplication and forming of inverses, a (non-Abelian) group.

Let $(G, .)$ and $(G', \times)$ be groups. A map $f : G \to G'$ is called a (group) *homomorphism* if for each $a, b \in G$, $f(a.b) = f(a) \times f(b)$ (or simply $f(ab) = f(a)f(b)$ if no confusion can arise). For such a map, if e is the identity of G and $a \in G$, $f(e)$ is the identity of G' and $f(a^{-1})$ is the inverse of $f(a)$ in G'. If $f : G \to G'$ is a homomorphism and f is bijective, f is called a (group) *isomorphism* of G onto G' and the inverse map f^{-1} is then necessarily an isomorphism $G' \to G$. In this situation G and G' are said to be *isomorphic* groups. If $f : G \to G'$ is a homomorphism between groups G and G', the subset $K \equiv \{g \in G : f(g) = e\}$, where e is the identity of G', is easily seen to be a subgroup of G called the *kernel* of f. If $H < G$ and if $g \in G$ the subset $H'_g \equiv \{g^{-1}hg : h \in H\}$ of G is easily seen to be a subgroup of G which is isomorphic to H under the isomorphism $h \to g^{-1}hg$. The subgroups H and H' are said to be *conjugate* .

If S is a subset of G, the family of all finite products of those members of G which are either the identity of G, a member of S or the inverse of a member of S is a subgroup of G containing S called the subgroup of G *generated* by S. It is, in fact, the intersection of all of the subgroups of G containing S and hence, in an obvious sense, is the smallest subgroup of G containing S. If S contains only one member, the resulting subgroup of G is called *cyclic*. If, for $n \in \mathbb{N}$, $G_1, ..., G_n$ are groups the direct product $G_1 \times .... \times G_n$ together with the binary operation $(a_1, ..., a_n).(b_1, ..., b_n) = (a_1 b_1, ..., a_n b_n)$ for $a_i, b_i \in G_i$ $(1 \leq i \leq n)$ is a group called the *product* of the groups $G_1, ..., G_n$.

Let G be a group and $H < G$. For $g \in G$, the subset $gH \equiv \{gh : h \in H\}$ is called a *left coset* of H in G. (One can similarly define a *right coset*.) This gives a partition of G arising from the equivalence relation given for $a, b \in G$ by $a \sim b \Leftrightarrow a^{-1}b \in H$. Then a and b are in the same left coset if and only if $a^{-1}b \in H$ or, equivalently, $aH = bH$. The collection of left cosets of G in G is denoted by $L(G, H)$. It is remarked that right and left cosets do not necessarily give the same partition of G, but if they do, that is, for each $g \in G$, $gH = Hg'$ for some $g' \in G$, $g = hg'$ for some $h \in H$, and so $h^{-1}g = g'$ and $Hg' = Hg = gH$. Then $Hg = gH$ for each $g \in G$ and H is a special kind of subgroup of G called a *normal* subgroup of G. It then easily follows that H is a normal subgroup of G if and only if $g^{-1}Hg \equiv \{g^{-1}hg : h \in H\} = H \ \forall g \in G$ and, of course, if G is Abelian, any subgroup H of G is normal. The kernel of a homomorphism $f : G \to G'$ between groups G and G', which was seen above to be a subgroup of G, is easily checked always to be a normal subgroup of G. It is now easily shown that if $H < G$ is normal, the collection of left cosets of H in G forms a group according to the product relation $aH.bH = a.bH$, for each $a, b \in G$, where $(aH)^{-1} = a^{-1}H$ and where the identity member is H. This group is called the *quotient group* of G by H and is denoted G/H. For H a normal subgroup of G there is a natural map $f : G \to G/H$ given by $g \to gH$ and which is easily seen to be a homomorphism, called the *natural homomorphism*, from G to G/H and its kernel is H. Slightly more generally, if $f : G \to G'$ is a homomorphism with kernel K, then G/K is isomorphic to $f(G)$, this isomorphism being given by $gK \to f(g)$ for $g \in G$. In addition, if f is *onto*, G/K and G' are isomorphic.

1.3   Vector Spaces and Linear Transformations

A *field* is a triple $(F, +, .)$ where F is a non-empty set and $+$ and $.$ are binary operations on F with the properties; (i) $(F, +)$ is an abelian group (with identity denoted by 0), (ii) $(F \setminus \{0\}, .)$ is an abelian group (with identity denoted by 1) and (iii) the operation $.$ is distributive over $+$, that is, $a.(b+c) = a.b + a.c$. The operations $+$ and $.$ are usually referred to as *addition* and *multiplication* with 0 and 1 the *additive and multiplicative identities*. The resulting inverses are written as $-a$ and (for $a \neq 0$) a^{-1}. The sets $\mathbb{R}$ and $\mathbb{C}$ above are obvious examples and are the only ones required here. Because of axiom (ii) $1 \neq 0$ and thus F contains at least two distinct members.

A *vector space* V over a field F consists of an abelian group $(V, \oplus)$ and a field $(F, +, .)$ with multiplicative identity 1 and additive identity 0 together with an operation $\odot$ of members of F on members of V with the properties that for $a, b \in F$ and $\mathbf{u}, \mathbf{v} \in V$, $a \odot v \in V$ and

(V1) $(a + b) \odot \mathbf{v} = (a \odot \mathbf{v}) \oplus (b \odot \mathbf{v})$,

(V2) $a \odot (\mathbf{u} \oplus \mathbf{v}) = (a \odot \mathbf{u}) \oplus (a \odot \mathbf{v})$,
(V3) $(a \cdot b) \odot \mathbf{v} = a \odot (b \odot \mathbf{v})$,
(V4) $1 \odot \mathbf{v} = \mathbf{v}$.

In practice, and where no confusion may arise, one usually writes $+$ for each of $+$ and $\oplus$ above and omits the symbols . and $\odot$. The members of V are called *vectors*, those of F *scalars* and $\odot$ *scalar multiplication*. Here only vector spaces over the real field $\mathbb{R}$ (*real vector spaces*) and the complex field $\mathbb{C}$ (*complex vector spaces*) will be needed. The identity member of V is labelled $\mathbf{0}$ and referred to as the *zero vector* of V and, using a minus sign to denote the additive inverses in each of the groups $(F+)$ and V, the above axioms then easily lead to (*i*) $a\mathbf{0} = \mathbf{0}$, (*ii*) $0\mathbf{v} = \mathbf{0}$, (*iii*) $(-a)\mathbf{v} = -(a\mathbf{v})$ and (*iv*) if $\mathbf{v} \neq \mathbf{0}$ then $a\mathbf{v} = \mathbf{0} \Rightarrow a = 0$. If $V = \mathbb{R}^n$ (respectively, $\mathbb{C}^n$) and $F = \mathbb{R}$ (respectively, $\mathbb{C}$) together with the standard component-wise operations one arrives at the usual vector space structures on $\mathbb{R}^n$ (respectively, $\mathbb{C}^n$). If $W \subset V$ is such that the naturally induced operations on W from those of V (and F) cause W to be a vector space over F then W is called a *subspace* of V. In fact, if a subset $W \subset V$ satisfies the property that for each $a, b \in F$ and $\mathbf{u}, \mathbf{v} \in W$ the member $a\mathbf{u} + b\mathbf{v} \in V$ is also in W then W is a subspace of V. For any vector space $\{\mathbf{0}\}$ is a subspace of it called the *trivial* subspace.

Let U and V be vector spaces over the same field F and let $f : U \to V$ be a map. Then f is called *linear* (or a *homomorphism*) between the vector spaces U and V if for each $a \in F$ and $\mathbf{u}, \mathbf{v} \in U$

(L1) $f(\mathbf{u} + \mathbf{v}) = f(\mathbf{u}) + f(\mathbf{v})$,
(L2) $f(a\mathbf{u}) = af(\mathbf{u})$.

Alternatively, if for each $a, b \in F$ and $\mathbf{u}, \mathbf{v} \in U$, $f(a\mathbf{u} + b\mathbf{v}) = af(\mathbf{u}) + bf(\mathbf{v})$ then f is linear, and conversely. If $f : U \to V$ is bijective it is easily checked that the inverse map $f^{-1} : V \to U$ is necessarily linear and f is then called a vector space *isomorphism* between U and V and U and V are called *isomorphic* vector spaces. Again if $f : U \to V$ is linear the subset $f(U) \equiv \{f(\mathbf{u}) : \mathbf{u} \in U\}$ is easily checked to be a subspace of V called the *range space* of f, denoted rgf whilst the subset $\{\mathbf{u} : f(\mathbf{u}) = \mathbf{0}\}$ is easily checked to be a subspace of U called the *kernel* of f, denoted $kerf$. If U, V, W are vector spaces over the field F and if $f : U \to V$ and $g : V \to W$ are linear then the map $g \circ f : U \to W$ is clearly linear and so isomorphism of vector spaces is an equivalence relation.

If V is a vector space over a field F, $\mathbf{u}_1, ... \mathbf{u}_n \in V$ and $a_1, ..., a_n \in F$ for $n \in \mathbb{N}$, the member $\sum_{i=1}^{n} a_i \mathbf{u_i}$ of V is called a *linear combination* of $\mathbf{u}_1, ..., \mathbf{u}_n$ (over F). If $\emptyset \neq S \subset V$ the set of all linear combinations of finite subsets of S is a subspace of V called the *span* of S, denoted by $Sp(S)$, (also called the subspace of V *spanned by* S) with S called a *spanning set* for $Sp(S)$. A non-empty subset $S \subset V$ is called *linearly independent* (over F) if given any $\mathbf{v}_1, ..., \mathbf{v}_n \in S$ the only solution of the equation $\sum_{i=1}^{n} a_i \mathbf{v_i} = \mathbf{0}$ $(a_1, ..., a_n \in F)$

is $a_1 = ... = a_n = 0$. Otherwise S is called *linearly dependent* (over F). It is true that any vector space V admits a linearly independent, spanning set and such a set is called a *basis* for V. In fact if S_1 is a linearly independent subset of V and S_2 is a spanning set for V with $S_1 \subset S_2$ there exists a basis S for V with $S_1 \subset S \subset S_2$. A vector space V over F which admits a *finite* spanning set is called *finite-dimensional* (and otherwise, *infinite-dimensional*) and the number of members in each basis for V is the *same positive integer* n called the *dimension* of V, written dim$V$, and V is called *n-dimensional*. Then any subspace of a finite-dimensional vector space V of dimension n is finite-dimensional and any subset S of V containing n members is a spanning set for V if and only if it is linearly independent (and then S constitutes a basis for V). In such a case and given that $\mathbf{u_1}, ...\mathbf{u_n} \in V$ is a basis for V, any $\mathbf{v} \in V$ may be written as $\mathbf{v} = \sum_{i=1}^{n} a_i \mathbf{u_i}$ of V where the $a_1, ..., a_n \in F$, called the *components* of $\mathbf{v}$ in this basis, are uniquely determined by $\mathbf{v}$ and the basis $\mathbf{u_1}, ...\mathbf{u_n}$. It easily follows that V is then, in an obvious way, isomorphic to F^n with the usual componentwise operations under an isomorphism which is dependent on the chosen basis. For finite-dimensional vector spaces U, V over a field F any linear map $f : U \to V$ is uniquely determined by its action on a basis of U by linearity. The dimensions of rgf and $kerf$ are called the *rank* and *nullity* of f, respectively, and it is easily checked that their sum equals dim$U$. The trivial subspace has, by definition, dimension zero. If V is finite-dimensional over F and U is a subspace of V with dim$U$ = dim$V$ then $U = V$.

Let $W_1,...,W_n$ ($n \in \mathbb{N}$) be subspaces of a finite-dimensional vector space V. Suppose that each $\mathbf{v} \in V$ can be written in *exactly one way* as $\mathbf{v} = \mathbf{w_1}+\cdots+\mathbf{w_n}$ ($\mathbf{w_i} \in W_i$). Then V is called the (internal) *direct sum* of $W_1, ..., W_n$. One could also view this construction by regarding $W_1,...,W_n$ as individual finite-dimensional vector spaces over the same field F, forming the Cartesian product $W_1 \times ... \times W_n$ and defining vector space addition and scalar multiplication by members of F in the usual component-wise fashion. This gives a vector space isomorphic in an obvious way to the original V and is called the (external) *direct sum* of $W_1, ..., W_n$, denoted by $W_1 + ... + W_n$. Usually one uses the term *direct sum* or just *(vector space) sum* for either of these constructions and, it is noted, dim$V$ =dimW_1 + ... + dimW_n. Slightly more generally if $U \subset V$ and $W \subset V$ are subspaces of V then $U \cap W$ is also a subspace of V as is $Sp(U \cup W)$ (which is the same as the subspace of V consisting of all members of V of the form $\mathbf{u} + \mathbf{w}$ for $\mathbf{u} \in U$ and $\mathbf{w} \in W$ but is only the direct sum of U and W if $U \cap W = \{\mathbf{0}\}$). Then an easily proved standard result gives the *dimension formula*

$$\dim Sp(U \cup W) + \dim(U \cap W) = \dim U + \dim W. \qquad (1.1)$$

Let U and V be finite-dimensional vector spaces over a field F with dimensions m and n, respectively, and let $f : U \to V$ be a linear map. Let $\{\mathbf{u_i}\} \equiv \{\mathbf{u_1}, ..., \mathbf{u_m}\}$ and $\{\mathbf{v_j}\} \equiv \{\mathbf{v_1}, ..., \mathbf{v_n}\}$ be respective bases for U and V. Then one can write $f(\mathbf{u_i}) = \sum_{j=1}^{n} a_{ij} \mathbf{v_j}$ where the $a_{ij} \in F$ give the $m \times n$ *matrix representation* $A \equiv (a_{ij})$ of f with respect to the above bases. If one makes a change of basis $\mathbf{u}_i \to \mathbf{u}'_i = \sum_{k=1}^m s_{ik}\mathbf{u}_k$ in U and a change of basis $\mathbf{v}_j \to \mathbf{v}'_j = \sum_{l=1}^n t_{jl}\mathbf{v}_l$ in V where the matrices $S \equiv (s_{ik})$ and $T \equiv (t_{jl})$ are *non-singular* $m \times m$ and $n \times n$ matrices, respectively, with entries in F the representative matrix of f with respect to the new bases is the $m \times n$ matrix SAT^{-1}. A special case of this occurs when $U = V$ and $\dim U$ ($=\dim V) = n$. In this case one can choose the same basis $\{\mathbf{u}_i\}$ in U and V and then $f(\mathbf{u_i}) = \sum_{j=1}^n a_{ij}\mathbf{u_j}$ so that f is represented with respect to this basis by the $n \times n$ matrix $A \equiv (a_{ij})$. Under a change of basis in U given by $\mathbf{u}_i \to \mathbf{u}'_i = \sum_{k=1}^n p_{ik}\mathbf{u}_k$ for some non-singular $n \times n$ matrix $P = (p_{ij})$, the matrix of f in the new basis is PAP^{-1}. Two $n \times n$ matrices A and B with entries in F are called *similar* (over F) if $B = PAP^{-1}$ for some non-singular $n \times n$ matrix P with entries in F and this relationship of similarity is an equivalence relation on such matrices. Thus matrices in the same equivalence class represent the same linear transformation on U in different bases. This allows the search for "simple" (canonical) representations (forms) for f to be made by seeking "convenient" bases for f in U. For all such choices of bases for U one achieves all possible forms for the representative matrix A and so this technique also allows for a search for canonical "forms" for an $n \times n$ matrix. This will be exploited later.

1.4 Dual Spaces and Bilinear Forms

Let V be an n-dimensional vector space over F. Define the set $\overset{*}{V} \equiv L(V, F)$ where $L(V, F)$ is the set of all linear maps $V \to F$ with F regarded as a 1-dimensional vector space over F. Then $\overset{*}{V}$ is easily checked to be a vector space over F with vector addition and scalar multiplication defined, for $f, g \in \overset{*}{V}$, $a \in F$ and $\mathbf{v} \in V$ by $(f + g)(\mathbf{v}) = f(\mathbf{v}) + g(\mathbf{v})$ and $(a \cdot f)(\mathbf{v}) = af(\mathbf{v})$ and is called the *dual (space)* of V over F. Now if $\{\mathbf{v_i}\}$ is a basis for V and if $a_1, ..., a_n \in F$ there exists exactly one $\mathbf{w} \in \overset{*}{V}$ such that $\mathbf{w}(\mathbf{v_i}) = a_i$ for $1 \leqslant i \leqslant n$ and so there is a uniquely determined set $\mathbf{w_1}...\mathbf{w_n} \in \overset{*}{V}$ such that $\mathbf{w_i}(\mathbf{v_j}) = \delta_{ij}$ for each $1 \leqslant i, j, \leqslant n$. This latter set is clearly a basis for $\overset{*}{V}$ called the *dual basis* of $\{\mathbf{v_i}\}$ and hence $\dim \overset{*}{V} = n$ and V and $\overset{*}{V}$ are each isomorphic to F^n and hence to each other. However, the isomorphism $V \to \overset{*}{V}$ uniquely defined by $\mathbf{v_i} \to \mathbf{w_i}$ for each i depends on the basis chosen for V and is not *natural* (in the usual mathematical sense of this word).

With V as above, one may construct in a similar fashion the dual of $\overset{*}{V}$, denoted $\overset{**}{V}$, and which is isomorphic to V. In this case the linear map $f : V \to \overset{**}{V}$ given for $\mathbf{v} \in V$ by $f(\mathbf{v})(\mathbf{w}) = \mathbf{w}(\mathbf{v})$ for each $\mathbf{w} \in \overset{*}{V}$ is basis independent and bijective and is a *natural* isomorphism. The vector spaces V and $\overset{**}{V}$ are thus naturally isomorphic and are often identified, as will be useful later.

Let U and V be vector spaces over F of dimension m and n, respectively. and define the $(m+n)$-dimensional vector space $W \equiv U + V$ over F. A map $f : W \to F$ is called a *bilinear form* if for $\mathbf{u}, \mathbf{u_1}, \mathbf{u_2} \in U$, $\mathbf{v}, \mathbf{v_1}, \mathbf{v_2} \in V$ and $a_1, a_2 \in F$ one has "linearity in each argument" (bilinearity), that is,

$$f(a_1\mathbf{u_1} + a_2\mathbf{u_2}, \mathbf{v}) = a_1 f(\mathbf{u_1}, \mathbf{v}) + a_2 f(\mathbf{u_2}, \mathbf{v}) \tag{1.2}$$

and

$$f(\mathbf{u}, a_1\mathbf{v_1} + a_2\mathbf{v_2}) = a_1 f(\mathbf{u}, \mathbf{v_1}) + a_2 f(\mathbf{u}, \mathbf{v_2}). \tag{1.3}$$

For bilinear forms f_1 and f_2 one can define the bilinear form $(a_1 f_1 + a_2 f_2)(\mathbf{u}, \mathbf{v}) = a_1 f_1(\mathbf{u}, \mathbf{v}) + a_2 f_2(\mathbf{u}, \mathbf{v})$ and so the set of all bilinear forms on W is itself a vector space over F. Given bases $\{\mathbf{u_i}\}$ for U and $\{\mathbf{v_i}\}$ for V and a matrix $A = (a_{ij})$ $(1 \le i \le m, 1 \le j \le n)$ with entries in F there is exactly one bilinear form f on W satisfying $f(\mathbf{u_i}, \mathbf{v_j}) = a_{ij}$ and A is referred to as *the matrix of f* with respect to the bases $\{\mathbf{u_i}\}$ and $\{\mathbf{v_i}\}$. Thus the bilinear forms f_{pq} associated, as above, with the arrays $a_{ij} = \delta_{ip}\delta_{jq}$ for each p, q with $1 \le p \le m$, $1 \le q \le n$, give a basis for the vector space of bilinear forms on W and hence the latter has dimension mn. Thus if $u = \sum_{i=1}^{m} x_i\mathbf{u_i} \in U$ and $v = \sum_{i=1}^{n} y_i\mathbf{v_i} \in V$, $f(\mathbf{u}, \mathbf{v}) = \sum_{i=1}^{m}\sum_{i=1}^{n} a_{ij}x_i y_j$. Under changes of bases in U and V given by $\mathbf{u}_i \to \mathbf{u}'_i = \sum_{k=1}^{m} s_{ik}\mathbf{u_k}$ in U and $\mathbf{v}_j \to \mathbf{v}'_j = \sum_{l=1}^{n} t_{jl}\mathbf{v_l}$ in V for non-singular matrices $S \equiv (s_{ik})$ and $T \equiv (t_{jl})$, the matrix of f changes to SAT^T where T^T denotes the *transpose* of T. It is often the case that $U = V$ with, say, their common dimension equal to n. A bilinear form f on $V + V$ is usually called a *bilinear form* on V and with a basis $\{\mathbf{v_i}\}$ chosen for V the matrix $A = a_{ij}$ for f is now given by $a_{ij} = f(\mathbf{v_i}, \mathbf{v_j})$ Under a change of basis given by $\mathbf{v}_j \to \mathbf{v}'_j = \sum_{l=1}^{n} t_{jl}\mathbf{v_l}$ for some non-singular matrix T the matrix of f changes to TAT^T. A bilinear form on V is then called *symmetric* if $f(\mathbf{u}, \mathbf{v}) = f(\mathbf{v}, \mathbf{u})$, $\forall \mathbf{u}, \mathbf{v} \in V$, equivalently, A is a symmetric matrix, and *non-degenerate* if $f(\mathbf{u}, \mathbf{v}) = 0$, $\forall \mathbf{v} \in V \Rightarrow \mathbf{u} = 0$, equivalently, A is non-singular, these definitions being independent of the basis chosen for V.

A bilinear form f on an n-dimensional vector space V over $\mathbb{R}$ is called a *real bilinear form*. In this case the associated map $f : V \to \mathbb{R}$ given by $\mathbf{v} \to f(\mathbf{v}, \mathbf{v})$ is called a *(real) quadratic form* (on V). If A is the matrix representing f with respect to the basis $\{\mathbf{v_i}\}$ of V and $\mathbf{v} = \sum_{i=1}^{n} x_i\mathbf{v_i} \in V$ this quadratic form is the map $\mathbf{v} \to \sum_{i=1}^{n} a_{ij}x^i x^j$. Thus only the symmetric part of the original bilinear form f (that is, of A) matters here. (There is a simple one-to-one relationship between real *symmetric* bilinear forms on V and real quadratic forms on V and a given real quadratic form uniquely determines its associated symmetric real bilinear form.) A quadratic form is then called *non-degenerate* if its associated symmetric bilinear form is non-degenerate. Under a change of basis in V given by $\mathbf{v}_j \to \mathbf{v}'_j = \sum_{l=1}^{n} s_{jl}\mathbf{v_l}$, the matrix representing the real quadratic form f changes according to $A \to SAS^T$. One calls two real, symmetric $n \times n$ matrices A and B *congruent* if $A = SBS^T$ for some real non-singular matrix S. This gives an equivalence relation on such matrices and one is thus lead to seek conditions which characterise the equivalence classes. This is *Sylvester's law of inertia* which states that for a given real symmetric $n \times n$ matrix A there exists a real non-singular $n \times n$ matrix S such that

$$SAS^T = diag(1, ..., 1, -1, ..., -1, 0, ..., 0) \qquad (1.4)$$

where, for $a, b, ..., c \in \mathbb{R}$, $diag(a, b, ..., c)$ denotes a diagonal matrix with zeros everywhere except on the diagonal where the entries are $a, b, ..., c$, and where, in (1.4), there are r entries 1, s entries -1, t entries 0 and r, s, t are non-negative integers with $r + s + t = n$. The right-hand side of (1.4) is called the *Sylvester canonical form* or the *Sylvester matrix* for A, and A has rank $r + s$. The triple (r, s, t) characterises the equivalence class of A and is called the *signature* of A. It is often written in the form $(+, ..., +, -1, ..., -1, 0, ...0)$ with the obvious number of each of the entries. Special cases are $t = 0$ (non-degenerate), $s = t = 0$ (*positive definite signature*), $r = t = 0$ (*negative definite signature*), $t = 0, s = 1, r \geq 1$ (or $t = 0, r = 1, s \geq 1$) (Lorentz signature) and $t = 0, r = s \neq 0$ (neutral signature).

In this book an *inner product* on an n-dimensional vector space V over $\mathbb{R}$ is a symmetric, non-degenerate, bilinear form $f : V + V \to \mathbb{R}$ and then V is referred to as an *inner product space*. (Sometimes the definition of an inner product, when applied to real vector spaces, requires a positive definite signature. This condition will not be enforced here.) An inner product on V is called a *metric* on V and it is either of positive (or negative) definite signature (sometimes called *Euclidean*) or it is not (and is then sometimes called *indefinite*). If $\mathbf{u}, \mathbf{v} \in V$, $f(\mathbf{u}, \mathbf{v})$ is called the *inner product* of $\mathbf{u}, \mathbf{v}$. If $\mathbf{u}, \mathbf{v}$ are non-zero vectors in V they are called orthogonal if $f(\mathbf{u}, \mathbf{v}) = 0$. From now on, unless explicitly stated to the contrary, $f(\mathbf{u}, \mathbf{v})$ will be written $\mathbf{u} \cdot \mathbf{v}$. A vector $\mathbf{v} \in V$ is called a *unit* vector if $\mathbf{v} \cdot \mathbf{v} = \pm 1$ and a *null* vector if it is not the zero vector and $\mathbf{v} \cdot \mathbf{v} = 0$. A basis for V is called *orthonormal* if it consists of mutually orthogonal unit vectors and, of course, the arrangement of signs for the unit vectors must be consistent with the Sylvester canonical form.

Two subspaces U, W of an inner product space V are called *orthogonal* if for any $\mathbf{u} \in U$ and any $\mathbf{w} \in W$, $\mathbf{u}$ and $\mathbf{w}$ are orthogonal, $\mathbf{u} \cdot \mathbf{w} = 0$. For a subspace U of V one can define its *orthogonal complement* $U^\perp = \{\mathbf{v} \in V : \mathbf{v} \cdot \mathbf{u} = 0, \forall \mathbf{u} \in U\}$. For positive (or negative) definite signature U and $U^\perp$ are always "complementary" in the sense that $U \cap U^\perp = \{\mathbf{0}\}$ and the span of their union equals V but this result can fail for indefinite signatures. However, if $\dim V = n$ and $\dim U = m$ then $\dim U^\perp = n - m$ follows from the theory of linear equations and it is clear that $U \subset (U^\perp)^\perp$. But then $\dim(U^\perp)^\perp = n - \dim U^\perp = m = \dim U$. So $(U^\perp)^\perp = U$ always.

Now let $V_1, ..., V_m$ be finite-dimensional vector spaces over the field F such that $\dim V_i = n_i$ ($1 \leqslant i \leqslant m$). A *multilinear map* (or *form*) f on $V \equiv V_1 + \cdots + V_m$ is a map $f : V \to F$ which is linear in each of its arguments (as for a bilinear form). The set of all such maps is then, with the obvious operations, a vector space over F. Then if $\{\mathbf{e}_i^1\}, ..., \{\mathbf{e}_j^m\}$ are bases for $V_1, ..., V_m$, respectively, and with corresponding dual bases $\{{}^*\mathbf{e}_i^1\}, ..., \{{}^*\mathbf{e}_j^m\}$ the multilinear maps $V \to F$ denoted by ${}^*\mathbf{e}_i^1 \otimes ... \otimes {}^*\mathbf{e}_j^m$ and defined by

$$
{}^*\mathbf{e}_i^1 \otimes ... \otimes {}^*\mathbf{e}_j^m(\mathbf{e}_a^1, ..., \mathbf{e}_b^m) = {}^*\mathbf{e}_i^1(\mathbf{e}_a^1)...{}^*\mathbf{e}_j^m(\mathbf{e}_b^m) = \delta_{ia}...\delta_{jb} \qquad (1.5)
$$

and extended by linearity to V give a basis for the vector space of all multilinear maps $V \to F$ and hence this latter vector space has dimension $n_1 n_2...n_m$.

1.5   Eigen-Structure, Jordan Canonical Forms and Segre Types

Let V be an n-dimensional vector space over the field F where $F = \mathbb{R}$ or $\mathbb{C}$ and let f be a linear map $f : V \to V$. A non-zero vector $\mathbf{v} \in V$ is called an *eigenvector* of f if $f(\mathbf{v}) = \lambda\mathbf{v}$ for $\lambda \in F$ and then λ is called the *eigenvalue* of (f associated with) $\mathbf{v}$. The terms *characteristic vector* and *characteristic value*, respectively, are also sometimes used. Each non-zero member of the 1-dimensional subspace of V spanned by $\mathbf{v}$ (the *direction* determined by $\mathbf{v}$) is then also an eigenvector with eigenvalue λ and this direction is referred to as the *eigendirection* determined by $\mathbf{v}$. If $\{\mathbf{e}_i\}$ is a basis for V, $A = a_{ij}$, the matrix representing f in this basis and $\mathbf{v} = \sum_{i=1}^{n} v_i \mathbf{e}_i$ then the eigenvector/eigenvalue condition above is $\sum_i^n v_i a_{ij} = \lambda v_j$, or in matrix language, $\mathbf{v}A = \lambda\mathbf{v}$. In this case the v_i and λ are called the *(components of the) eigenvector and associated eigenvalue* of A. It follows that $\lambda \in F$ is an eigenvalue of f (or A) if and only if $x = \lambda$ satisfies the equation $\det(A - xI_n) = 0$ where I_n is the unit $n \times n$ matrix. The left-hand side of this last equation is the *characteristic polynomial* of A and is of order n with coefficients in F whilst the equation itself is called the *characteristic equation* of A. Recalling section 1.3, it is remarked that the above concepts are independent of any bases, that is, similar matrices have the same characteristic polynomial. For any eigenvalue λ, the number of times the factor $(x - \lambda)$ appears in the characteristic equation is called the *multiplicity* of λ. If λ is an eigenvalue of f (or A) the set of all eigenvectors of f with eigenvalue λ together with the zero vector is a subspace of V called the λ-*eigenspace* of f. Its dimension may not equal the multiplicity of λ as will be made clear later. If $ker f$ is not trivial it is the $0-$eigenspace of f and if $\dim ker f = m$ then f has rank $n - m$.

Now let U be a subspace of V and suppose the linear map $f : V \to V$ satisfies $f(U) \subset U$. Then U is called an *invariant* subspace of (for) f. Thus any eigendirection and any eigenspace of any linear map $V \to V$ is invariant for that map. However, a non-zero member of an invariant subspace of f need not be an eigenvector of f. In fact an invariant subspace of f may not contain any eigenvectors of f. This remark and the concept of an invariant subspace will be important later.

The linear map f described above is, of course, uniquely determined by its action on a basis for V. A simple description of f can thus be achieved by a judicious choice of such a basis. If a basis for V consisting of eigenvectors exists (and, in general, it does not) the matrix representing f in this basis is, conveniently, a diagonal matrix with diagonal entries consisting of the eigenvalues of f and then f (or A) is said to be *diagonalisable (over F)*. Thus if the characteristic equation of f admits n distinct solutions in F the associated eigenvectors are easily checked to be independent and constitute a basis of eigenvectors for f which is then diagonalisable over F. Even if the characteristic equation does not admit n distinct solutions but factorises over F into n linear factors the map f will still be diagonalisable if the dimension of each eigenspace equals the multiplicity of its associated eigenvalue. Failing this, but retaining the factorising of the characteristic polynomial into n linear factors over F, one could seek some "almost" diagonal form for f as its "canonical form". This process leads to the Jordan (canonical) form for f and is rather useful. It will be discussed next. However, it depends on the characteristic polynomial of f factoring into n linear factors over F and this can only be guaranteed if F is an *algebraically closed* field, for example, if $F = \mathbb{C}$. If $F = \mathbb{R}$ such a factoring may not exist.

So suppose that V is an n-dimensional *complex* vector space and $f : V \to V$ is linear and admits a basis of eigenvectors. The matrix representing f in this basis is diagonal with the diagonal entries equal to the eigenvalues of f in some order. The Jordan theorem solves the problem when such a basis may not exist. This leads to the following generalisation (the *Jordan canonical form*) for a linear map $f : V \to V$ and the algebraically closed nature of $\mathbb{C}$ is crucial here. Suppose the distinct eigenvalues of f are $\lambda_1, ..., \lambda_r \in \mathbb{C}$ $(1 \leqslant r \leqslant n)$. If f is diagonalisable it can be viewed as decomposing V into a direct sum of subspaces $V_1 + \cdots + V_r$ where V_i $(1 \leq i \leq r)$ is the λ_i-eigenspace of f and $\sum_{i=1}^{r} \dim V_i = n$. In this case the multiplicity of each eigenvalue equals the dimension of the corresponding eigenspace. Now suppose f is not diagonalisable. Then the characteristic polynomial of f factorises into n linear factors over $\mathbb{C}$, with distinct eigenvalues $\lambda_1, ..., \lambda_r$ $(1 \leqslant r < n)$ with respective multiplicities $m_1, ..., m_r$ $(\sum_{i=1}^{r} m_i = n)$. It can then be shown that V may be decomposed as $V = V_1 + \cdots + V_r$ where $\dim V_i = m_i$ and each V_i is an invariant subspace of V under f "associated" with the eigenvalue λ_i which contains, but is not necessarily equal to, the λ_i-eigenspace. Further, one may choose a basis for V_i on which the restriction of f has representative matrix with λ_i in each diagonal position, some arrangements of zeros and ones along the superdiagonal and zeros elsewhere. Choosing such a basis for each V_i one obtains a *Jordan basis* for V and a representative matrix for f of the form (dots denote zeros)

$$A = \begin{bmatrix} A_1 & \cdots & \cdots \\ \cdots & A_2 & \cdots \\ \cdots & \cdots & \cdots \\ \cdots & \cdots & A_r \end{bmatrix}$$

(A basis scalable to a Jordan basis is still called a Jordan basis). Then, for a particular eigenvalue λ_i, the matrix representing f in the above basis when restricted to V_i is A_i above and is given by

$$A_i = \begin{bmatrix} B_{i1} & \cdots & \cdots \\ \cdots & B_{i2} & \cdots \\ \cdots & \cdots & \cdots \\ \cdots & \cdots & B_{ik(i)} \end{bmatrix}$$

where each matrix B_{ij} is a $p_{ij} \times p_{ij}$ matrix with λ_i in each diagonal position, a "1" in every superdiagonal place and zeros elsewhere and the order is usually chosen so that $p_{i1} \geq \cdots \geq p_{ik(i)}$. It is called a *basic Jordan block*. With an ordering established for the eigenvalues $\lambda_1, ..., \lambda_r$ the canonical form for A is uniquely determined (and $m_i = p_{i1} + \cdots + p_{ik(i)}$). This information is usually collectively called the *Jordan canonical form* for f and the A is the *Jordan matrix* for f. The symbol

$$\{(p_{11}, ..., p_{1k(1)})(p_{21}, ..., p_{2k(2)})\cdots(p_{r1}, ..., p_{rk(r)})\} \tag{1.6}$$

is referred to as the *Segre type, Segre symbol* or *Segre characteristic* of f and carries with it the information in the Jordan canonical form. The characteristic polynomial of f is then

$$(-1)^n (x - \lambda_1)^{m_1} (x - \lambda_2)^{m_2} ... (x - \lambda_r)^{m_r}. \tag{1.7}$$

It is known from the *Cayley-Hamilton theorem* that every $n \times n$ real or complex matrix A satisfies its own characteristic equation. Thus there exists a polynomial of least degree m $(1 \leq m \leq n)$ which is satisfied by A. If it is agreed that this polynomial is monic then it is unique and is called the *minimal polynomial* of A. It can be shown that any two (similar) matrices have the same minimal polynomial and thus one has a *minimal polynomial* for the map f above. It can also be shown that for this map the minimal polynomial is given by

$$(x - \lambda_1)^{p_{11}} (x - \lambda_2)^{p_{21}} ... (x - \lambda_r)^{p_{r1}} \tag{1.8}$$

that is, the power to which $(x - \lambda_i)$ is raised is the largest integer amongst the set $\{p_{i1}, ... p_{ik(i)}\}$ and thus the minimal polynomial divides the characteristic polynomial.

For each i the polynomials $(x - \lambda_i)^{p_{ij}}$ in (1.7) are called the *elementary divisors* associated with the eigenvalue λ_i. Recalling the above ordering on the p_{ij} such an elementary divisor is called *simple* if $p_{ij} = 1$ and *non-simple of order p_{ij}* if $p_{ij} > 1$. In the above Jordan form, each V_i is an invariant subspace for f with matrix A_i and within this subspace each basic Jordan block gives rise to an invariant subspace of V_i.

It is useful to note here that, for example, if a 4×4 matrix C is a basic Jordan block with eigenvalue λ and if the Jordan basis members are

$\mathbf{u} = (1,0,0,0)$, $\mathbf{w} = (0,1,0,0)$, $\mathbf{r} = (0,0,1,0)$ and $\mathbf{v} = (0,0,0,1)$ then the following pattern ensues

$$C = \begin{bmatrix} \lambda & 1 & 0 & 0 \\ 0 & \lambda & 1 & 0 \\ 0 & 0 & \lambda & 1 \\ 0 & 0 & 0 & \lambda \end{bmatrix}$$

and

$$f(\mathbf{v}) = \lambda\mathbf{v}, \qquad f(\mathbf{r}) = \lambda\mathbf{r} + \mathbf{v},$$
$$f(\mathbf{w}) = \lambda\mathbf{w} + \mathbf{r}, \qquad f(\mathbf{u}) = \lambda\mathbf{u} + \mathbf{w}. \tag{1.9}$$

An eigenvalue λ of f is called *non-degenerate* if the associated λ-eigenspace is 1-dimensional and otherwise *degenerate*. Equal eigenvalues are indicated by enclosing the integers associated with the (equal) eigenvalues inside round brackets in the Segre type for f. A non-degenerate eigenvalue λ could be associated with a simple or a non-simple elementary divisor.

It is remarked that the above Jordan theory demanded that the characteristic polynomial factorised into n linear factors over the appropriate field and that this was satisfied since a vector space over $\mathbb{C}$ was considered. If one is really interested in a vector space over $\mathbb{R}$ techniques are available for this (for example, the so-called rational canonical form). This book will be mainly concerned with real vector spaces but a more direct approach to its eigenstructure will now be described.

So let V be an n-dimensional *real* vector space V. One can now describe a technique relating V to an associated complex vector space and usually referred to as "complexifying" V. Then the above theory for complex vector spaces may be used. This construction will allow for scalar multiplication of members of V by members of $\mathbb{C}$. Starting from the real vector space V, first construct the real vector space $V + V$ and for $\mathbf{u}, \mathbf{v} \in V$ define a linear map on $V + V$ by $c : (\mathbf{u}, \mathbf{v}) \to (-\mathbf{v}, \mathbf{u})$ so that $c \circ c$ is the negative of the identity map on $V + V$ (thought of as "multiplication by i"). Then one can think of $V + V$ as a complex vector space (the *complexification* of V) with multiplication by members of $\mathbb{C}$ defined for $a, b \in \mathbb{R}$ and $\mathbf{u}, \mathbf{v} \in V$ by

$$(a + ib)(\mathbf{u}, \mathbf{v}) = a(\mathbf{u}, \mathbf{v}) + bc(\mathbf{u}, \mathbf{v}) = a(\mathbf{u}, \mathbf{v}) + b(-\mathbf{v}, \mathbf{u}) = (a\mathbf{u} - b\mathbf{v}, a\mathbf{v} + b\mathbf{u}) \tag{1.10}$$

as is easily checked. Thus V, which was an n-dimensional vector space over $\mathbb{R}$, is converted to a vector space over $\mathbb{C}$. One thinks of $(\mathbf{u}, \mathbf{v})$ as $\mathbf{u} + i\mathbf{v}$. Then if $\mathbf{e_i}$ ($1 \leqslant i \leqslant n$) is a basis for V the members $(\mathbf{e_i}, \mathbf{0})$ constitute a basis for $V + V$ when the latter is taken over the field $\mathbb{C}$ and so the complex vector space $V + V$ is also n-dimensional. This follows since if $(\mathbf{u}, \mathbf{v}) \in V + V$ with $\mathbf{u} = \sum_{i=1}^{n} u_i \mathbf{e_i}$ and $\mathbf{v} = \sum_{i=1}^{n} v_i \mathbf{e_i}$ for $u_i, v_i \in \mathbb{R}$ then from the above definition of complex multiplication for V

$$(\mathbf{u}, \mathbf{v}) = \sum_{i=1}^{n} (u_i + iv_i)(\mathbf{e_i}, \mathbf{0}). \tag{1.11}$$

Now let $f : V \to V$ be a linear map on V and extend it to a linear map on $V + V$ by defining $\widetilde{f} : (V + V) \to (V + V)$ by $\widetilde{f}((\mathbf{u}, \mathbf{v})) = (f(\mathbf{u}), f(\mathbf{v}))$. Then if $f(\mathbf{e_i}) = \sum_{i=1}^{n} a_{ij}\mathbf{e_j}$, $\widetilde{f}(\mathbf{e_i}, \mathbf{0}) = a_{ij}(\mathbf{e_j}, \mathbf{0})$ and so the map $f : V \to V$ has the same (real) representative matrix (a_{ij}) as the extended map $\widetilde{f} : (V + V) \to (V + V)$ for any such basis. Then if $f(\mathbf{v}) = \lambda \mathbf{v}$ so that $\mathbf{v}$ is an eigenvector of f with eigenvalue $\lambda \in \mathbb{R}$, $\widetilde{f}(\mathbf{v}.\mathbf{0}) = (\lambda \mathbf{v}, \mathbf{0})$ and $(\mathbf{v}, \mathbf{0})$ is an eigenvector of $\widetilde{f}$ with (real) eigenvalue λ. However, if $(\mathbf{u}, \mathbf{v})$ is an eigenvector of $\widetilde{f}$ with eigenvalue $a + ib$ $(a, b \in \mathbb{R}, b \neq 0)$, $\widetilde{f}(\mathbf{u}, \mathbf{v}) = (a + ib)(\mathbf{u}, \mathbf{v})$ and so $f(\mathbf{u}) = a\mathbf{u} - b\mathbf{v}$ and $f(\mathbf{v}) = a\mathbf{v} + b\mathbf{u}$. Similarly $\widetilde{f}(\mathbf{u}, -\mathbf{v}) = (a\mathbf{u} - b\mathbf{v}, -a\mathbf{v} - b\mathbf{u}) = (a - ib)(\mathbf{u} - \mathbf{v})$ and so $(\mathbf{u}, -\mathbf{v})$ is an eigenvector of $\widetilde{f}$ with eigenvalue $a - ib$. Thus the eigenvectors and eigenvalues for $\widetilde{f}$ come, as expected (since the a_{ij} are real and hence the characteristic polynomial has real coefficients), in complex conjugate pairs. It follows that the "real" and "imaginary" parts, ($\mathbf{u}$ and $\mathbf{v}$), of a complex eigenvector $(\mathbf{u}, \mathbf{v})$ of $\widetilde{f}$ span a 2-dimensional invariant subspace (for f) of V and which, for $b \neq 0$, contains no eigenvectors of f. This result will be important later. One thinks of the above results, informally, as the statement that $\mathbf{u} + i\mathbf{v}$ is a "complex" eigenvector of f with "complex" eigenvalue $a + ib$. The reference to the map $\widetilde{f}$ will usually be dropped and one will speak of "complex" eigenvectors and "complex" eigenvalues of f. This summary of the eigenstructure of f on V and $\widetilde{f}$ on $V + V$ is all that is required for this book. It is noted (and easily checked) here for future use that any (real) invariant $2-$space of the real linear map $f : V \to V$ either contains one or two independent real eigenvectors or a conjugate pair of complex eigenvectors (in the above sense).

An important classical result arises at this point. Let V be an n-dimensional real vector space, let $f : V \to V$ be a linear map on V and $h : V + V \to \mathbb{R}$ a *positive definite* metric on V. Suppose that f is *self-adjoint* (with respect to h), that is, $h(\mathbf{u}, f(\mathbf{v})) = h(f(\mathbf{u}), \mathbf{v})$ for each $\mathbf{u}, \mathbf{v} \in V$. Then in any basis $\{\mathbf{u_i}\}$ the product matrix AH is symmetric, where A and H are the matrices representing f and h in this basis. Let P be a real non-singular matrix effecting the transformation from the above basis to the basis in which h takes its Sylvester canonical form I_n. The matrix representing f in this new basis is then $Q = PAP^{-1}$ and can be checked to be symmetric (and conversely, the symmetry of Q implies the symmetry of AH). A well-known classical result now says that the characteristic polynomial of the *real symmetric matrix* Q factorises into n real factors and that f admits a basis of eigenvectors which may be chosen orthonormal with respect to the metric represented by the matrix (δ_{ij}) in this basis (the Sylvester form for h on this basis). This result, that f is diagonalisable, is usually referred to as the *principal axes theorem* for a real symmetric matrix . However, it should be noted that, given the self-adjoint assumption on f, this result depends on the positive definite nature of h. It fails if this is not the case since the characteristic equation may admit non-real solutions and/or there may not exist a basis of eigenvectors for f.

In this book the following convention will be made when writing a Segre symbol {*abc...*}. If the entry *a*, say, refers to a *real* eigenvalue it will be written as a positive integer and which gives the order of the associated elementary divisor. If an entry refers to a complex eigenvalue arising from a simple elementary divisor it will be written as z and since this book is mostly concerned only with maps whose characteristic polynomials have real coefficients $\bar{z}$ will also occur in the Segre symbol. If an eigenvalue is complex and arises from a non-simple elementary divisor it will be written as an integer (≥ 2) equal to the order of the elementary divisor and the fact that it is complex will be specified separately. As stated earlier, equal eigenvalues are enclosed inside round brackets in the Segre symbol.

1.6 Lie Algebras

Let V be a vector space over F ($F = \mathbb{R}$ or $\mathbb{C}$). One wishes to impose a type of multiplication between members of V. Suppose there is a binary operation on V represented by $(\mathbf{u}, \mathbf{v}) \to \mathbf{uv}$ where, for $\mathbf{u}, \mathbf{v} \in V$, $\mathbf{uv}$ represents this product, and which satisfies for $\mathbf{u}, \mathbf{v}, \mathbf{w} \in V$ and $a \in F$

(**LA1**) $(\mathbf{u} + \mathbf{v})\mathbf{w} = \mathbf{uw} + \mathbf{vw}$,
(**LA2**) $\mathbf{u}(\mathbf{v} + \mathbf{w}) = \mathbf{uv} + \mathbf{uw}$,
(**LA3**) $a(\mathbf{uv}) = (a\mathbf{u})\mathbf{v} = \mathbf{u}(a\mathbf{v})$,
(**LA4**) $(\mathbf{uv}) = -(\mathbf{vu})$,
(**LA5**) $\mathbf{u}(\mathbf{vw}) + \mathbf{v}(\mathbf{wu}) + \mathbf{w}(\mathbf{uv}) = \{\mathbf{0}\}$.

Then V with this operation (the *Lie product*) is called a *Lie algebra* over F. The last condition above is the *Jacobi identity* and it follows from the fourth that $\mathbf{uu} = 0$ for each $\mathbf{u} \in V$. If U and V are Lie algebras and $f : U \to V$ a vector space homomorphism (respectively, an isomorphism) such that, in an obvious notation, $f(\mathbf{uv}) = f(\mathbf{u})f(\mathbf{v})$ then f is a *Lie algebra homomorphism* (respectively, *isomorphism*) and then U and V are *Lie algebra homomorphic* (respectively, *isomorphic*). If V is a Lie algebra and $U \subset V$ is a subspace of V such that U, with the induced Lie product from V, is a Lie algebra, then U is called a *Lie subalgebra* of V. For each $\mathbf{w} \in V$, $0\mathbf{w} = \mathbf{0}$ from **LA1** (put $\mathbf{v} = -\mathbf{u}$). If $f : U \to V$ is a Lie algebra homomorphism, as above, the range space $f(U)$ of f is a subalgebra of V and the kernel of f is a subalgebra of U, as is easily checked. If $\mathbf{uv} = \mathbf{0}$, $\mathbf{u}$ and $\mathbf{v}$ are said to *commute* and if $\mathbf{uv} = \mathbf{0}$ for any $\mathbf{u}, \mathbf{v} \in V$, V is called *Abelian*. If V and W are Lie algebras the vector space direct sum $V + W$ may be given the following structure of a Lie algebra. Let $\mathbf{v}, \mathbf{v}' \in V$ and $\mathbf{w}, \mathbf{w}' \in W$ and using the products in V and W, define the Lie product on $V + W$ given by $(\mathbf{v}, \mathbf{w})(\mathbf{v}', \mathbf{w}') \equiv (\mathbf{vv}', \mathbf{ww}')$. Then $V + W$

is a Lie algebra (the *product algebra* of V and W) and any member of the subspace $V + \{\mathbf{0}\}$ of $V + W$ (which is Lie isomorphic to V) commutes with any member of the subspace $\{\mathbf{0}\} + W$ of $V + W$ (which is Lie isomorphic to W) since $(\mathbf{v}, \mathbf{0})(\mathbf{0}, \mathbf{w}') = (\mathbf{0}, \mathbf{0})$. The projection maps from $V + W$ to V and W are easily checked to be Lie algebra homomorphisms. The set $M_n\mathbb{R}$ of all real $n \times n$ matrices is an n^2-dimensional real vector space and if $A, B \in M_n\mathbb{R}$ the product operation which sends $(A, B) \to AB - BA$ gives $M_n\mathbb{R}$ the structure of a Lie algebra.

1.7   Topology

Let X be any set. The idea here is to put a structure on X which allows one to make sense of concepts such as "nearness", "limit", "convergence", "continuity", etc. Most geometrical intuition relies on such concepts and precise proofs regarding these concepts require such a structure to be laid down axiomatically. This leads to the idea of a topological structure for X. Thus with X arbitrary let $\mathcal{T}$ be a family of subsets of X satisfying the following conditions:

T1 The empty set $\emptyset$ is a member of $\mathcal{T}$,

T2 X is a member of $\mathcal{T}$,

T3 The union of an arbitrary family of members of $\mathcal{T}$ is itself a member of $\mathcal{T}$, and

T4 The intersection of a *finite* family of members of $\mathcal{T}$ is itself a member of $\mathcal{T}$.

The pair $(X, \mathcal{T})$ is then called a *topological space*, $\mathcal{T}$ is called a *topology* for X and the members of $\mathcal{T}$ are called *open* (sub)sets of (or in) X (or said to be members of, or *open* in, $\mathcal{T}$) . It is easily checked that in **T4**, if one replaces the finite family of members by any two members, the resulting axioms are equivalent to the originals. A subset $F \subset X$ is called *closed* (in X) if its complement $X \setminus F$ in X is open, that is, if $X \setminus F \in \mathcal{T}$. Thus $\emptyset$ and X are closed subsets. Also it is easily checked from de Morgan's laws that any arbitrary intersection, and any finite union, of closed subsets of X is closed in X. One may easily write the above four axioms in an equivalent way in terms of closed subsets of X by using the de Morgan laws.

The set $\mathbb{R}$ has a *standard topology* in which a subset $U \subset \mathbb{R}$ is open if, for any $p \in U$, there is an open interval $I \equiv (a, b)$ of $\mathbb{R}$ $(a, b \in \mathbb{R}, a < b)$ such that $p \in I \subset U$. It follows that such intervals (a, b) are open subsets of $\mathbb{R}$ and the intervals of the form $[a, b]$ are closed subsets of $\mathbb{R}$. This topology on $\mathbb{R}$ will always be understood. For any set X, if $\mathcal{T}$ is defined to be the collection of *all* subsets of X, then $\mathcal{T}$ is a topology for X called the *discrete topology* for X whilst if X and $\emptyset$ are the only members of $\mathcal{T}$ again $\mathcal{T}$ is a topology for X called the *indiscrete topology* for X. More examples can be given after a little more topology is described.

Let $(X, \mathcal{T})$ be a topological space and let $p \in X$. A subset $N \subset X$ containing p is called a *neighbourhood* of p if there exists $U \in \mathcal{T}$ with $p \in U \subset N$. Thus any open set containing p is a neighbourhood of p but a neighbourhood N of p need not be open (and is open if and only if it is a neighbourhood of *each* of its points). If $A \subset X$, define the *interior* of A, denoted int$A$, as the set of points (of A) for which A is a neighbourhood. Then int$A$ is an open set and is the largest open subset of A in the sense that if U is open and $U \subset A$ then $U \subset$ int$A$. Further for $A \subset X$, a point $p \in X$ is a *limit point* of A if every open subset containing p intersects A in some point other than (possibly) p. Then A is closed if and only if it contains each of its limit points. If one defines the *closure* $\overline{A}$ of A to be the union of A with the set of all of its limit points then $\overline{A}$ is closed and, in the sense given above, is the smallest closed set containing A. It follows from these definitions that A is open if and only if $A =$ int$A$ and that A is closed if and only if $A = \overline{A}$. A point $p \in X$ is a *boundary point* of a subset $A \subset X$ if for any open neighbourhood U of p, $U \cap A$ *and* $U \cap (X \setminus A)$ are not empty. Then the *boundary* $\delta(A)$ of A is the collection of all boundary points of A. It is easily checked that, for any $A \subset X$, $\delta(A) = \overline{A} \cap \overline{(X \setminus A)}$ and is thus a closed subset of X, that $\delta(A) = \delta(X \setminus A)$ and that a subset of X is closed if and only if it contains all its boundary points.

A sense of topological "size" (large and small) will be required later and this can be described now. A subset $A \subset X$ is called *dense* in X if $X = \overline{A}$ and *nowhere dense* in X if int$A = \emptyset$. The statement int$A = \emptyset$ is equivalent to the statement that A contains no non-empty open subsets. One can think of a subset A of X as being "topologically large" in X if it is open and dense in X and "topological small" in X if it is closed and nowhere dense in X. Now A is dense if and only if its complement in X has empty interior and so the complement of an open dense subset is closed and nowhere dense, and vice versa. Any finite subset of $\mathbb{R}$ is closed and nowhere dense in $\mathbb{R}$ whilst the subset $\mathbb{Q}$ of $\mathbb{R}$, which is neither open nor closed, is dense in $\mathbb{R}$ as also is its complement $\mathbb{R} \setminus \mathbb{Q}$ of irrational numbers. It can be checked that a finite union of closed, nowhere dense subsets of X is closed and nowhere dense in X and hence that the intersection of finitely many open dense subsets of X is itself open and dense in X. To see this, let F and F' be closed and nowhere dense in X and let $\emptyset \neq U \subset F \cup F'$ be open in X so that $U \subset F$ and $U \subset F'$ are each false. Then $U \cap (X \setminus F)$ is an open subset of F' and is hence empty. Similarly, $U \cap (X \setminus F')$ is empty and so one achieves the contradiction $U = \emptyset$. An induction argument completes the proof. The second part follows from de Morgan's laws.

Let $(X, \mathcal{T})$ and $(Y, \mathcal{T}')$ be topological spaces and let $f : X \to Y$ be a map. Then f is said to be *continuous at* $p \in X$ if $f^{-1}(N)$ is a neighbourhood of p (in $\mathcal{T}$) whenever N is a neighbourhood of $f(p)$ (in $\mathcal{T}'$). The map f is then said to be *continuous* if it is continuous at each $p \in X$ and this is equivalent to the statement that the inverse image under f of any open set in $\mathcal{T}'$ is open in $\mathcal{T}$. If $(X, \mathcal{T})$, $(Y, \mathcal{T}')$ and $(Z, \mathcal{T}'')$ are topological spaces and $f : X \to Y$ and $g : Y \to Z$ are continuous then $g \circ f : X \to Z$ is continuous. If, in the above, f is continuous, a bijection and is such that $f^{-1} : Y \to X$ is continuous, then f is called a *homeomorphism* and $(X, \mathcal{T})$ and $(Y, \mathcal{T}')$ are said to be *homeomorphic* topological spaces.

In the actual construction of a topology for a set the following technique is sometimes useful. Let $(X, \mathcal{T})$ be a topological space and let $\mathcal{B}$ be a subset of $\mathcal{T}$ such that each $U \in \mathcal{T}$ is a union of members of $\mathcal{B}$. Then $\mathcal{B}$ is called a *base* (or *basis*) for $\mathcal{T}$ and $\mathcal{T}$ is said to be *generated* by $\mathcal{B}$. Thus $\mathcal{B}$ is a base for $\mathcal{T}$ if and only if for each $p \in X$ and open set U containing p there exists $B \in \mathcal{B}$ with $p \in B \subset U$. On the other hand let $\mathcal{B}$ be a family of subsets of X. Under what conditions on $\mathcal{B}$ does it become a base for *some* topology on X? If X equals the union of the members of $\mathcal{B}$ and if when $B_1, B_2 \in \mathcal{B}$ and $p \in B_1 \cap B_2$ there exists $B \in \mathcal{B}$ and $p \in B \subset B_1 \cap B_2$ then $\mathcal{B}$ is a base for *some* topology on X. The idea of a basis for a topology can be simplified even further. Let X be any set and $\mathcal{B}'$ be a collection of subsets of X whose union equals X. Let $\mathcal{B}$ be the set of all *finite* intersections of members of $\mathcal{B}'$. Then $\mathcal{B}$ is a basis for *some* topology $\mathcal{T}$ on X. The collection $\mathcal{B}'$ is called a *subbase* (or *subbasis*) for $\mathcal{T}$ and is said to generate $\mathcal{T}$. For a topological space $(X, \mathcal{T})$, let $p \in X$ and let $\mathcal{D}$ be a family of open sets each containing p such that for any open set U in X with $p \in U$, there exists $D \in \mathcal{D}$ and $p \in D \subset U$. Then $\mathcal{D}$ is called a *local base* (or *local basis*) at p. In fact these concepts are easily related since a collection of subsets $\mathcal{B}$ of X is a basis for $\mathcal{T}$ if and only if for each $p \in X$ the family $\mathcal{B}_p = \{B \in \mathcal{B} : p \in B\}$ is a local basis at p. The concepts of basis and local basis simplify the idea of continuity because it is easily checked that a map $f : X \to Y$ between topological spaces X, Y is continuous if and only if the inverse image under f of each member of a base (or subbase) of Y is open in X.

For the set $\mathbb{R}$ and $a, b \in \mathbb{R}, a < b$, the intervals of the form (a, b) give a base for the standard topology whilst the intervals of the form $(-\infty, a)$ and (a, ∞) for each $a \in \mathbb{R}$ together constitute a subbase for this topology.

Let $(X, \mathcal{T})$ be a topological space and let $A \subset X$ be any subset of X. Then A inherits a natural topology from the topology $\mathcal{T}$ on X as follows. Define a collection $\overline{\mathcal{T}}$ of subsets of A by $\overline{\mathcal{T}} = \{U' \subset A : U' = A \cap U, U \in \mathcal{T}\}$. Then $\overline{\mathcal{T}}$ is easily checked to be a topology for A called the *subspace* or *relative* or *inherited* topology on A from $(X, \mathcal{T})$, and $(A, \overline{\mathcal{T}})$ is called a *topological subspace* of $(X, \mathcal{T})$. The usual inclusion map $i : A \to X$ is then continuous with respect to $\mathcal{T}$ and $\overline{\mathcal{T}}$. Of course, if A and B are subsets of X with $A \subset B$ then B inherits a topology from X and A inherits a topology from X and also one from the topology inherited by B from X. Fortunately these are the same topology. Thus the open subsets of $(A, \overline{\mathcal{T}})$ are the intersections with A of the open subsets of X (and it is easily checked that this statement is true if "open" is replaced by "closed"). If $(X, \mathcal{T})$ and $(Y, \mathcal{T}')$ are topological spaces and $f : X \to Y$ is a continuous map then, in the above notation, the restriction $f_{|A} : A \to Y$ is continuous as a map from $(A, \overline{\mathcal{T}})$ to $(Y, \mathcal{T}')$. One may also show that the continuity of $f : X \to Y$ above is equivalent to the continuity of the map $\tilde{f} : X \to f(X)$ defined by $\tilde{f}(p) = f(p)$ when $f(X)$ has subspace topology from $(Y, \mathcal{T}')$.

Let $(X_1, \mathcal{T}_1), ..., (X_n \mathcal{T}_n)$ be a finite collection of topological spaces and let $X \equiv X_1 \times ... \times X_n$. Consider the family of subsets of X of the form $U_1 \times ... \times U_n$ where each $U_i \in \mathcal{T}_i$. It is easily checked that this is a base for a topology $\mathcal{T}$ on X called the *product topology* for X (and then $(X, \mathcal{T})$ is referred to simply as the *product* of the above topological spaces). It then follows that the *projection maps* $p_i : X \to X_i$ given by $(x_1, ..., x_n) \to x_i$ are continuous. Only such *finite* products will be considered here. Using the set $\mathbb{R}$ (with its standard topology given earlier) one can construct the set $\mathbb{R}^n$ for $n \in \mathbb{N}$ and give it the product topology as above to get the *standard topology* on $\mathbb{R}^n$. Further, identifying the set of complex numbers $\mathbb{C}$ as $\mathbb{R}^2$ in the usual way, one can get the *standard topology* on $\mathbb{C}$ and then the *standard topology* on $\mathbb{C}^n$. These will always be understood. It easily follows that the collection of all finite products of open intervals in $\mathbb{R}$ give a base for the above standard topology on $\mathbb{R}^n$. A very slight modification of this argument gives a base for the standard topology on $\mathbb{C}^n$.

A topological space X is called *first countable* (or said to satisfy the *first axiom of countability*) if X admits a countable local base at each of its points. A topological space X is called *second countable* (or said to satisfy the *second axiom of countability*) if X admits a countable base for its topology. Since those members of a base for X, containing $p \in X$, is a local base at p for X, every second countable space is first countable, but not conversely. Since it is easily checked that the open intervals of $\mathbb{R}$ of the form (a, b) with $a, b \in \mathbb{Q}$ and hence with centre point in $\mathbb{Q}$ form a base for $\mathbb{R}$, this latter topological space is second and hence first countable (since it is known that $\mathbb{Q}$ and $\mathbb{Q}^n$ are denumerable for any positive integer n). It follows by taking finite products of these intervals, as described above, that $\mathbb{R}^n$ (and by a similar argument $\mathbb{C}^n$) are first and second countable.

Let $(X, \mathcal{T})$ be a topological space and let $\sim$ be an equivalence relation on X with quotient set $X/\sim$ and natural projection $\mu : X \to X/\sim$. The family $\{U \subset X/\sim : \mu^{-1}U \in \mathcal{T}\}$ can easily be shown to be a topology on $X/\sim$ called the *quotient topology* for $X/\sim$ (and the resultant topological space is the *quotient space of X by $\sim$*) and with this topology the map μ is continuous. Thus, for example, let $X \equiv [0, 1]$ with $[0, 1]$ having the subspace topology inherited from the usual topology on $\mathbb{R}$ and let $X/\sim$ consist of the all the subsets $\{x\}$ for $x \in (0, 1)$ together with the subset $\{0, 1\}$. Then the resulting quotient topology on $X/\sim$ is that of the unit circle in $\mathbb{R}^2$ inherited from the usual topology on $\mathbb{R}^2$. Another example arises as follows. Define an equivalence relation $\sim$ on the non-zero members of $\mathbb{R}^n$ by $u \sim v$ if and only if $u = av$ for $0 \neq a \in \mathbb{R}$. The resulting quotient space of the non-zero members of $\mathbb{R}^n$ is denoted by $P^{n-1}\mathbb{R}$ and called the *real projective space* of dimension $n - 1$. Let X be a topological space, $\sim$ an equivalence relation on X with natural projection $\mu : X \to X/\sim$ and $f : X \to Y$ a map from X to a topological space Y. Then f is said to *respect* $\sim$ if it is constant on each equivalence class under $\sim$. In this case there is a continuous map $f' : X/\sim \to Y$ such that $f = f' \circ \mu$.

Let $(X, \mathcal{T})$ be a topological space and $\mathcal{B}$ a collection of subsets of X. If the union of the members of $\mathcal{B}$ equals X, $\mathcal{B}$ is called a *covering* of X and if, in addition, each member of $\mathcal{B}$ is an open subset of X, $\mathcal{B}$ is called an *open covering*. If some subfamily $\mathcal{B}'$ of a covering $\mathcal{B}$ of X is also a covering of X it is called a *subcovering* of $\mathcal{B}$. Another form of "topological smallness" can now be described. A topological space $(X, \mathcal{T})$ is called *compact* if every open covering of X contains a *finite* subcovering. A subspace of $(X, \mathcal{T})$ is called *compact* if it is compact with its subspace topology. The topological spaces $\mathbb{R}$, $\mathbb{R}^n$, $\mathbb{C}$ and $\mathbb{C}^n$ with their standard topologies are *not* compact but $P^n\mathbb{R}$ is compact. A subspace of a compact space is not necessarily compact (for example, the closed interval $[a, b]$ of $\mathbb{R}$ can be shown to be compact but the open interval (a, b) is not) but *any* closed subspace of a compact topological space *is* compact. Two important results can be given here. First, suppose $(X, \mathcal{T})$ and $(Y, \mathcal{T}')$ are topological spaces with $(X, \mathcal{T})$ compact and let $f : X \to Y$ be continuous. Then $f(X)$ is a compact subspace of $(Y, \mathcal{T}')$. Second, if $(X, \mathcal{T})$ is compact and $f : X \to \mathbb{R}$ is continuous (when $\mathbb{R}$ has its standard topology) then the function f is *bounded* (that is, $f(X)$ is a bounded (and compact) subset of $\mathbb{R}$) and attains its bounds (that is, $\exists y, z \in X$ such that $f(y) = sup[f(X)]$ and $f(z) = inf[f(X)]$. If, in addition, f is a *positive* function (that is, $f(x) > 0 , \forall x \in X$) then f is *bounded away from zero* (that is, $\exists \epsilon \in \mathbb{R}, \epsilon > 0$ such that $f(x) \geq \epsilon \, \forall x \in X$). It is true that the (finite) product of non-empty topological spaces is compact if and only if the individual topological spaces are compact. [The continuity of the projection maps and a remark above easily give part of the proof of this result.]

Some special types of topological spaces can now be described. A topological space $(X, \mathcal{T})$ is called *Hausdorff* if given any two *distinct* members $x, y \in X , \exists$ disjoint open subsets U, V of X (that is, $U \cap V = \emptyset$) with $x \in U, y \in V$. Thus $\mathbb{R}^n$ and $\mathbb{C}^n$ are Hausdorff. [In fact, all the topological spaces encountered in this book will be Hausdorff.] Standard results state that a subspace of a Hausdorff space is Hausdorff (in the subspace topology) and that a compact subspace of a Hausdorff space is closed. There is also a topological method of deciding if a topological space is all in "one piece". A topological space $(X, \mathcal{T})$ is called *connected* if whenever U, V are open subsets of X satisfying $U \cup V = X$ and $U \cap V = \emptyset$ then one of U, V is empty. In other words X is not the union of two disjoint non-empty open subsets. Otherwise it is called *disconnected*. A subspace of a topological space is *connected* (respectively, *disconnected*) if it is connected (respectively, disconnected) in the subspace topology. Thus $\mathbb{R}^n$ and $\mathbb{C}^n$ are connected as is any interval (a, b) or $[a, b]$ of $\mathbb{R}$. The subspaces $\mathbb{Q}$ and $(0, 1) \cup (2, 3)$ of $\mathbb{R}$ are clearly disconnected. A subspace $C \subset X$ is called a *component* of X if it is "maximally connected", that is, C is connected and if D is connected and $C \subset D$ then $C = D$. It can then be shown to follow that any component of a topological space $(X, \mathcal{T})$ is a closed (but not necessarily open) subset and that X can be written as a disjoint union of its components. If $(X, \mathcal{T})$ and $(Y, \mathcal{T}')$ are topological spaces with $(X, \mathcal{T})$ connected and with $f : X \to Y$ continuous then $f(X)$ is a connected subspace of $(Y, \mathcal{T}')$. Also the topological product of a finite number of non-empty topological spaces is connected if and only if the individual topological spaces are connected. Thus $\mathbb{R}^n$ and $\mathbb{C}^n$ are connected. There is another concept of connectedness for a topological space X. A *path* (or *curve*) in X is a continuous map c from some closed interval $[a, b]$ of $\mathbb{R}$ (with subspace topology from $\mathbb{R}$) to X. The points $c(a)$ and $c(b)$ are, respectively, the *initial* (or starting, or beginning) and *final* (or end) points of c and c is sometimes said to be *from $c(a)$ to $c(b)$*. A topological space is called *path-connected* if given any two distinct points $x, y \in X$ there exists a path c in X from x to y. As a consequence one may introduce the idea of a *path component* of X. Although it is easy to show that a path-connected topological space is necessarily connected, the converse is false. There is a special class of topological spaces for which connectedness and path-connectedness are equivalent. Suppose X is *locally path-connected*, that is, for each $x \in X$ and neighbourhood V of x there exists an open subset U in M, with $x \in U \subset V$, which is path-connected in its subspace topology (thus X admits a base of path connected subsets). Then for such a topological space connectedness is equivalent to path-connectedness and the decomposition of X into components described above is also a decomposition of X into path components and each of these components is open and closed in X. [It is remarked at this point that most of this text deals with manifolds and these will be introduced in the next chapter. A manifold will be seen to have a "natural" topology which is locally path-connected and hence, for manifolds, connectedness and path-connectedness are equivalent conditions.]

Let $p, q \in X$ and let c_1 and c_2 be paths from p to q so that, with a, b as above, $c_1(a) = c_2(a) = p$ and $c_1(b) = c_2(b) = q$. Then c_1 and c_2 are called *homotopic* if there exists a continuous (*homotopy*) map $G : [a, b] \times [a, b] \to X$ such that $G(t, a) = c_1(t)$, $G(t, b) = c_2(t)$, $G(a, t) = c_1(a) (= c_2(a))$ and $G(b, t) = c_1(b) (= c_2(b))$ for each $t \in [a, b]$. The relation of being homotopic for paths from p to q can be checked to be an equivalence relation. A path c is called *closed* at p if its initial and final points are equal (to p). A path c is called a *constant* or a *null* path at $x \in X$ if $c(t) = x$ for each t. Finally if every closed path in X is homotopic to a constant path (sometimes said to be *homotopic to zero*) X is called *simply-connected*. If each $p \in X$ admits an open neighbourhood U such that, with its induced topology, U is simply connected and X is called *locally simply connected*.

Now suppose that X and X' are Hausdorff, connected, locally simply connected and locally path connected (hence path connected) topological spaces and let $\pi' : X' \to X$ be continuous and surjective. Then X' is called a *covering space for X with covering map π'* if for each $p \in X$ there exists a connected open neighbourhood U of p such that each component of $\pi'^{-1}U$ is homeomorphic to U under the appropriate restriction of π'. It turns out that if X

is a topological space with the above restrictions, it must admit a covering space X' which is simply connected. Such a covering space X' for X is called a *universal covering space for X* and is unique in the sense that if X'' is another universal covering space for X with covering map π'' there exists a homeomorphism $f : X' \to X''$ such that $\pi'' \circ f = \pi'$.

Here it is convenient to make some remarks on what might be called the "rank theorems". Such theorems occur in many guises and a few can be described here. First, if $(X, \mathcal{T})$ is a topological space and $f : X \to \mathbb{R}$ is continuous and is such that $f(x) \neq 0$ for some $x \in X$, then $\exists$ an open neighbourhood U of x in X such that $f(y) \neq 0$ for each $y \in U$. Second let $M_n\mathbb{R}$ denote the set of all real $n \times n$ matrices This can be given a natural topology in the following way. The set $M_n\mathbb{R}$ may be put into a natural bijective correspondence with $\mathbb{R}^{n^2}$ by enumerating the entries of such a matrix $A = (a_{ij})$ as $a_{11}, ..., a_{1n}, a_{21}, ..., ...a_{nn}$. Then one can put the standard topology on $\mathbb{R}^{n^2}$ and hence on $M_n\mathbb{R}$. Then, for example, the determinant function $M_n\mathbb{R} \to \mathbb{R}$ is a continuous map and thus, from the above result, the subset $GL(n, \mathbb{R})$ of all non-singular members of $M_n\mathbb{R}$ is an open subset of $M_n\mathbb{R}$. More generally let X be a topological space and $f : X \to M_n\mathbb{R}$ a continuous map. If $x_0 \in X$ and $f(x_0)$ is a matrix of rank $p \leqslant n$ then there exists an open subset $U \subset X$ containing x_0 such that $f(x)$ has rank $\geq p$ for each $x \in U$. Another example, which is a corollary of this one, is that if X is a topological space and $f : X \to \mathbb{R}^n \times \cdots \times \mathbb{R}^n$ (m times) is a continuous map then if $x_0 \in X$ and $f(x_0)$ consists of m vectors in $\mathbb{R}^n$ spanning a p-dimensional subspace of $\mathbb{R}^n$ then there exists an open subset U of X containing x_0 such that for $x \in U$, the members of $f(x)$ span a subspace of $\mathbb{R}^n$ of dimension $\geq p$.

1.8   Euclidean Geometry

In this final section a brief digression will be made in order to introduce Euclidean geometry in a modern setting. Although not strictly needed it is a useful foil for what is to come. The original "Elements" of Euclid were given about 2,300 years ago and collected together the works of many people. Apart from giving rise to the study of geometry, it initiated the axiomatic method in mathematics encouraging the laying down of certain primitive (unquestioned) "features" of the study and then imposing conditions, also unquestioned, on them (axioms). One then proceeds strictly logically with no further input from intuition. Euclid did not always stick to his own rules but the clarity and power of his method laid the basis of modern mathematics. David Hilbert [8] joined in the spirit of Euclid and set down his procedures in a precise axiomatic form. It is worth discussing this briefly (without proofs) to appreciate its beauty and to point out the difference between Euclidean geometry and the differential geometry to be described later. Although Hilbert's original work was concerned with 3-dimensional geometry it is sufficient for present purposes to reduce this to 2-dimensions to simplify the situation.

Hilbert starts with three undefined (primitive) non-empty sets, $\mathcal{P}$ (thought of as "points"), $\mathcal{L}$ (thought of as "lines") and an *incidence relation* $\mathcal{I} \subset \mathcal{P} \times \mathcal{L}$ with $(p, L) \in \mathcal{I}$ thought of as "p is incident with L" or "p is on L". More will be needed later. If $(p, L) \in \mathcal{I}$ one writes $p \circ L$. If $p, q, r \in \mathcal{P}$ and $\exists L \in \mathcal{L}$ with $p \circ L$, $q \circ L$ and $r \circ L$, the set $\{p, q, r\}$ is called *collinear*. These are controlled by five groups of axioms: (1) axioms of incidence, (2) axioms of betweenness, (3) axioms of congruence (length and angle), (4) the completeness axiom and finally (5) the parallel axiom.

There are three *axioms of incidence* and which are fairly explicit.

$I(1)$ For each distinct pair $p, q \in \mathcal{P}, \exists$ a unique $L \in \mathcal{L}$ such that $p \circ L$ and $q \circ L$. One can thus write $L = pq$.

$I(2)$ For each $L \in \mathcal{L}, \exists$ at least two distinct $p, q \in \mathcal{P}$ with $p \circ L$ and $q \circ L$.

$I(3)$ There exists a subset $\{p, q, r\}$ of $\mathcal{P}$ which is not collinear.

It is easily checked that one can (and will) uniquely identify a member $L \in \mathcal{L}$ with those members of $\mathcal{P}$ incident with it.

Next there are four *axioms of betweenness* and they are based on another primitive set $B \subset \mathcal{P} \times \mathcal{P} \times \mathcal{P}$. If $(p, q, r) \in B$ one writes $p - q - r$ and says "q is between p and r".

$B(1)$ If $p - q - r$ then p, q, r are distinct, collinear members of $\mathcal{P}$ and $r - q - p$.

$B(2)$ given distinct $b, d \in \mathcal{P}, \exists a, c, e \in \mathcal{P}$ such that $a - b - d$, $b - c - d$ and $b - d - e$ hold.

$B(3)$ if $a, b, c \in \mathcal{P}$ are distinct and collinear then *exactly one* of these points is between the other two.

For the final axiom of betweenness some further definitions are needed. The *segment* $[a, b] \equiv \{p \in \mathcal{P} : p = a$ or $p = b$ or $a - p - b\}$. The points a, b are called its *endpoints*. The *ray* $\overrightarrow{ab} = [a, b] \cup \{p : a - b - p\}$ and this ray is said to *emanate* from a. If $a - b - c$ the ray $\overrightarrow{bc}$ is said to be *opposite* to the ray $\overrightarrow{ba}$. It then follows from the betweenness axioms that $\overrightarrow{ba}$ is opposite to $\overrightarrow{bc}$, that $[a, b] \subset \overrightarrow{ab}$, that $\overrightarrow{ab} \neq \overrightarrow{ba}$, that $\overrightarrow{ba} \neq \overrightarrow{bc}$, that $\overrightarrow{ab} \cap \overrightarrow{ba} = [ab]$ and that $\overrightarrow{ab} \cup \overrightarrow{ba} = ab$. Now let $L \in \mathcal{L}$ and let $a, b \in \mathcal{P}$ such that neither a nor b is on L. Then a, b are said to be *on the same side of* L if either $a = b$ or $[ab] \cap L = \emptyset$. Otherwise a, b are said to be *on opposite sides of* L. The final betweenness axiom comes in two parts and is given by

$B(4)$ (The *plane separation* axiom.) For any line L and points a, b, c *not* on L:

(a) if a, b are on the same side of L and b, c are on the same side of L then a, c are on the same side of L,

(b) if a, b are on opposite sides of L and b, c are on opposite sides of L then a, c are on the same side of L.

One can now define a *triangle* as follows: Let $a, b, c \in \mathcal{P}$ be non-collinear. Then the triangle $\triangle_{abc} \equiv [ab] \cup [bc] \cup [ca]$. One can then prove *Pasch's theorem* which is: *If $\triangle_{abc}$ is any triangle and L any line distinct from ab but intersecting $[ab]$ in a point d between a and b then L also intersects $[ac]$ or $[bc]$. If $c \notin L$ then L does not intersect both $[ac]$ and $[bc]$.*

In fact, given the axioms I(1)...I(3), B(1)...B(3), axiom B(4) is *equivalent to* the statement of Pasch's theorem and the latter is sometimes given as an alternative axiom to B(4). The betweenness axioms force plane separation onto the model and remove "circular" lines from it.

Next there are three *axioms of length congruence*:

$C(1)$ The collection of all segments admits an *equivalence relation*, denoted by $\sim$ and if $a, b, c, d \in \mathcal{P}$ and $[ab] \sim [cd]$ one says that $[ab]$ and $[cd]$ are *congruent*.

$C(2)$ If a, b are distinct members of $\mathcal{P}$ and if $a' \in \mathcal{P}$ then on any ray emanating from $a', \exists$ a unique $b' \in \mathcal{P}$ such that $[ab] \sim [a'b']$.

$C(3)$ If $a, b, c, a', b', c' \in \mathcal{P}$ with $a - b - c$ and $a' - b' - c'$ and if $[ab] \sim [a'b']$ and $[bc] \sim [b'c']$ then $[ac] \sim [a'c']$.

One can now introduce a concept of length into the geometry. Let $a, b, c, d \in \mathcal{P}$ and write $[ab] < [cd]$ if $\exists e \in \mathcal{P}$ such that $c - e - d$ and $[ab] \sim [ce]$. Then it can be shown that if $a, b, c, d, e, f \in \mathcal{P}$,

(i) exactly one of the following holds; $[ab] < [cd]$, $[ab] \sim [cd]$ or $[cd] < [ab]$,

(ii) $[ab] < [cd]$ and $[cd] \sim [ef]$ imply $[ab] < [ef]$,

(iii) $[ab] < [cd]$ and $[ab] \sim [ef]$ imply $[ef] < [cd]$,

(iv) $[ab] < [cd]$ and $[cd] < [ef]$ imply $[ab] < [ef]$.

Thus one can introduce the idea of a *free segment* as an *equivalence class* of segments as given above. If A and B are free segments, one writes $A < B$ if $\exists [ab] \in A$ and $[cd] \in B$ such that $[ab] < [cd]$ (which makes sense by (ii) and (iii) above). Also if A, B, C are free segments one can define an *addition* on them by writing $A + B = C$ if $\exists [cd] \in C$ and $e \in \mathcal{P}$ with $c - e - d$ and $[ce] \in A$ and $[ed] \in B$. A *multiplication* on free segments can be defined, inductively, by $1A = A$ and $nA = (n-1)A + A$ $(n \in \mathbb{N})$. From this one can show that if $[ab]$ is a segment $\exists$ a unique $c \in \mathcal{P}$ such that $a - c - b$ and $[ac] = [cb]$. Then c is the *midpoint* of $[ab]$ and so for any free segment $A, \exists$ a free segment B such that $A = B + B$ and so one may write $B = \frac{1}{2}A$. Thus given a segment one may "multiply" it by numbers of the form $\frac{m}{2^n}$, with m, n in the set of positive integers $\mathbb{N}$. Thus a concept of length requires only the choice of a "unit segment". The numbers of the form $\frac{m}{2^n}$ with $m, n \in \mathbb{N}$ are called *dyadic numbers* and satisfy the condition that they constitute a subset of $\mathbb{Q}$ and are denumerable.

One can now introduce (briefly) angles and congruence axioms for angles. Let $a, b, c \in \mathcal{P}$ be distinct and giving *non-opposite* rays $\vec{ab}$ and $\vec{ac}$ emanating from a. An *angle* is such a point a together with the rays $\vec{ab}$ and $\vec{ac}$ and it is denoted by $\widehat{bac}$ or $\widehat{cab}$ (showing that one intends the order in which the rays are given to be irrelevant).

There are three congruence axioms for angles.

$C(4)$ The collection of all angles admits an equivalence relation, also denoted by $\sim$. If $\widehat{abc} \sim \widehat{xyz}$ say that $\widehat{abc}$ and $\widehat{xyz}$ are *congruent*.

$C(5)$ Given any angle $\widehat{abc}$, any $b' \in \mathcal{P}$ and any ray $\vec{b'a'}$ emanating from b', $\exists$ a unique ray $\vec{b'c'}$ emanating from b' *on a given side of* the line $a'b'$ (that is the points $\vec{b'c'} \setminus \{b'\}$ are on that side of the line $a'b'$) such that $\widehat{abc} \sim \widehat{a'b'c'}$.

$C(6)$ Let $\triangle_{abc}$ and $\triangle_{a'b'c'}$ be any two triangles with $[ab] \sim [a'b']$, $[ac] \sim [a'c']$ and $\widehat{bac} \sim \widehat{b'a'c'}$. Then $[bc] \sim [b'c']$, $\widehat{abc} \sim \widehat{a'b'c'}$ and $\widehat{acb} \sim \widehat{a'c'b'}$.

One can then define a concept of addition for angles, the idea of a *free angle* as an equivalence class under the relation $\sim$ and a *measure* of free angles.

The next axiom concerns a type of "completeness" of the lines in $\mathcal{L}$, that is, the requirement that such lines have "no points missing". For $L \in \mathcal{L}$ and distinct $o, a \in L$ consider the ray $\vec{oa}$ and let $p, q \in \vec{oa}$. Say $p \leq q$ if $[op] \leq [oq]$ and define $o < r, \forall r \in \vec{oa} \setminus \{o\}$. The *Completeness Axiom* states that for any ray $\vec{oc}$ on any line oc, if $\vec{oc} \equiv A \cup B$ with $A \neq \emptyset$, $B \neq \emptyset$, $A \cap B = \emptyset$ and for each $p \in A$ and $q \in B$, $p < q$, there exists $x \in \vec{oc}$ such that if $a \in A$, and $b \in B$ and $a \neq x \neq b$ then $a < x < b$. The point x is necessarily unique. It can now be shown that each $L \in \mathcal{L}$ is essentially a copy of the real line.

The axioms laid down so far are $I(1),...I(3)$, $B(1),...B(4)$, $C(1),...C(6)$ and the completeness axiom. They define what is sometimes called *Neutral or Absolute (Plane) Geometry*. It can be shown from these axioms that if $p \in \mathcal{P}$ and if $L \in \mathcal{L}$ with p not on L, *there exists* a line $L' \in \mathcal{L}$ such that p is on L' and $L \cap L' = \emptyset$, that is, L and L' are *parallel*. The final axiom is the so-called *parallel axiom*. It states that if $p \in \mathcal{P}$ and if $L \in \mathcal{L}$ with p not on L, there exists a *unique* line $L' \in \mathcal{L}$ such that p is on L' and L and L' are *parallel*.

With these axioms it may be shown that one may define distance (up to a unit of length) between two members of $\mathcal{P}$ and also a measure of angle (again up to units) and, in particular, the concept of a right angle. Then one may set up Cartesian-type coordinates on $\mathcal{P}$ to put it into a bijective correspondence with the set $\mathbb{R}^2$, the set $\mathcal{L}$ becoming the usual collection of straight lines in $\mathbb{R}^2$. Thus this model is just the "usual" Euclidean plane, and is the only model which satisfies these axioms, that is, this axiom system is *categorical*. [If one does not choose the parallel axiom as above it, in fact, follows that there are infinitely many choices for the line L' through p parallel to L and this gives rise to another single model which is the (non-Euclidean, plane) geometry of Lobachevsky and Bolyai.]

A salient point here is that Euclidean geometry is "homogeneous" in that it is "the same" everywhere. [One can, with a little effort, express this in terms of global bijective "symmetry" maps on $\mathcal{P}$, a flavour of which is given in chapter 6]. It also allows "movement without change" within $\mathcal{P}$ in that one may move regions of $\mathcal{P}$ about without change of "shape" or "size" (rigid motions) and, using the concept of parallel lines, make statements like "this direction at p is the same as that direction at q" for any $p, q \in \mathcal{P}$ (parallel displacement). Further the whole of $\mathcal{P}$ can also be described in the single coordinate system $\mathbb{R}^2$. In the study of more general geometries none of these luxuries is necessarily present. In these latter structures the geometry is allowed to "change" from region to region (given, that is, that "change" makes sense) and may not be coordinatisable, globally, as Euclid's is. Measurement of length will turn out to require a geometrical structure called a metric to be postulated at the outset and preservation of "direction" requires a structure called a connection. The "natural" idea of "moving" a segment of a line in Euclid's geometry "without changing its direction or its length" suggests imposing the structure of a *compatible* metric and connection pair. The concepts of metric, connection and compatible metric-connection pair will be very important in what is to follow and have a natural setting in a mathematical structure called a manifold. Such objects are covered in the next two chapters.

Chapter 2

Manifold Theory

2.1 Manifolds

A manifold M is a mathematical construction which endows a set $M \neq \emptyset$ with local coordinates, that is, each point of M is contained in a subset of M which "looks like" $\mathbb{R}^n$ for some $n \in \mathbb{N}$, that is, the subset looks like a local coordinate system. There is no requirement that the whole of M should look like $\mathbb{R}^n$, that is, this feature is local. In addition, should two of these coordinate systems overlap and hence lead to a coordinate change on their intersection, some degree of differentiability is imposed on this coordinate change. Some authors prefer to start by declaring M to be a certain type of topological space and then proceed to impose the local coordinate structure. This has the advantage that (topological) continuity is available from the outset but the disadvantage that more topological conditions may be assumed than is necessary. In this book M will be initially assumed to be nothing more than a bare set upon which only the important features of the local coordinates are imposed together with the differentiability (in $\mathbb{R}^n$) requirements mentioned above. A natural topology will then be shown to arise from this. Thus some knowledge of calculus on $\mathbb{R}^n$ will be assumed with, of course, the standard topology on $\mathbb{R}^n$ being given. As mentioned earlier only a brief resume of the subject is given but, it is hoped, with all the salient points included and definitions given since manifold theory is discussed in many books of which [9, 10, 11] are recommended and [12, 13, 14, 15] give useful summaries.

Starting from the non-empty set M a bijective map x from some subset $U \subset M$ onto an open subset of $\mathbb{R}^n$ is called an *(n-dimensional) chart* of M (and thus M is an infinite set). The projection maps $p_i : \mathbb{R}^n \to \mathbb{R}$ then allow the *ith coordinate functions* $x^i = p_i \circ x : U \to \mathbb{R}$ to be defined. The set U is then called the *chart* or *coordinate domain* of x and if $p \in U$ the $n-$tuple $(x^1(p), ..., x^n(p))$ is referred to as the coordinates of p in the chart U. The set U is also referred to as a *coordinate neighbourhood* of any of its points. A chart is called *global* if its domain is M. A collection A of charts of M whose domains form a covering of M is called a *(smooth) atlas* for M if for any two charts x and y in A with respective domains U and V such that $U \cap V \neq \emptyset$, $x(U \cap V)$ and $y(U \cap V)$ are open subsets of $\mathbb{R}^n$ and the bijective map $y \circ x^{-1} : x(U \cap V) \to y(U \cap V)$ and its inverse $x \circ y^{-1}$ are *smooth* (C^∞) maps (*coordinate transformations*) between open subsets of $\mathbb{R}^n$. Thus a smooth atlas gives a local coordinatisation on the whole of M together with the associated smooth coordinate transformations. Extra charts (and their domains) could be added, of course, but only subject to the above restrictions so that the original and added charts together still constitute a smooth atlas for M and two smooth atlases are called *equivalent* if their union (in an obvious sense) is also a smooth atlas for M. One could thus define a *complete (smooth) atlas* for M as one which is not properly contained in any other smooth atlas and this latter atlas is then unique. Any smooth atlas is then said to determine a *smooth, n-dimensional structure* on M and with this structure M is called an *n-dimensional smooth manifold* and one writes $\dim M = n$.

Some useful examples of manifolds are now given.

(*i*) Consider the set $\mathbb{R}^n$. Here the identity map immediately gives a global chart (the *identity chart*) for the *standard n-dimensional smooth manifold structure on* $\mathbb{R}^n$.

(*ii*) Let V be an n-dimensional real vector space and let $\{\mathbf{e}_i\}$ $(1 \leqslant i \leqslant n)$ be a basis for V. Any $\mathbf{v} \in V$ may be written uniquely as a linear combination of the members of $\{\mathbf{e}_i\}$ with components v_i. The map $x : \mathbf{v} \to (v_1, ..., v_n)$ is a global chart for V and hence V becomes an n-dimensional smooth manifold. If a different basis is used, the work in chapter 1 shows that the coordinate transformations resulting from any two bases are smooth and hence different bases give rise to equivalent atlases.

(*iii*) The set $M_n\mathbb{R}$ can be given an n^2-dimensional manifold structure by constructing an n^2-dimensional global chart x whose action maps $A = (a_{ij}) \in M_n\mathbb{R}$ to the n^2-tuple $(a_{11}, ..., a_{1n}, a_{21}, ..., a_{2n}, ..., a_{nn})$. Similarly the subsets $S(n, \mathbb{R})$ (symmetric real matrices) and $Sk(n, \mathbb{R})$ (skew-symmetric real matrices) of $M_n\mathbb{R}$ can, by using the global charts $x : (a_{ij}) \to (a_{11}, ..., a_{1n}, a_{22}, ..., a_{2n}, ..., a_{nn})$ and $y : (a_{ij}) \to (a_{12}, ..., a_{1n}, a_{23}, ..., a_{2n}, ...,$ $a_{(n-1)n})$, be given the structure of $\frac{1}{2}n(n+1)$- and $\frac{1}{2}n(n-1)$-dimensional smooth manifolds, respectively.

(*iv*) If M_1 and M_2 are smooth manifolds of dimensions n_1 and n_2, respectively, the set $M_1 \times M_2$ can be given the structure of a smooth $(n_1 + n_2)$-dimensional manifold by constructing an atlas of charts on $M_1 \times M_2$ consisting of maps of the form $x_1 \times x_2$ where x_1 and x_2 are charts belonging to atlases for M_1 and M_2 with domains U_1 and U_2, respectively, where $x_1 \times x_2 : U_1 \times U_2 \to \mathbb{R}^{(n_1+n_2)}$ is the map $(p, q) \to (x_1(p), x_2(q))$. The smoothness of the resulting coordinate transformations is easily checked and the resulting manifold $M_1 \times M_2$ is called the *manifold product* of M_1 and M_2. This construction easily extends to finitely many manifolds.

(*v*) The subset S^n of $\mathbb{R}^{n+1}$ $(n \geq 1)$ may also be shown to be an n-dimensional smooth manifold. This manifold does not possess a global chart for topological reasons.

Let M and M' be smooth manifolds with $\dim M = n$ and $\dim M' = n'$ and let $f : M \to M'$ be a map. One first needs to make sense of the differentiability or otherwise of f. Suppose $p \in M$ and choose charts x in M and x' in M' whose domains contain p and $f(p)$, respectively. Then the function $F \equiv x' \circ f \circ x^{-1}$, called the *coordinate representative of f with respect to the charts x and x'*, has domain an open subset of $\mathbb{R}^n$ and range in $\mathbb{R}^{n'}$. Thus one has succeeded in representing f by a function F on some open subset of $\mathbb{R}^n$ by use of the chart maps on M and M'. It may be thought that a degree of differentiability for f may be defined by assuming it to be that of F (since such a notion is well-defined for F on open subsets of $\mathbb{R}^n$) but it must be checked that this is independent of the charts x and x' chosen. This is easily verified by noting that, if different charts y and y' are chosen containing p and $f(p)$, respectively, the coordinate representative for f is now $F' \equiv y' \circ f \circ y^{-1}$. But then F' restricts to the map $(y' \circ x'^{-1}) \circ F \circ (x \circ y^{-1})$ on the obvious intersection of domains and which is an open subset of $\mathbb{R}^n$. Consider the assumption that F is smooth. Then so is F' and one may unambiguously define f to be *smooth* if, say, F is, since the coordinate transformations are smooth . The coordinate functions x^i are then, from this approach, easily seen to be smooth functions from their coordinate domains to $\mathbb{R}$. It is noted that if a function f is not defined on the whole of M it is assumed that it is defined at least on some chart domain of M and so the representative F is defined on some open subset of $\mathbb{R}^n$ from which one may make sense of the smoothness of F and hence that of f on its domain. If $f : M \to M'$ is smooth and M'' is another smooth manifold with $g : M' \to M''$ smooth then the map $g \circ f : M \to M''$ is smooth. If M and M' are smooth manifolds of dimension n and n', respectively, and $f : M \to M'$ is a bijective map such that f and f^{-1} are smooth then f and f^{-1} are called *diffeomorphisms* and M and M' are said to be *diffeomorphic*. Here, since the coordinate representatives of f and f^{-1} give rise to smooth, bijective maps between open subsets of $\mathbb{R}^n$ and $\mathbb{R}^{n'}$, these representatives are continuous maps (since continuity makes sense for maps between these open sets). Thus these open subsets of $\mathbb{R}^n$ and $\mathbb{R}'^n$ are homeomorphic and this can be shown to imply $n = n'$. So if M and M' are diffeomorphic, $\dim M = \dim M'$. [Here it is remarked, first, that one requires the smoothness of *both* f and f^{-1} and second that the concept of continuity for maps *between manifolds* has not yet been defined since no topology for a manifold has yet been specified.]

2.2 The Manifold Topology

Let M be a smooth manifold with $\dim M = n$ and let $\mathcal{B}$ be the set of *all* chart domains of a *complete* atlas for M. Then if x is a chart of M with domain U and if $V \subset U$ is such that $x(V)$ is open in $\mathbb{R}^n$ then the restriction of x to V is also a chart of M with domain V. Also if x and y are charts of M

with domains U and V, respectively, and with $U \cap V \neq \emptyset$, then $x(U \cap V)$ is an open subset of $\mathbb{R}^n$ and so the intersection of two chart domains of M is a chart domain of M. It now follows that $\mathcal{B}$ is a *base for a topology on M* and this topology is called the *manifold topology* for M. Thus any chart x of M with domain U is then a homeomorphism from U to an open subset of $\mathbb{R}^n$ where each has subspace topology in an obvious way and then for manifolds M and M' a smooth map $f : M \to M'$ is now necessarily continuous with respect to the manifold topologies. The same applies if f is defined only on some open subset of M. In example (iv) above the manifold topology on $M_1 \times M_2$ is now easily seen to coincide with the product topology from the manifold topologies on M_1 and M_2 and the projection maps are seen to be smooth and continuous.

It may be expected that, since a manifold is "locally like" $\mathbb{R}^n$, some topological properties of a manifold arise directly from the topological properties of $\mathbb{R}^n$ (but some do not). Thus it can be shown that (i) a manifold is first countable but is second countable if and only if it admits a countable atlas, (ii) a manifold is not necessarily Hausdorff, (iii) a manifold is locally path-connected and is hence connected if and only if it is path-connected (and note that a path here is a continuous map in the manifold topology [see chapter 1]) and (iv) every component of a manifold M is open and closed in M. The first of these points reveals that the discussions of limit points, convergence and continuity can be achieved using sequences and does not require the use of nets or filters (see, for example, [3]). The third point requires a little explanation before proceeding. Path-connectedness has been defined in terms of *continuous* paths between points. However, it is true that, for a manifold, this definition is equivalent to a similar definition in terms of *smooth* paths. More precisely, for $p, q \in M$, a (continuous) path between p and q has been defined as a continuous map $c : [a, b] \to M$ with $a, b \in \mathbb{R}$ and $a < b$, $c(a) = p$ and $c(b) = q$. A smooth path from p to q is a smooth map $c : I \to M$ where I is an open interval of $\mathbb{R}$ and where $\exists a, b \in I$ with $a < b$, $c(a) = p$ and $c(b) = q$. A *piecewise-smooth path* in M from p to q is then a map c' on $[a, b]$ into M with $a, b \in \mathbb{R}$ and $a < b$ and with $c'(a) = p$ and $c'(b) = q$ such that one may divide $[a, b]$ into finitely many closed (sub)intervals $[a, s_1], [s_1, s_2]...$ $[s_{m-1}, b]$ with $a < s_1 < ... < s_{m-1} < b$ on each one of which c' agrees with a smooth map from an open interval in $\mathbb{R}$ containing that subinterval. It is then easily shown that a piecewise-smooth path is continuous and for $p, q \in M$ the relation $p \sim q \Leftrightarrow$ there exists a piecewise-smooth path from p to q is an equivalence relation on M whose equivalence classes are open and closed in M and hence, if M is connected in the manifold topology, there is only one equivalence class, equal to M. Thus if one can find a continuous path between any two points of M (that is, if M is path-connected) then M is connected (chapter 1) and one can also find a piecewise-smooth path between them. The converse to this also follows and so one may state the concept of path connectedness in terms of either continuous paths or piecewise-smooth paths. But since M is a manifold, its system of charts shows that it is locally path-connected and hence connectedness and path-connectedness are equivalent statements for M (see chapter 1). Finally it can be shown that if one can find a piecewise-smooth path between p and q one can also find a smooth path from p to q (and clearly conversely). Thus, for a manifold, (topological) path-connectedness can be equivalently defined in terms of smooth paths. This means that a "natural manifold" approach to connectedness using smooth paths is sufficient. This is especially useful in the applications of manifold theory to general relativity theory where an observer's link to the rest of the universe and hence to physical observations is (usually) interpreted in terms of information transmitted along certain types of paths.

In this book, *all manifolds will henceforth be assumed smooth, connected, Hausdorff and second countable.* The assumption of connectedness will be seen to be convenient and natural whilst the Hausdorff and second countable restrictions are made for technical reasons.

2.3 Vectors, Tensors and Their Associated Bundles

Let M be an n-dimensional manifold. At each $p \in M$ there is a certain finite-dimensional vector space of fundamental importance. This will now be defined, and from it the concepts of vectors and tensors at p will follow. However, for a real-valued smooth map g whose domain is some open subset of M (and using the identity chart for $\mathbb{R}$) one must first define a derivative of g. Now let x be a chart of M whose domain U includes p and is contained in the domain of g and let G be the representative for g in this chart so that, using the identity chart in $\mathbb{R}$, $g = G \circ x$. Then G is a smooth map from some open subset of $\mathbb{R}^n$ to $\mathbb{R}$. So one can define the functions $g_{,a}$ (sometimes written $\frac{\partial g}{\partial x^a}$ and called the *partial derivatives* of g in this chart) by

$$g_{,a} \equiv \frac{\partial G}{\partial x^a} \circ x \qquad (1 \leqslant a \leqslant n). \tag{2.1}$$

Now let $F(p)$ denote the family of all smooth real-valued functions whose (open) domains each include p. A *derivation* on M is a map $L : F(p) \to \mathbb{R}$ satisfying the two conditions

$$L(af + bg) = aL(f) + bL(g), \qquad L(fg) = fL(g) + gL(f), \tag{2.2}$$

where $a, b \in \mathbb{R}$, $f, g \in F(p)$ and with the sum and product of members of $F(p)$ defined on the obvious (non-empty and open) domain intersections. Now in the above chart the n maps $(\frac{\partial}{\partial x^a})(p) : F(p) \to \mathbb{R}$ given by $g \to g_{,a}(p)$ are then derivations on $F(p)$. The family of all derivations on $F(p)$ can be shown to be an n-dimensional real vector space for which the $(\frac{\partial}{\partial x^a})(p)$ form a basis and this vector space (attached to $p \in M$) is called the *tangent space* to M at p, denoted by $T_p M$. Its members are called *tangent vectors* (or just *vectors*) at p and could be thought of as "vector arrows" at p. For $0 \neq \mathbf{v} \in T_pM$ the 1-dimensional subspace of T_pM spanned by $\mathbf{v}$ is called the *direction determined* (or *spanned*) by $\mathbf{v}$. Thus if $\mathbf{v} \in T_pM$ one may write in this chart $\mathbf{v} = \sum_{a=1}^{n} v^a \frac{\partial}{\partial x^a}(p)$, $(v^a \in \mathbb{R})$ and $\mathbf{v}(g)$ is interpreted as the "directional derivative of g in the direction $\mathbf{v}$". The v^a are the *components* of $\mathbf{v}$ at p in the chart x and are uniquely determined by $\mathbf{v}$, p and x. Thus regarding $\mathbf{v}$ as a derivation at p and since the coordinate functions are in $F(p)$, $\mathbf{v}(x^a) = v^a$. If y is another chart of M at p with domain V containing p one has $\mathbf{v} = \sum_{a=1}^{n} v'^a \frac{\partial}{\partial y^a}(p)$ where the $v'^a \in \mathbb{R}$ are the components of $\mathbf{v}$ in the chart y. From this one may easily calculate the relationship between the v^a and the v'^a as

$$v'^a = \mathbf{v}(y^a) = \sum_{b=1}^{n} v^b \left(\frac{\partial}{\partial x^b}\right)_p (y^a) = \sum_{b=1}^{n} \left(\frac{\partial y^a}{\partial x^b}\right)_p v^b \tag{2.3}$$

which gives the transformation law for the components of $\mathbf{v}$ in the charts x (v^a) and y (v'^a).

The collection of all tangent vectors at all points of M is denoted by TM, called the *tangent bundle* of M and is defined by $TM \equiv \bigcup_{p \in M} T_pM$. This leads to the useful map $\pi : TM \to M$ defined by $\pi(\mathbf{v}) = p$ if $\mathbf{v} \in T_pM$ and which attaches tangent vectors in TM to the point of M from whence they came and so $\pi^{-1}\{p\} = T_pM$. The set TM can be given a manifold structure by noting that TM is the union of sets of the form $\pi^{-1}U$ where U is some chart domain of M with chart x. The write each $\mathbf{v} \in \pi^{-1}U$ as $\mathbf{v} = \sum_{a=1}^{n} v^a (\frac{\partial}{\partial x^a})_p$ where $\pi(\mathbf{v}) = p$ to get an injective map $\pi^{-1}U \to \mathbb{R}^{2n}$ given by $\mathbf{v} \to (x^1(p), ..., x^n(p), v^1, ..., v^n)$ whose range is the open subset of $x(U) \times \mathbb{R}^n$ of $\mathbb{R}^{2n}$. This gives a chart for TM for each chart of M and the collection of all such charts is easily shown to be an atlas for TM giving the latter a smooth manifold structure of dimension $2n$ and for which π is a smooth map $TM \to M$. The collection of the zero vectors in T_pM for each $p \in M$ is a subset of TM called the *zero section* of TM.

For manifolds M and M' and with $p \in M$ and $p' \in M'$ consider the product manifold $M \times M'$ and the point $(p, p') \in M \times M'$. The tangent space to $M \times M'$ at (p, p') can, as intuitively expected, be shown to be isomorphic, as a vector space, to the vector space sum of T_pM and $T_{p'}M'$ [9].

Now consider the dual space $\overset{*}{T}_pM$ of T_pM and let x be a chart of M whose domain contains p so that $\{\frac{\partial}{\partial x^a}(p)\}$ is a basis for T_pM. Let $\{dx^a\}_p$ denote the corresponding dual basis in $\overset{*}{T}_pM$ so that $(dx^a)_p(\frac{\partial}{\partial x^b})_p = \delta_b^a$. Then each $\mathbf{w} \in \overset{*}{T}_pM$ may be written as $\mathbf{w} = \sum_{a=1}^{n} w_a (dx^a)_p$ where the $w_a \in \mathbb{R}$ are the components of $\mathbf{w}$ in the basis $(dx^a)(p)$ of the chart x. The real n-dimensional vector space $\overset{*}{T}_pM$ is called the *cotangent space* to M at p. Its members are called *cotangent vectors* (or just *covectors* or *1-forms*) at p. One can similarly construct the *cotangent bundle* $\overset{*}{T}M \equiv \bigcup_{p \in M} \overset{*}{T}_pM$ which may be given the structure of a $2n$-dimensional smooth manifold with obvious smooth projection $\overset{*}{T}M \to M$ also denoted by π. Just as for tangent vectors, if x and y are charts whose domains contain p and $\mathbf{w} \in \overset{*}{T}_pM$, then one has, at p, $\mathbf{w} = \sum_{a=1}^{n} w_a(dx^a)_p = \sum_{a=1}^{n} w'_a(dy^a)_p$. Now from the previous part one has $(\frac{\partial}{\partial x^a})_p = A_a^b(\frac{\partial}{\partial y^b})_p$ $(A_a^b \in \mathbb{R})$ and applying to the functions y^b one sees that $(\frac{\partial}{\partial x^a})_p = (\frac{\partial y^b}{\partial x^a})_p(\frac{\partial}{\partial y^b})_p$. Then one derives the transformation law for the components of $\mathbf{w}$ with respect to the charts x and y as

$$w_a = \mathbf{w}(\frac{\partial}{\partial x^a})_p = \sum_{b=1}^{n}(\frac{\partial y^b}{\partial x^a})_p w'_b. \tag{2.4}$$

Members of T_pM are sometimes called *contravariant vectors* and members of $\overset{*}{T}_pM$ *covariant vectors*. Although T_pM and $\overset{*}{T}_pM$ are each isomorphic to $\mathbb{R}^n$ as vector spaces, there is no natural isomorphism between them unless some other structure on M can provide it. This will appear later in the form of a metric.

The above notions may be generalised by considering the vector space of all multilinear maps on $V \equiv T_pM + \cdots + T_pM + \overset{*}{T}_pM + \cdots + \overset{*}{T}_pM$ with s copies of T_pM and r copies of $\overset{*}{T}_pM$ for non-negative integers r, s. This real vector space of dimension n^{r+s} is called the *vector space of tensors of type* (r, s) at p, and is denoted by $T_s^r M_p$ where the slight deviation from the above terminology is for notational ease. Such tensors are said to be of *order* $r + s$. One then proceeds as above. The tensor space of type $(0,1)$ corresponds to $\overset{*}{T}_pM$ whereas type $(1,0)$ corresponds to the dual of $\overset{*}{T}_pM$ which is, in a natural way, T_pM under the isomorphism which associates with $(\frac{\partial}{\partial x^a})_p$ a member $\mathbf{e_a}$ in the dual of $\overset{*}{T}_pM$ satisfying, at p, $\mathbf{e_a}(dx^b)_p = (dx^b)_p(\frac{\partial}{\partial x^a})_p = \delta_b^a$ (see section 1.4). One normally also uses the symbol $(\frac{\partial}{\partial x^a})_p$ for $\mathbf{e_a}$ giving $(\frac{\partial}{\partial x^a})_p(dx^b)_p = \delta_b^a$, at p. One can, just as before, form the *bundle of tensors of type* (r, s) on M, denoted by $T_s^r M$ and this can be given the structure of an $(n^{r+s} + n)-$ dimensional smooth manifold by following procedures similar to those for the tangent and cotangent bundles and then the projection map onto M (denoted also by π) is smooth. If $t \in T_s^r M_p$ it may be written in the above basis as

$$t = \sum_{a_1,\ldots,b_s=1}^{n} t_{b_1\ldots b_s}^{a_1\ldots a_r}(dx^{b_1})_p \otimes \cdots \otimes (dx^{b_s})_p \otimes (\frac{\partial}{\partial x^{a_1}})_p \otimes \cdots \otimes (\frac{\partial}{\partial x^{a_r}})_p \tag{2.5}$$

where the $t_{b_1\ldots b_s}^{a_1\ldots a_r}$ are the components of t in the above chart x and are given by

$$t_{b_1\ldots b_s}^{a_1\ldots a_r} = t((\frac{\partial}{\partial x^{b_1}})_p, \cdots, (\frac{\partial}{\partial x^{b_s}})_p, (dx^{a_1})_p, \cdots, (dx^{a_r})_p). \tag{2.6}$$

For any other chart y, the coordinates of t are $t'^{c_1 \ldots c_r}_{d_1 \ldots d_s}$ and are given by

$$t'^{c_1 \ldots c_r}_{d_1 \ldots d_s} = \sum_{a_1, \ldots, b_s = 1}^{n} (\frac{\partial y^{c_1}}{\partial x^{a_1}})_p \cdots (\frac{\partial y^{c_r}}{\partial x^{a_r}})_p (\frac{\partial x^{b_1}}{\partial y^{d_1}})_p \cdots (\frac{\partial x^{b_s}}{\partial y^{d_s}})_p t^{a_1 \ldots a_r}_{b_1 \ldots b_s}. \qquad (2.7)$$

Addition of tensors of the same type is carried out within their vector space structure by adding their associated components given in the same coordinate system. Scalar multiplication of a tensor by a real number is similarly done. Sometimes an index in the upper position is referred to as a *contravariant* index and one in the lower position as a *covariant* index.

A tensor T at $p \in M$ of type $(0, 2)$ (of order 2) is called *symmetric* (respectively, *skew-symmetric*) if for each $\mathbf{u}, \mathbf{v} \in T_pM$, $T(\mathbf{u}, \mathbf{v}) = T(\mathbf{v}, \mathbf{u})$ (respectively, $T(\mathbf{u}, \mathbf{v}) = -T(\mathbf{v}, \mathbf{u})$). Thus, in any chart domain, $T_{ab} = T_{ba}$ (respectively, $T_{ab} = -T_{ba}$). Similar comments apply to tensors of type $(2, 0)$.

2.4   Vector and Tensor Fields

A *vector field* on M is the attachment to each point of M of a member of T_pM, that is, it is a map $X : M \to TM$ such that $X(p) \in T_pM$ for each $p \in M$. Such a map is called a *section* of TM. One usually requires such vector fields to be smooth and so a *smooth vector field* on M is a vector field such that the map $X : M \to TM$, which is a map between manifolds, is a smooth section of TM. It is easily checked that this is equivalent to the statement that for each $p \in M$ and chart x with chart domain U containing p the components of X in this chart are smooth functions $U \to \mathbb{R}$. Another equivalent statement is that if $V \subset M$ is an open subset and $f : V \to \mathbb{R}$ is smooth, the function $Xf : V \to \mathbb{R}$ given for $p \in V$ by $Xf(p) = X(p)(f)$ is smooth. A vector field defined on the whole of M is called *global*. However a vector field may only be defined on some open subset of M and the above definitions of smoothness are easily modified in this case. It follows that the *coordinate vector fields* $\frac{\partial}{\partial x^a}$ defined on the chart domain U by $\frac{\partial}{\partial x^a}(p) = (\frac{\partial}{\partial x^a})_p$ ($p \in U$) are smooth vector fields (on U). Recalling the above remarks about addition and scalar multiplication in T_pM one can add global, smooth vector fields together and scalar multiply them by real numbers so that the set of all global, smooth vector fields becomes a vector space over $\mathbb{R}$. The vector space of such vector fields is denoted by $\mathcal{F}$. One can similarly draw these conclusions about smooth vector fields on a given open subset $U \subset M$. It is remarked that independent members of $\mathcal{F}$, when evaluated at $p \in M$, do not necessarily give independent members of T_pM (for example the vector fields on the manifold $\mathbb{R}^2$ with components $X = (x, 1)$ and $Y = (1, y)$ in the usual global chart with coordinates x, y on $\mathbb{R}^2$ are independent in $\mathcal{F}$ but give dependent vectors in $T_{(1,1)}\mathbb{R}^2$.

The concept of a *smooth covector field* (or *smooth 1-form field*) on M or on some open subset $U \subset M$, that is, a smooth section of the cotangent bundle $\overset{*}{T}M$ restricted to U, can be defined in a similar way to that for vector fields. Thus a 1-form (covector) field w on U is smooth if and only if for any open subset $V \subset U$ and any smooth vector field X defined on V the map $w(X) : V \to \mathbb{R}$ given by $p \to w(p)(X(p))$ $(p \in V)$ is smooth. Then this process is easily extended to *smooth tensor fields* (smooth sections of the tensor bundles $T_s^r M$) either on M or restricted to some open subset of M with the idea of smooth components in each chart of M being the most useful one for this book. Tensor fields defined on the whole of M are called *global*. One remark may be useful here. Let $f : M \to \mathbb{R}$ be smooth and construct a smooth covector field df on M (not to be confused with another use of the symbol "d" above) by its action on any smooth vector field on M (or some open subset of M) and defined by $df(X) = X(f)$. Then on some chart x with domain U one has $df = w_a dx^a$ on U and then an application of this to the smooth vector fields $\frac{\partial}{\partial x^a}$ on U reveals that $w_a = \frac{\partial f}{\partial x^a}$ and so, on U, $df = (\frac{\partial f}{\partial x^a})dx^a$. Thus one sees the origin of the (traditional) use of the symbol "d" earlier.

If t and t' are two smooth tensor fields of type (r, s) and (p, q), respectively, on M or on some open subset $U \subset M$ one may define their *tensor product* $t \otimes t'$ as that (smooth) tensor field of type $(r + p, s + q)$ on M (or U) such that its (smooth) components in any chart x are given by

$$(t \otimes t')^{a_1 \ldots a_{r+p}}_{b_1 \ldots b_{s+q}} = t^{a_1 \ldots a_r}_{b_1 \ldots b_s} t'^{a_{r+1} \ldots a_{r+p}}_{b_{s+1} \ldots b_{s+q}}. \tag{2.8}$$

If t is a smooth type, (r, s) $(rs \neq 0)$ tensor on M (or some open subset $U \subset M$) the *contraction* of t over the indices a_p and b_q is the type $(r - 1, s - 1)$ tensor $\bar{t}$ with components

$$\bar{t}^{a_1 \ldots a_{p-1} a_{p+1} \ldots a_r}_{b_1 \ldots b_{q-1} b_{q+1} \ldots b_s} = \sum_{k=1}^{n} t^{a_1 \ldots a_{p-1} k a_{p+1} \ldots a_r}_{b_1 \ldots b_{q-1} k b_{q+1} \ldots b_s}. \tag{2.9}$$

One can similarly contract one tensor field t with another t' over a pair of specified indices by forming the tensor product $t \otimes t'$ and contacting according to (2.9) over those indices. A smooth, real-valued function on M is, in this sense, sometimes regarded as a smooth tensor on M of type $(0, 0)$.

Given smooth vector fields X and Y on M (or some open subset $U \subset M$) one may construct another smooth vector field $[X, Y]$ called the *Lie bracket* of X and Y and defined by its action on a smooth function f defined on some appropriate open subset of M by

$$[X, Y](f) = X(Y(f)) - Y(X(f)). \tag{2.10}$$

If X and Y have components X^a and Y^a in some chart x of M then the components of $[X, Y]$ are $\sum_{b=1}^{n} (Y^a{}_{,b} X^b - X^a{}_{,b} Y^b)$, where a comma denotes the usual partial derivative, $Y^a{}_{,b} \equiv \frac{\partial Y^a}{\partial x^b}$, in the chart coordinates. It is easily checked that for smooth, global vector fields X, Y, Z on M (or on some open subset of M), $[X, Y] = -[Y, X]$ and that the Jacobi identity $[X, [Y, Z]] + [Y, [Z, X]] + [Z, [X, Y]] = 0$ holds and so the set of all smooth, global vector fields on M (or on some open subset of M) is a Lie algebra under the binary operation of the Lie bracket.

At this point it is convenient to introduce the *Einstein summation convention* which says that a twice repeated index, one contravariant and one covariant, is automatically summed over its range without the necessity of using the summation symbol. It will always be stated clearly if this convention is to be temporarily suspended.

2.5   Derived Maps and Pullbacks

Let M and M' be manifolds, $\phi : M \to M'$ be a smooth map and let $p \in M$ and $p' = \phi(p)$. For $\mathbf{v} \in T_p M$ define $\phi_{*p} \mathbf{v} \in T_{p'} M'$ by $\phi_{*p} \mathbf{v}(f) = \mathbf{v}(f \circ \phi)$ for any smooth real-valued function f defined on some open neighbourhood V of p' (and so the domain of $f \circ \phi$ includes the open neighbourhood $\phi^{-1} V$ of p). The map ϕ_{*p} is thus a linear map $T_p M \to T_{p'} M'$ between vector spaces. Choosing charts x about p and y about p', and noting that ϕ_{*p} is completely determined by its action on a basis of $T_p M$, one finds for this action (and recalling the Einstein summation convention)

$$(\frac{\partial}{\partial x^a})_p \to (\frac{\partial(y^b \circ \phi)}{\partial x^a})_p (\frac{\partial}{\partial y^b})_{p'}. \qquad (2.11)$$

The matrix $(\frac{\partial(y^b \circ \phi)}{\partial x^a})_p$ is the *Jacobian* of the coordinate representative $y \circ \phi \circ x^{-1}$ of ϕ at p and its rank, which is clearly independent of the charts x and y, is called the *rank of ϕ at p*. The map ϕ_{*p} is called the *derived linear function of ϕ* or the *differential of ϕ*, at p. Thus as ϕ "moves" $p \to p'$, ϕ_{*p} "moves" $T_p M \to T_{p'} M'$. If M'' is another smooth manifold and $\psi : M' \to M''$ a smooth map, then $\psi \circ \phi : M \to M''$ is smooth and one finds $(\psi \circ \phi)_{*p} = \psi_{*p'} \circ \phi_{*p}$. The map ϕ also leads to the natural smooth map $\phi_* : TM \to TM'$ between tangent bundles, called the *differential* of ϕ and given by $\mathbf{v} \to \phi_* \mathbf{v} = \phi_{*p} \mathbf{v}$ for $\mathbf{v} \in T_p M$ and then $(\psi \circ \phi)_* = \psi_* \circ \phi_*$. Also if t is a type $(0, s)$ tensor at $p' = \phi(p)$ one may define a type $(0, s)$ tensor $\phi_p^* t$ at p, called the *pullback* of t under ϕ by

$$\phi_p^* t(\mathbf{v}_1, ..., \mathbf{v}_s) = t(\phi_{*p}(\mathbf{v}_1), ... \phi_{*p}(\mathbf{v}_s)) \qquad (2.12)$$

for $\mathbf{v}_1, ..., \mathbf{v}_s \in T_p M$. Recalling the earlier definition of a tensor this construction allows one to generalise the above differential ϕ_{*p} of ϕ by defining, for any $\mathbf{w} \in \overset{*}{T}_{p'} M'$ and $\mathbf{v} \in T_p M$, $\phi_{*p} \mathbf{v}(\mathbf{w}) = \mathbf{v}(\phi_p^* \mathbf{w})$. This is equivalent to the original definition but has the advantage that it may be generalised in order to move type $(r, 0)$ tensors at p to type $(r, 0)$ tensors at p'. Thus if t is a type

$(r, 0)$ tensor at p one defines the *pushforward* $\phi_{*p}t$ of t (under ϕ) at p' by

$$\phi_{*p}t(\mathbf{w}_1, ..., \mathbf{w}_r) = t(\phi_p^*\mathbf{w}_1, ..., \phi_p^*\mathbf{w}_r) \qquad (2.13)$$

for $\mathbf{w}_1, ..., \mathbf{w}_r \in \overset{*}{T}_{p'}M'$. It should be noted here that the above differential and pushforward maps each map type $(r, 0)$ tensors *at* $p \in M$ to type $(r, 0)$ tensors *at* $p' \in M'$. They cannot, in general, map vector or type $(r, 0)$ tensor *fields* on M to vector or type $(r, 0)$ *fields* on M'. For example, if X is a vector field on M one cannot necessarily map it to a vector field on M' by operating with ϕ_* on each $X(p) \in T_pM$ since there may exist $p, q \in M$ with $\phi(p) = \phi(q)$ but $\phi_{*p}(X(p)) \neq \phi_{*q}(X(q))$. Similar problems arise for general "pushforwards". However, the pullback can be used to map (that is, to "pullback") tensor *fields* of type $(0, s)$ on M' to similar tensor *fields* on M. This is clear from the definitions above (naively, since although a function between manifolds may not be injective, it is always, by definition, "single valued"). If, on the other hand, $\phi : M \to M'$ is a smooth *diffeomorphism* one can achieve more since one may utilise the smooth diffeomorphism $\phi^{-1} : M' \to M$ in addition to ϕ. Let t be any smooth tensor field on M' of type (r, s). Then one can define a smooth type (r, s) tensor field ϕ^*t on M, called the *pullback* of t under ϕ, by

$$\phi^*t(p)(\mathbf{v}_1, ..., \mathbf{v}_s, \mathbf{w}_1, ..., \mathbf{w}_r) = t(p')(\phi_{p*}\mathbf{v}_1, ..., \phi_{p*}\mathbf{v}_s, \phi_{p'}^{-1*}\mathbf{w}_1, ..., \phi_{p'}^{-1*}\mathbf{w}_r)$$

for $\mathbf{v}_1, ..., \mathbf{v}_s \in T_pM$ and $\mathbf{w}_1, ..., \mathbf{w}_r \in \overset{*}{T}_pM$ and $p' = \phi(p)$. Also if X is a global, smooth vector field on M one may define a global, smooth vector field on M' by attaching the vector $\phi_{*p}(X(p))$ to $\phi(p) \in M'$.

2.6   Integral Curves of Vector Fields

Let M be an n-dimensional manifold. As given earlier a (smooth) path or curve in M is a smooth map $c : I \to M$ where I is an *open* interval in $\mathbb{R}$. Let t be the identity chart on I (from $\mathbb{R}$) and suppose $p \in M$ and x a chart for M whose domain contains p and which intersects the range of c non-trivially. Let $c^a = x^a \circ c$ be the coordinate representative of c in x whose domain is some open subset of I and let $\frac{\partial}{\partial t}$ be the vector field corresponding to the chart t on $\mathbb{R}$. Define the map $\dot{c} \equiv c_* \circ \frac{\partial}{\partial t}$ so that $\dot{c} : I \to TM$ is a path in TM which associates with each $t_0 \in I$ a vector at $c(t_0)$ in M, that is, a member of $T_{c(t_0)}M$. The function t is usually called the *parameter* of c (and $\{c(t) : t \in I\}$ the set of points of M "on the path c") and this latter vector at $c(t_0)$ is the classical *tangent vector* (or is said to be *tangent*) to the path c at $c(t_0)$ (or at t_0) since

$$\dot{c}(t_0) \equiv (c_* \circ \frac{\partial}{\partial t})(t_0) = c_*(\frac{\partial}{\partial t})_{t_0} = (\frac{d(x^a \circ c)}{dt})_{t_0}(\frac{\partial}{\partial x^a})_{c(t_0)}. \qquad (2.14)$$

Now suppose that X is a vector field on M. A path c in M is called an *integral curve* of X if its tangent vector at any point of the path equals the value of X there, that is, if for each t_0 in the domain of c, $\dot{c}(t_0) = X(c(t_0))$ or, equivalently, $\dot{c} = X \circ c$. If the domain of c contains $0 \in \mathbb{R}$, c is said to *start* from $p \equiv c(0)$. Thus, if within a chart domain of M, $X = X^a \frac{\partial}{\partial x^a}$ for smooth component functions X^a of X, c is an integral curve of X if

$$\frac{d}{dt}[c^a(t)] = \frac{d}{dt}[x^a(c(t))] = X^a(x^1(c(t)), ..., x^n(c(t))). \qquad (2.15)$$

If $f : J \to I$ is a bijective map with J, another open interval of $\mathbb{R}$, with f and f^{-1} smooth and f having nowhere zero derivative, then $c' = c \circ f$ is also a path (with the same range in M as c but different parameter) called a *reparametrisation* of c. The tangent vectors to c and c' at the same point $p = c(t_0) = c'(t_0')$ differ only by a non-zero scaling, as is easily checked. One may relate the pushforward of a vector field and an integral curve as follows. If M and M' are smooth manifolds, $f : M \to M'$ a smooth map and $c : I \to M$ a smooth path in M then $f \circ c : I \to M'$ is a smooth path in M' and if $\mathbf{v} \in T_pM$ is tangent to c at $p \in M$, $f_{*p}\mathbf{v}$ is tangent to $f \circ c$ at $f(p) \in M'$. This follows since $(f \circ c)\dot{}(t_0) = f_{*c(t_0)}(c_*(\frac{\partial}{\partial t})_{t_0})$.

Let X be a smooth vector field on M and let c_1 and c_2 be integral curves of X with domains I_1 and I_2, respectively, each of which includes the member 0 and which both start from the same point $p \in M$. Since M is assumed to be Hausdorff, c_1 and c_2 can be shown to coincide on $I_1 \cap I_2$ [9]. Thus the union of the domains of all the integral curves of X starting from p is an open interval of $\mathbb{R}$ containing 0 and on which is defined an integral curve of X called the *maximal integral curve of X* starting from p. If the domain of this maximal integral curve through any $p \in M$ is $\mathbb{R}$, X is called *complete*.

There is a very important result regarding integral curves of a smooth vector field. Let X be a smooth vector field on M. Then *given any $p' \in M$ there exists an open neighbourhood U of p' and an open interval I of $\mathbb{R}$ containing 0 such that there is an integral curve of X with domain I starting from any $p \in U$. Any other integral curve of X starting from p coincides with this curve on some neighbourhood of 0.*

2.7   Submanifolds and Quotient Manifolds

Let M and $\bar{M}$ be manifolds and let $f : \bar{M} \to M$ be a smooth map. Then f is called an *immersion* if at each $p \in \bar{M}$ f has rank equal to the dimension of $\bar{M}$. Thus f_{*p} is injective at each $p \in \bar{M}$ and so $\dim\bar{M} \leqslant \dim M$. Now let M be an n-dimensional smooth manifold, let $M' \subset M$ and let $i : M' \to M$ be the natural inclusion map. Then M' is a *submanifold* of M if M' can be given a smooth manifold structure such that the map i is an immersion. Intuitively, one insists on a smooth manifold structure on M' such that it is "contained in" M (through i) in a smooth way. This definition is not as strong as might be thought. In fact, for a given manifold M, different (that is non-diffeomorphic) manifold structures may be imposed on M' consistent with the above definition and these distinct structures may have different dimensions. Another problem arises from the following observation. The manifold structure given on M' will lead to a natural (manifold) topology on M' (section 2.2). But another topology emerges on M', this time when M' is regarded as a subspace of M when the latter has its manifold topology (section 1.7). The two topologies may differ [but the smoothness, hence continuity, of the map i ensures that the subspace topology on M' (from M), viewed in the usual way as a collection of subsets of M' is contained in the manifold topology on M']. A submanifold M' of M for which these two topologies coincide is called a *regular submanifold* of M. This notation is taken from [9] and is not universal. Other authors use the terms *immersed* and *embedded* submanifolds for what here are called submanifolds and regular submanifolds, respectively. Any topological property ascribed to a submanifold will always refer to its manifold topology.

Any non-empty open subset of an n-dimensional manifold M may be given a natural submanifold structure directly from that of M by restricting charts of M in an obvious way and is then called an *open submanifold* of M. It is necessarily a regular submanifold of M and has the same dimension as M. In fact, any submanifold of M whose manifold structure is n-dimensional is an open submanifold of M [9]. If U and V are open submanifolds of M and $f : U \to V$ is a smooth bijection between U and V with f^{-1} also smooth, f is called a *local diffeomorphism* of, or on, M. An open submanifold of $M_n\mathbb{R}$ arises from those matrices with non-zero determinant, that is, $GL(n, \mathbb{R})$. Also the subsets $S(n, \mathbb{R})$ and $Sk(n, \mathbb{R})$ of symmetric and skew-symmetric matrices in $M_n\mathbb{R}$ with the manifold structures given to them earlier can be checked to be regular submanifolds of $M_n\mathbb{R}$. If V is an n-dimensional real vector space and W an m-dimensional subspace of V each with their manifold structures of dimension n and m, respectively, given earlier, then W is a regular submanifold of V. More interestingly, if a subset M' of a smooth manifold M admits the structure of a *regular* submanifold of M of dimension n' then M' admits no other (non-diffeomorphic) submanifold structures of this dimension and no other (non-diffeomorphic) regular submanifold structures of *any* dimension.

A potential problem arises here. Suppose M_1 and M_2 are smooth manifolds with submanifolds M_1' and M_2', respectively, and let $f : M_1 \to M_2$ be smooth. Then the restriction of f to M_1' is a smooth map $M_1' \to M_2$ as is seen by composing with the (smooth) inclusion map. But if the range $f(M_1)$ is contained in M_2' then the consequent map $f : M_1 \to M_2'$ may not be smooth (but *is* smooth if M_2' is a regular submanifold of M_2).

A submanifold (respectively, regular submanifold) of a submanifold (respectively, regular submanifold) is a submanifold (respectively, regular submanifold) and further if $M' \subset M$ is a *regular* submanifold of M and $M'' \subset M$ is a submanifold of M with $M'' \subset M'$ then M'' is a submanifold of M' [9].

It is noted here that, in the above notation, any smooth tensor of type $(0, r)$ on M may be pulled back to a similar one on M' using the inclusion map i. If M' is an m-dimensional submanifold of M, with natural inclusion i and if $p \in M'$, $i_{*p} : T_pM' \to T_pM$ is an injective map whose range is an m-dimensional subspace of T_pM called the *subspace of T_pM tangent to M'* and its members are said to be *tangent to M'* at p.

Let M and M' be smooth manifolds of dimension n and n', respectively, $(n > n')$ and $f : M \to M'$ a smooth map such that f has rank n' at each point of the subset $f^{-1}\{p'\} \equiv \{p \in M : f(p) = p'\} \equiv \bar{M} \subset M$ for $p' \in M'$. Then $\bar{M}$ can be given the structure of a regular submanifold of M of dimension $n - n'$ and $\bar{M}$ is a closed subset of M. The special case of this when $M' = \mathbb{R}$ turns out to be useful and in this case $\bar{M}$ is sometimes referred to as an $(n - 1)$-dimensional *hypersurface* of M (or a *level surface* of f in M).

Let M and M' be smooth manifolds of dimension n and n', respectively, and $f : M \to M'$ be a smooth map. Then f is called a *submersion* if at each $p \in M$ the rank of f equals $\dim M'$. Thus $\dim M' \leqslant \dim M$ and f_{*p} is surjective for each $p \in M$. As an example note that when $\mathbb{R}^n$ and $\mathbb{R}$ have their usual manifold structures the projection maps $p_i : \mathbb{R}^n \to \mathbb{R}$ are submersions. Next let $\sim$ be an equivalence relation on M with $M/\sim$ denoting the associated quotient set and $\mu : M \to M/\sim$ the natural projection. If $M/\sim$ can be given the structure of a smooth manifold such that μ is a submersion then $M/\sim$ is called a *quotient manifold* of M. The manifold topology on $M/\sim$ is the same as the quotient topology on $M/\sim$ (chapter 1) arising from the manifold topology on M and if $M/\sim$ admits the structure of a quotient manifold of M it does so in only one way. If M and M' are smooth manifolds, $\sim$ an equivalence relation on M such that $M/\sim$ is a quotient manifold of M, μ the natural projection and $f : M \to M'$ a smooth map which respects $\sim$ then there exists a map $f' : M/\sim \to M'$ such that $f = f' \circ \mu$ and where f' is smooth. Further, the rank of f at $p \in M$ equals the rank of f' at $\mu(p)$ [9]. As an example the real projective space $P^{n-1}\mathbb{R}$ of dimension $n - 1$ discussed in chapter 1 is an $(n-1)$-dimensional quotient manifold of the non-zero members of $\mathbb{R}^n$ its members being directions (in $\mathbb{R}^n$).

As another example of this define an m-frame in $\mathbb{R}^n$ $(m, n \in \mathbb{N}, m < n)$ as an ordered set of m independent members of $\mathbb{R}^n$. The collection of all such m-frames in $\mathbb{R}^n$ is denoted by $V(m, \mathbb{R}^n)$ and is in bijective correspondence with the open subset of the set $M_{n \times m}\mathbb{R}^n$ of $n \times m$ real matrices of rank m and is hence an open submanifold of the mn-dimensional manifold $M_{n \times m}\mathbb{R}^n$. The set $V(m, \mathbb{R}^n)$ is thus a manifold called the *Stiefel manifold of m-frames in $\mathbb{R}^n$*. If one then maps each m-frame onto to the m-dimensional subspace of $\mathbb{R}^n$ spanned by its members (and denoting the collection of all such subspaces by $G(m, \mathbb{R}^n)$) one gets a surjective map $f : V(m, \mathbb{R}^n) \to G(m, \mathbb{R}^n)$. The set $G(m, \mathbb{R}^n)$ can be shown to have a natural manifold structure of dimension $m(n - m)$ and is called a *Grassmann manifold*. (More details of this, in the important special case when $n = 4, m = 2$, will be given in the next chapter.) Further, the map f can then be shown to be a submersion.

Now, for m-frames p and q define an equivalence relation $\sim$ on $V(m, \mathbb{R}^n)$ by $p \sim q \Leftrightarrow f(p) = f(q)$. The map f respects $\sim$ and thus $G(m, \mathbb{R}^n)$ is diffeomorphic to a quotient manifold of $V(m, \mathbb{R}^n)$. The manifold $G(m, \mathbb{R}^n)$ can be shown to be Hausdorff, second countable, compact and diffeomorphic to $G(n - m, \mathbb{R}^n)$. [Second countability follows since a quotient manifold of a second countable manifold is necessarily second countable, and since $V(m, \mathbb{R}^n)$ is second countable (because $M_{n \times m}\mathbb{R}$ is).]

2.8 Distributions

There is a generalisation of the idea of an integral curve for a smooth vector field. Suppose M is an n-dimensional manifold and $m \in \mathbb{N}$ is fixed with $m < n$ and that for each $p \in M$ there is allocated an m-dimensional subspace D_p of T_pM. Suppose also that for any $p' \in M$ there exists an open neighbourhood U of p' and m smooth vector fields $X_1, ..., X_m$ defined on U such that $X_1(p), ..., X_m(p)$ span D_p for each $p \in U$. Then the map $D : p \to D_p$ is called an *m-dimensional (smooth) distribution on M*. Since the integer m is fixed, this distribution is sometimes said to be "in the sense of Frobenius" to distinguish it from more general distributions where the dimension of D_p (that is, m) may vary over M. One might ask if, for distributions, there are analogues of the integral curves encountered for smooth vector fields, that is, are there "local submanifolds" to which the above local vector fields $X_1, ..., X_m$ are tangent. To answer this question let D be a distribution on M and call a submanifold M' of M an *integral manifold of D* if the inclusion (immersion) map $i : M' \to M$ has the property that the range of the map $i_{p*} : T_pM' \to T_pM$ is exactly D_p for each $p \in M'$, that is, $D(p)$ is the subspace of T_pM tangent to M'. If such an M' exists it is necessarily m-dimensional since i is an immersion. The distribution D is then called *integrable* if each $p \in M$ is contained in an integral manifold of D. A smooth vector field X defined on some non-empty open subset of M is said to *belong to D* if $X(p) \in D_p$ for each p in the domain of X. An important theorem due to Frobenius then states that, under the above conditions, D is integrable if and only if $[X, Y]$ belongs to D whenever X and Y belong to D, where X and Y are any smooth vector fields defined on some non-empty open subset of M (the *involutive* condition for D). A 1-dimensional distribution is sometimes referred to as a *direction field* on M and is necessarily integrable. It should not be confused with the term "direction".

A generalised distribution on M can now be defined and this will be done in terms of a family of smooth vector fields on M and which is sufficient for present purposes. Let S be a non-trivial, real, finite-dimensional Lie algebra of global, smooth vector fields on M under the usual addition and scalar multiplication of vector fields, and under the Lie bracket operation. For each

$p \in M$, define the subspace $S_p = \{X(p) : X \in S\}$ of T_pM and then consider the map $p \to S_p$ on M.

This map is called the *generalised distribution on M determined by S*. It is not necessarily a distribution in the sense of Frobenius since $\dim S_p$ may not be constant over M. However, starting with the family of smooth vector fields S ensures that a "smoothness" is built into such a generalised distribution. A submanifold M' of M is called an *integral manifold of S* (or of the generalised distribution determined by S) if for each $p' \in M'$ the subspace of $T_{p'}M$ tangent to M' equals $S_{p'}$. Further such an integral manifold M' is called a *maximal integral manifold of S* if it is a connected integral manifold of S which is not properly contained in any other connected integral manifold of S. It can then be shown [23], [24], [25] that there exists a unique maximal integral manifold of S through any $p \in M$ for which $\dim S_p \geq 1$.

2.9   Linear Connections and Curvature

A very important concept in differential geometry is the idea of a connection on an n-dimensional manifold M. Not all manifolds admit a connection but those which do are amongst the most important manifolds of study. On a general manifold, one has no natural means of providing a link between the tensor spaces at different points of M and this a connection does in the following way. One requires a method of "moving", say, $\mathbf{v} \in T_pM$ along some smooth (or piecewise smooth) path c in M from p to $p' \in M$ to a member $\mathbf{v}' \in T_{p'}M$. It is recalled that M is assumed connected and hence path-connected. Given $\mathbf{v} \in T_pM$ and c it is required that $\mathbf{v}' \in T_{p'}M$ is *uniquely determined* by $\mathbf{v}$ and c. Such a means of vector or tensor "transfer" is called *parallel transfer (or transport)* and provides a standard of "no change" as the vector or tensor at p is moved along the path from p to p' although it does, in general, depend on the path chosen. This (possible) path dependency distinguishes it from the usual (path independent) parallel transfer familiar from Euclidean geometry and in this lies the genesis of the curvature tensor, to be dealt with later. This standard of no change allows a coordinate independent differentiation using a kind of Newton quotient familiar from real analysis. Here a process will be followed which starts with this idea of a derivative, leaving the parallel transfer concept to follow as a "zero derivative".

A *smooth (linear) connection* ∇ on M is a map which associates with two smooth vector fields X and Y defined on open subsets U and V of M, respectively, a third smooth vector field denoted by $\nabla_X Y$ defined on $U \cap V$ such that for smooth real-valued functions f, g, a smooth vector field Z (each defined on an appropriate subset of M), $a, b \in \mathbb{R}$ and with all appropriate domains assumed open and non-empty, one has

$$\nabla_Z(aX + bY) = a\nabla_Z X + b\nabla_Z Y \tag{2.16}$$

$$\nabla_{fX+gY} Z = f\nabla_X Z + g\nabla_Y Z$$

$$\nabla_X(fY) = f\nabla_X Y + (X(f))Y.$$

It is convenient to define $\nabla_X f \equiv X(f)$ so that these formulae all take a Leibniz form. One calls $\nabla_X Y$ the *covariant derivative of the vector field Y along the vector field X*. This allows the development of the concept of the covariant derivative of a vector field Y along the integral curves of the vector field X. If x is a chart in M with domain U then using the coordinate vector fields $\frac{\partial}{\partial x^a}$ one has, since $\nabla_X Y$ is a vector field on U,

$$\nabla_{\frac{\partial}{\partial x^b}}\left(\frac{\partial}{\partial x^c}\right) = \Gamma^a_{bc}\frac{\partial}{\partial x^a} \tag{2.17}$$

for smooth functions Γ^a_{bc} on U called the *coefficients of the connection* or the *connection coefficients* associated with ∇. These coefficients are not the components of any tensor; in fact using transformations like (2.3) one finds that upon a change of charts $x \to x'$, so that the coefficients of the connection in the chart x' are Γ'^a_{bc},

$$\Gamma'^a_{bc} = \frac{\partial x^e}{\partial x'^b}\frac{\partial x^f}{\partial x'^c}\frac{\partial x'^a}{\partial x^d}\Gamma^d_{ef} + \frac{\partial^2 x^d}{\partial x^{b'}\partial x^{c'}}\frac{\partial x'^a}{\partial x^d}. \tag{2.18}$$

In this book only *symmetric* connections will be considered which means that the condition $\Gamma^a_{bc} = \Gamma^a_{cb}$ will be imposed at each point of M and which is easily seen from (2.18) to be independent of the coordinates used. The condition that ∇ be symmetric can be expressed, using (2.16) and (2.17), in the form

$$\nabla_X Y - \nabla_Y X - [X, Y] = 0. \tag{2.19}$$

Let $c : I \to M$ be a smooth path in M with I an open interval in $\mathbb{R}$ such that c is injective with $c(I)$ contained in some chart domain U of M. Let $X(t) = \dot{c}(t)$ be the tangent vector at $c(t)$ and let $Y(t) = Y(c(t)) \in T_{c(t)}M$ be a collection of vectors along c which are *smooth on c* in the sense that $Y(t)(f)$ is a smooth function $I \to \mathbb{R}$ for each smooth function $f : I \to \mathbb{R}$. Then it can be shown that there exist smooth vector fields X and Y on U which restrict to $X(t)$ and $Y(t)$ on $c(I)$ [11]. Then $Y(t)$ is said to be *covariantly constant* or *parallel* along c if $\nabla_X Y = 0$ along c. Now $\nabla_X Y$ in the coordinates on U is, from (2.16) and (2.17),

$$\left(X^a Y^b \Gamma^c_{ab} + X^a\frac{\partial Y^c}{\partial x^a}\right)\frac{\partial}{\partial x^c} \tag{2.20}$$

where $X^a = \frac{dx^a}{dt}$ and Y^a are the components of Y in U and so on c one gets

$$\frac{dY^c}{dt} + \Gamma^c_{ab}\frac{dx^a}{dt}Y^b = 0. \tag{2.21}$$

This involves only the values of X and Y on the path c and, in fact, is independent of the choices of X and Y on U as long as they restrict to the given $X(t)$ and $Y(t)$ on c. Now the first-order differential equation (2.21) ensures that if $p, p' \in c(I)$ and with $Y(t) = Y(c(t))$ assumed smooth on c, $Y(p')$ is uniquely determined by $Y(p)$ and c and the map $T_p M \to T_{p'} M$ which arises is a path-dependent vector space isomorphism called the *parallel transfer or transport* (of $T_p M$) along c from p to p'. If $c(I)$ is not contained in a single chart domain then, since I and hence $c(I)$ are compact (chapter 1), one may cover $c(I)$ with a finite number of coordinate domains of M and perform the above procedure on the restriction of c to each of these domains. This gives a linear isomorphism, called *parallel transport or transfer* between $T_p M$ and $T_{p'} M$ for any $p, p' \in M$ (see, e.g. [12]).

Now consider the situation when $Y(t) = X(t)$, that is, $Y(t)$ is the tangent vector to c at $c(t)$. Thus one is insisting that the tangent vector to c is parallely transferred along c. Then (2.21) gives $\nabla_X X = 0$ which is

$$\frac{d^2 x^a}{dt^2} + \Gamma^a_{bc} \frac{dx^b}{dt} \frac{dx^c}{dt} = 0 \qquad (2.22)$$

where $x^a \equiv x^a \circ c$. In this form c is called an *affinely parametrised geodesic* (for ∇) and t is referred to as an *affine parameter* for c (and is determined up to a linear transformation of t). Under a general parameter change (a reparametrisation of c) and retaining the symbol t for the new parameter (2.22) becomes

$$\frac{d^2 x^a}{dt^2} + \Gamma^a_{bc} \frac{dx^b}{dt} \frac{dx^c}{dt} = \lambda(t) \frac{dx^a}{dt} \qquad (2.23)$$

for some smooth function λ, and reflects the fact that the parallel transport of the tangent vector to c along c is, at each point of c, proportional to the tangent vector at that point, that is, $\nabla_X X = \lambda X$. In this sense (2.23) is the most general form of the *geodesic equation* for c. Given (2.23) it is easily checked that a reparametrisation of c may be used to achieve an affine parameter and (2.22). Any map $c : I \to M$ satisfying (2.23) is simply called a *geodesic* (and sometimes this term is also used to denote the subset $c(I)$ of M).

An affinely parametrised geodesic $c : I \to M$ for some open interval I in $\mathbb{R}$ is called *maximal* if it cannot be extended to an affinely parametrised geodesic on an interval properly containing I and *complete* if $I = \mathbb{R}$. If every affinely parametrised geodesic of ∇ on M is complete the connection ∇ (or M, if ∇ is understood) is called *geodesically complete* (or just *complete*).

Theorem 2.1 *Let M be an n-dimensional manifold with smooth, symmetric connection ∇ and let $p \in M$ and $0 \neq \mathbf{v} \in T_p M$.*

(i) *There is a unique affinely parametrised maximal geodesic c (whose domain contains 0) such that $c(0) = p$ and $\dot{c}(0) = \mathbf{v}$.*

(ii) *If c' is an affinely parametrised geodesic satisfying $c'(0) = p$ and $\dot{c}'(0) = \mathbf{v}$ then c' is defined on some open subinterval of the domain of c above and agrees with it there.*

(iii) If c' is an affinely parametrised geodesic (whose domain contains 0) satisfying $c'(0) = p$ and $\dot{c}'(0) = \lambda\mathbf{v}$ for $0 \neq \lambda \in \mathbb{R}$ then c' is a reparametrisation of c in (i) on the intersection of their domains.

These results give existence and uniqueness theorems for geodesics and *(iii)* shows that, roughly speaking, the range $c(I)$ of a geodesic path in M is determined by its *starting point* (say $c(0)$) and *initial direction* (that is, the 1-dimensional subspace of $T_{c(0)}M$ spanned by $\mathbf{v}(0)$).

Now let X, Y, Z be smooth vector fields whose domains are open subsets of M and define a smooth vector field on the intersection of these domains (assumed non-empty) by

$$\bar{R}(X,Y)Z \equiv \nabla_X(\nabla_Y Z) - \nabla_Y(\nabla_X Z) - \nabla_{[X,Y]}Z(= -\bar{R}(Y,X)Z). \quad (2.24)$$

The function $\bar{R}$ is called the *curvature structure* arising from ∇ and is, in an obvious sense, $\mathbb{R}$-linear in its arguments (from (2.16)) and, in addition, if f, g, h are smooth real-valued functions on the appropriate domains, satisfies

$$\bar{R}(fX, gY)hZ = fgh\bar{R}(X,Y)Z. \quad (2.25)$$

It can be shown that if $p \in M$, $\mathbf{u}, \mathbf{v}, \mathbf{w} \in T_pM$ and if X, Y and Z are vector fields on some open neighbourhood of p satisfying $X(p) = \mathbf{u}$, $Y(p) = \mathbf{v}$ and $Z(p) = \mathbf{w}$ then $\bar{R}(X,Y)Z$, on evaluation at p, is independent of X, Y and Z providing that, on their evaluation at p, they give $\mathbf{u}, \mathbf{v}, \mathbf{w}$, respectively. Thus one may define a type $(1,3)$ tensor $Riem$ at p by

$$Riem(\mathbf{q}, \mathbf{u}, \mathbf{v}, \mathbf{w}) \equiv (\bar{R}(Y,Z)X)_p(\mathbf{q}) \quad (2.26)$$

for $\mathbf{q} \in \overset{*}{T}_pM$. Thus, on some chart x whose domain U contains p, one has

$$Riem = R^a{}_{bcd}\frac{\partial}{\partial x^a} \otimes dx^b \otimes dx^c \otimes dx^d. \quad (2.27)$$

Since $\bar{R}$ is smooth the components $R^a{}_{bcd}$ are smooth functions on U and are called the (components of the) *curvature tensor* on U arising from the connection ∇. The assumption that ∇ is symmetric together with the definition of $\bar{R}$ give

$$\bar{R}(X,Y)Z + \bar{R}(Y,Z)X + \bar{R}(Z,X)Y = 0 \quad (2.28)$$

for smooth vector fields X, Y, Z and using (2.27) one then gets a condition, equivalent to (2.28) in terms of components and which is

$$R^a{}_{bcd} + R^a{}_{cdb} + R^a{}_{dbc} = 0. \quad (2.29)$$

Then using (2.17) and the definitions of $\bar{R}$ and $Riem$ one gets a useful expression for the components of $R^a{}_{bcd}$ in terms of the connection coefficients Γ^a_{bc}

$$R^a{}_{bcd} = \Gamma^a_{db,c} - \Gamma^a_{cb,d} + \Gamma^e_{db}\Gamma^a_{ce} - \Gamma^e_{cb}\Gamma^a_{de} \quad (2.30)$$

where a comma denotes a partial derivative. From this one immediately sees that

$$R^a{}_{bcd} = -R^a{}_{bdc}. \tag{2.31}$$

Now let X, Y be smooth vector fields on some open chart domain U of M so that one can write $X = X^a \frac{\partial}{\partial x^a}$ and $Y = Y^a \frac{\partial}{\partial x^a}$ for smooth functions X^a and Y^a, on U. Then using (2.20) one finds for the smooth vector field $\nabla_Y X$ on U

$$\nabla_Y X = \left(\frac{\partial X^a}{\partial x^b} + \Gamma^a_{bc} X^c\right) Y^b \left(\frac{\partial}{\partial x^a}\right) \equiv (X^a{}_{;b} Y^b)\left(\frac{\partial}{\partial x^a}\right) \tag{2.32}$$

where $X^a{}_{;b} \equiv \frac{\partial X^a}{\partial x^b} + \Gamma^a_{bc} X^c$. But then $(X^a{}_{;b} Y^b)(p)$ are the components of a vector in $T_p M$ for any $p \in U$ and any vector $Y(p) \in T_p M$ and it easily follows after a short calculation using (2.3) and (2.4) and the arbitrariness of $Y(p)$ that $X^a{}_{;b}$ are the components of a smooth, type $(1,1)$ tensor field on U called the *covariant derivative* of the vector field X (with respect to ∇). It is then convenient to define, for a real-valued function f on U, $\nabla f = df$ where the "d" operator is as given in section 2.4.

One may extend the idea of a covariant derivative to arbitrary tensor fields defined on open subsets of M. In fact for any vector field X on an open subset U of M there is a unique operator ∇_X which maps a smooth tensor field of type (r,s) on U to a tensor field of the same type on U which coincides with $\nabla_X Y$ above when applied to a smooth vector field Y on U and which satisfies the following conditions for smooth tensor fields S and T on U of the same type, for a smooth real-valued function f on U and $a, b \in \mathbb{R}$

$$\nabla_X f = X(f),$$
$$\nabla_X (aS + bT) = a\nabla_X S + b\nabla_X T,$$
$$\nabla_X (S \otimes T) = \nabla_X S \otimes T + S \otimes \nabla_X T,$$
$$\nabla_X \text{ commutes with the contraction operator,}$$
$$\nabla_{fX} S = f\nabla_X S,$$
$$\nabla_{X+Y} Z = \nabla_X Z + \nabla_Y Z. \tag{2.33}$$

Then one extends the idea of the covariant derivative to tensor fields in a way similar to that done above for vector fields. If T is a smooth tensor field of type (r,s) on U one achieves a smooth tensor field ∇T of type $(r, s+1)$ on U given by

$$\nabla T \equiv T^{a_1 \ldots a_r}_{b_1 \ldots b_s; b} \frac{\partial}{\partial x^{a_1}} \otimes \ldots \otimes \frac{\partial}{\partial x^{a_r}} dx^{b_1} \ldots \otimes dx^{b_s} \otimes dx^b \tag{2.34}$$

where the components $T^{a_1 \ldots a_r}_{b_1 \ldots b_s; b}$ of ∇T are

$$T^{a_1 \ldots a_r}_{b_1 \ldots b_s; b} = \frac{\partial}{\partial x^b} T^{a_1 \ldots a_r}_{b_1 \ldots b_s} + \Gamma^{a_1}_{bc} T^{ca_2 \ldots a_r}_{b_1 \ldots b_s} + \cdots + \Gamma^{a_r}_{bc} T^{a_1 \ldots a_{r-1} c}_{b_1 \ldots b_s}$$
$$- \Gamma^c_{bb_1} T^{a_1 \ldots a_r}_{cb_2 \ldots b_s} - \cdots - \Gamma^c_{bb_s} T^{a_1 \ldots a_r}_{b_1 \ldots b_{s-1} c}. \tag{2.35}$$

If $X = X^a \frac{\partial}{\partial x^a}$ on U the components of $\nabla_X T$ in U are $T^{a_1 \ldots a_r}_{b_1 \ldots b_s; b} X^b$. The semi-colon symbol will always be used to denote a covariant derivative in this way. If $\nabla T = 0$ on U one says that T is *covariantly constant*, or *parallel*, on U and, in the case when $U = M$, this leads, as before, to the concept of parallel transfer of the various tensor spaces at $p \in M$ to $q \in M$ along any smooth or piecewise-smooth path from p to q and is a path-independent vector space isomorphism between them.

The concept of covariant differentiation gives rise to a number of important identities between the tensor concerned and the geometrical objects arising from ∇. For this the notation $T^{a \ldots b}_{c \ldots d; ef} \equiv (T^{a \ldots b}_{c \ldots d; e})_{; f}$ will be used. Thus for a vector field X, a 1-form field w and a type $(0, 2)$ tensor field T, one has the useful *Ricci identities* (recalling ∇ is always assumed symmetric)

$$X^a_{;bc} - X^a_{;cb} = X^d R^a_{dcb}, \tag{2.36}$$

$$w_{a;bc} - w_{a;cb} = w_d R^d_{abc}, \tag{2.37}$$

$$T_{ab;cd} - T_{ab;dc} = T_{eb} R^e_{acd} + T_{ae} R^e_{bcd}. \tag{2.38}$$

Suppose w is a smooth 1-form (covector) field on M. Then w is called *exact* (or a *global gradient*) if there exists a smooth function $f : M \to \mathbb{R}$ such that $w = df$ and *closed* if for each $p \in M$ there exists an open neighbourhood U of p and a smooth function $f : U \to \mathbb{R}$ such that $w = df$ on U, that is, a closed 1-form is *locally* a gradient (locally exact). The 1-form w is closed if and only if each $p \in M$ admits an open coordinate neighbourhood U on which $w_{a,b} = w_{b,a}$ and so, since the connection is assumed symmetric, this can be restated in terms of covariant derivatives as $w_{a;b} = w_{b;a}$.

There is another identity involving only the curvature tensor components and the coefficients of the connection. It is called the *Bianchi identity* and is

$$R^a_{bcd;e} + R^a_{bde;c} + R^a_{bec;d} = 0. \tag{2.39}$$

Another important tensor which derives naturally and simply from the (global, smooth) curvature tensor is the *Ricci tensor*. It is a global, smooth type $(0, 2)$ tensor obtained from *Riem* by a contraction and has coordinate components $R_{ab} \equiv R^c_{acb}$. Since the connection is symmetric it is easily checked that the Ricci tensor is a symmetric tensor, that is, $R_{ab} = R_{ba}$ in any coordinate system.

If ∇ is such that the associated curvature tensor *Riem* is identically zero on M then M is called *flat* and ∇ is called a *flat connection* on M. For the purposes of this book a manifold admitting a symmetric connection whose associated curvature tensor does not vanish over any non-empty open subset of M is called *non-flat*. Of course, a manifold is *not flat* if it does not satisfy the flat condition and this is not the same as non-flat.

2.10   Lie Groups and Lie Algebras

Let G be a set admitting a group structure with multiplication denoted by . and a topological structure. These structures may be imposed quite independently and need have no relationship to each other. But if the group and topological structures are chosen so that, with the given topology on G (and the product topology on $G \times G$), the map $G \times G \to G$ given by $(a, b) \to a.b$ *and* the map $G \to G$ given by $a \to a^{-1}$, for $a, b \in G$, are continuous, then G, together with its group and topological structures, is called a *topological group*.

With this as a foil, again let G be a group with multiplication . and suppose that G has the structure of a smooth n-dimensional manifold. Then $G \times G$ has a natural product manifold structure of dimension $2n$ arising from that on G. Suppose one insists as a compatibility requirement that the map $G \times G \to G$ given by $(a, b) \to a.b$ for $a, b \in G$, is smooth (and hence continuous). It then follows (from the manifold structure on G) that the map $G \to G$ given by $a \to a^{-1}$ is *necessarily smooth* (and hence continuous) and a diffeomorphism and then G, together with its group and manifold structures, is called an n-dimensional *Lie group*. It is clear that a Lie group is a topological group. (Henceforth the . in the group operations on G will be dropped, $a.b$ being written as ab, unless any confusion may arise.) A standard example of a Lie group which will be needed later is the set $GL(n, \mathbb{R})$ with its standard group and manifold structures, the latter as an n^2-dimensional open submanifold of $M_n\mathbb{R}$.

The manifold topology arising on a Lie group G can be shown to be Hausdorff and, if connected, it is necessarily second countable. If G is not connected the component containing the identity e of G, the *identity component*, is denoted by G_e and is a subgroup and an open subset of G, hence an open submanifold of G. Also G_e is generated (section 1.2) by any open subset of G_e containing e and hence if $G' \subset G_e$ is both a subgroup and an open subset of G, $G' = G_e$. As another example, it is easily checked that, with the obvious structures, $\mathbb{R}^n$ is a Lie group.

If G_1 and G_2 are Lie groups and $f : G_1 \to G_2$ is smooth and a group homomorphism, f is called a *Lie group homomorphism* and if f is also bijective and f^{-1} is smooth f is a *Lie group isomorphism* and G_1 and G_2 are said to be *Lie isomorphic*. If $G_1,...,G_n$ are Lie groups, then $G_1 \times ... \times G_n$ is also Lie group with the product group and manifold structures.

Now for a Lie group G if H is a subset of G which is a subgroup and a submanifold of G and, in addition, with these structures (and topology derived from its manifold structure), H *is a Lie group*, it is called a *Lie subgroup* of G of dimension equal to that of H as a submanifold of G. There is a possible complication here in that one may ask if the conditions that H be a subgroup and a submanifold of G are sufficient to claim Lie subgroup status for H? The problem arises in the following way. The map $G \times G \to G$ given by $(a, b) \to ab$ for $a, b \in G$, which is guaranteed to be smooth since G is a Lie group, restricts to a map $H \times H \to H$. But this last map may not be smooth (see section 2.7). It is, of course, smooth if H is a regular submanifold of G (section 2.7) but, in fact, H may not be such a submanifold. However, it is, in fact, true (but not obviously so) that if H is just a subgroup and a submanifold of G then it is a Lie subgroup of G [22]. Thus the identity component G_e of G is an open submanifold of G and is hence a regular submanifold and a Lie subgroup of G. A well-known theorem states that if G is a Lie group and $H \subset G$ is a subgroup of G which is a closed subset of G and does not inherit a discrete subspace topology from G then H admits a unique structure as a regular submanifold of G and is then a Lie subgroup of G [9].

Let M be any smooth manifold admitting a global vector field X and let $f : M \to M$ be a smooth diffeomorphism on M. Then X is called *f-invariant* if $f_* \circ X = X \circ f$, that is, if the pushforward of $X(p)$ with f equals $X(f(p))$. If X and Y are smooth, global f-invariant vector fields on M then it can be shown that $[X, Y]$ is also f-invariant [9]. Now suppose G is an n-dimensional Lie group with identity e. The maps L_a and R_a for each $a \in G$ defined by $L_a : g \to ag$ and $R_a : g \to ga$ are smooth diffeomorphisms on G called *left translations* and *right translations*, respectively. A global, smooth vector field X on G is called *left-invariant* (respectively, *right-invariant*) if it is L_a-invariant (respectively, R_a-invariant) for each $a \in G$. Now let $\mathbf{v} \in T_eG$ and define a global vector field X on G by $X(a) = L_{a*}(\mathbf{v})$ for each $a \in G$. It can be shown that X is smooth and left-invariant and that any left-invariant global, smooth, vector field X on G arises in this way from $X(e) \in T_eG$. Thus the collection of all left-invariant vector fields on G is a real vector space isomorphic to T_eG under the isomorphism $X(e) \to X$ and is n-dimensional. It can also be given the structure of an n-dimensional Lie algebra under the Lie bracket operation since if X, Y are left-invariant so is $[X, Y]$. Now suppose $\mathbf{u}, \mathbf{v} \in T_eG$ give rise to smooth, left-invariant vector fields X and Y on G, respectively, so that $X(e) = \mathbf{u}$ and $Y(e) = \mathbf{v}$. Then if one defines $[\mathbf{u}, \mathbf{v}] \equiv [X, Y](e)$ this Lie algebra structure is transferred to T_eG. The resulting Lie algebra is referred to as the *Lie algebra of G* and denoted by LG. Similar remarks apply also to right-invariant vector fields but give rise to a Lie algebra structure denoted by RG in which the bracket product differs only in sign from that on LG.

Recalling the earlier section on distributions let D be a d-dimensional distribution (in the sense of Frobenius) on G and call it *left-invariant* if for each $a, g \in G$, $D(ag) = L_{a*}D(g)$ where L_{a*} is applied in an obvious way to the subspace $D(g)$. Then D uniquely determines a subspace $D(e)$ of T_eG and every subspace $U \subset T_eG$ uniquely determines a left-invariant distribution D on M according to $D(a) = L_{a*}U$ for each $a \in G$ (and so $D(e) = U$). Now let U be any subspace of T_eG and let D be the associated left-invariant distribution on G (so that $D(e) = U$). Then, using the Frobenius theorem, one achieves the following important results.

(*i*) D is integrable if and only if U is a subalgebra of LG.

(*ii*) Let H be a Lie subgroup of G with Lie algebra LH and let $i : H \to G$ be the smooth inclusion map. Then $i_{*e} : LH \to LG$ is a Lie algebra isomorphism between LH and a subalgebra of LG and H is an integral manifold of the left-invariant distribution on G determined by the subspace $i_{*e}(T_eH)$.

(*iii*) Let U be a subalgebra of LG. Then there exists a unique connected Lie subgroup H of G such that, if i is the inclusion map $H \to G$, i_{*e} is a Lie algebra isomorphism between LH and U.

Thus there is a one-to-one relationship between the connected Lie subgroups of G and the subalgebras of LG.

2.11   The Exponential Map

Let G be a Lie group with identity e and let X be a smooth left-invariant vector field on G. Suppose c is an integral curve of X with domain I containing 0 and which starts at e. For $a \in G$ consider the smooth curve $c_a = L_a \circ c$ with domain I starting from $a \in G$. Then with the usual coordinate t on c

$$\dot{c}_a = c_{a*} \circ \frac{\partial}{\partial t} = L_{a*} \circ c_* \circ \frac{\partial}{\partial t} = L_{a*} \circ \dot{c}$$
$$= L_{a*}X \circ c = X \circ L_a \circ c = X \circ c_a. \tag{2.40}$$

So c_a is also an integral curve of X with the same domain I as c. Since $a \in G$ was arbitrary, it can be shown [9] that this is sufficient to make X a complete vector field. Hence each left-invariant smooth vector field is complete. So if $\mathbf{v} \in T_eG$ let X be the left-invariant vector field determined by $\mathbf{v}$, $X(e) = \mathbf{v}$, and let c be the maximal integral curve of X starting from e. Now one can define the *exponential map* for G, $exp : T_eG \to G$ given by $exp(\mathbf{v}) = c(1)$. The completeness of X means that this map is defined on the whole of T_eG and is smooth when T_eG has its standard manifold structure. The name "exponential" is suggested by the easily checked property that, for $\alpha, \beta \in \mathbb{R}$, $exp((\alpha+\beta)\mathbf{v}) = exp(\alpha\mathbf{v})exp(\beta\mathbf{v})$. Since the Lie group $GL(n, \mathbb{R})$ and certain of its subgroups will play an important role in what is to follow one can use it as an example now. The n^2-dimensional Lie group $GL(n, \mathbb{R})$ can be shown to have as its Lie algebra the vector space $M_n\mathbb{R}$ with the binary (Lie) operator on it given by $[A, B] = AB - BA$, the commutator of A and B. The exponential map is then given in a standard notation, for $A \in M_n\mathbb{R}$, by [9]

$$exp(tA) = I_n + \sum_{s=1}^{\infty} \frac{t^s}{s!} A^s. \tag{2.41}$$

If H is a Lie subgroup of G it is important to distinguish between the exponential map for H, denoted by exp_H with $exp_H : T_eH \to H$, and the above map exp for G, restricted to H. Fortunately there is no problem since if $i : H \to G$ is the natural inclusion map, then $exp \circ i_{*e} = i \circ exp_H$ [9] and so for $\mathbf{v} \in i_* T_e H$, $exp(t\mathbf{v})$ lies in H for each t. It is also useful to ask how efficient the exponential for LG is at generating G from T_eG. If $exp : T_eG \to G$ is surjective then G is called an *exponential* Lie group (and for this to be the case G is necessarily connected). However, this is not the case in general. But *if G is a connected Lie group it is generated by any open subset of G containing e (and the range of the map exp is just such a set [9]) and so every $g \in G$ is the product of finitely many members of G each of which is the exponential of some member of LG.*

2.12 Covering Manifolds

Let M be an n-dimensional manifold. Then M is locally path connected and locally simply connected in its manifold topology. Thus it makes sense to consider results regarding (topological) homotopy for M as a topological space (chapter 1). Of course, one would like these to be phrased in terms of smooth paths and it turns out that this can be done without much essential change. It is known that if c is a continuous closed path at p in M there exists a smooth closed path at p which is (continuously) homotopic to c whilst if c_1 and c_2 are smooth closed paths at p which are continuously homotopic to each other they are smoothly homotopic to each other (in an obvious sense from chapter 1 by taking the homotopy map to be smooth). Then let $\sim$ denote the equivalence relation of being continuously homotopic for continuous, closed paths at $p \in M$ and let $\approx$ denote the equivalence relation of being smoothly homotopic for smooth, closed paths at p. The M is (continuously) simply connected if any continuous, closed path c at p satisfies $c \sim c_p$ where c_p is the constant path at p and M is (smoothly) simply connected if any smooth, closed path at p satisfies $c \approx c_p$. Suppose M is continuously simply connected and let c be a smooth, closed path at p. Then c is continuous and $c \sim c_p$. But c_p is smooth and the earlier remarks show that $c \approx c_p$ so that M is smoothly simply connected. Now suppose M is smoothly simply connected and let c be a continuous, closed path at p. Then the above remarks reveal a smooth, closed path c' at p satisfying $c' \sim c$. But then the initial assumption on M gives $c' \approx c_p$. So $c \sim c_p$ and M is continuously simply connected.

Now let M be the manifold above with its manifold topology and let M' be a topological space. Suppose $\pi' : M' \to M$ is a topological covering map (chapter 1) so that M' is a (topological) covering space of the (topological) space M . Then, on remembering that M has a manifold structure it turns out [26] that there exists a unique manifold structure for M' whose associated topology equals the original given topology for M' and for which the map π' is smooth and is such that each $p \in M$ admits a connected open neighbourhood U such that each component of $\pi'^{-1}U$ is, with its open submanifold structure, diffeomorphic to the open submanifold U under the appropriate restriction of π'. Thus the dimensions of M and M' and the rank of π' at any $p \in M'$ are equal. This gives the definition of a smooth covering $\pi' : M' \to M$ and M' is called a *covering manifold* of M. If M' is simply connected (in either the continuous or smooth sense from the previous paragraph) it is called a *universal covering (manifold)* of M. Every manifold such as M admits a universal covering and the latter is unique in the sense that if M'' is also a universal covering of M with smooth covering π'' there exists a diffeomorphism $f : M' \to M''$ such that $\pi'' \circ f = \pi'$.

2.13   Holonomy Theory

The theory of holonomy groups is, in essence, a closer inspection of the connection on a manifold and a direct link to the curvature tensor. It will, of course, be more fruitful later when metrics are introduced. The work in this section is largely taken from [10].

Let M be an n-dimensional manifold (recalling that M is always assumed connected and hence path-connected) admitting a smooth, symmetric connection ∇, let $p \in M$ and let C_p denote the set of all piecewise-smooth (closed) paths starting and ending at p. If $c \in C_p$ there is an associated vector space isomorphism τ_c on T_pM obtained by parallelly transporting (section 2.9) each member of T_pM around c. Using a standard notation for combining and inverting paths one has for $c_1, c_2 \in C_p$, $\tau_{c^{-1}} = \tau_c^{-1}$ and $\tau_{c_1 \circ c_2} = \tau_{c_1} \cdot \tau_{c_2}$. Then the set $\{\tau_c : c \in C_p\}$ of all such isomorphisms is a subgroup of the group $GL(T_pM)$ of all isomorphisms of T_pM onto itself and called the *holonomy group of M at p*, denoted by Φ_p. Now since M is connected, and hence path connected, given any $p, p' \in M$ there exists a smooth curve from p to p' and then, using this curve to transfer closed curves at p to closed curves at p', in a standard way, one can easily see that the holonomy groups at each $p \in M$ are conjugate and hence isomorphic. Thus one can drop the reference to the point of M in Φ_p and refer to the *holonomy group of M* denoted by Φ. Repeating these operations but now using only paths which are homotopic to zero one similarly obtains the *restricted holonomy group of M* denoted by Φ^0. Thus Φ and Φ^0 are subgroups of $GL(T_pM) = GL(n, \mathbb{R})$ and are equal if M is simply connected.

It turns out that Φ and Φ^0 are Lie subgroups of $GL(n, \mathbb{R})$ and that Φ^0 is connected. In addition, Φ^0 is the identity component of Φ and so $\dim\Phi = \dim\Phi^0$. If M is simply connected, $\Phi = \Phi^0$ and then Φ is connected. The Lie algebras of Φ and Φ^0 are equal and isomorphic to a subalgebra of the Lie algebra of $GL(n, \mathbb{R})$ denoted by ϕ and called the *holonomy algebra of M* (of course, with respect to ∇).

Suppose $p \in M$ and let x be a chart whose domain contains p. In the coordinates x^a on U, and recalling the use of a semi-colon as a covariant derivative, one can compute the curvature tensor *Riem* and its covariant derivatives at p and then for any $X, Y, Z, \ldots \in T_p M$ compute the following matrices

$$R^a{}_{bcd} X^c Y^d, \qquad R^a{}_{bcd;e} X^c Y^d Z^e, \ldots \qquad (2.42)$$

It can then be shown that this set of matrices, for all $X, Y, Z, \ldots \in T_p M$ (and under matrix commutation), *spans a subalgebra of the Lie algebra* ϕ and hence only finitely many terms are required in the list (2.42). Recalling the symmetry properties of *Riem* the family (2.42) may be rewritten as $R^a{}_{bcd} F^{cd}$, $R^a{}_{bcd;e} H^{cde}, \ldots$ where $F, H, \ldots$ are arbitrary tensors at p satisfying $F^{ab} = -F^{ba}$, $H^{abc} = -H^{bac}, \ldots$ The subalgebra spanned by (2.42) is called the *infinitesimal holonomy algebra at* p, denoted by ϕ'_p, and is a subalgebra of ϕ. The unique connected Lie subgroup of $GL(n, \mathbb{R})$ that ϕ'_p gives rise to is called the *infinitesimal holonomy group (of M) at* p and denoted by Φ'_p. In general, ϕ'_p depends on p. If $\dim \Phi'_p$ is constant on M then $\Phi'_p = \Phi^0$ for each $p \in M$. Clearly ∇ is flat ($\Leftrightarrow$ *Riem* $\equiv 0$ on M) $\Leftrightarrow \Phi'_p$ is trivial for each $p \in M \Leftrightarrow \Phi^0$ is trivial.

Let $p \in M$ and let V be a non-trivial subspace of $T_p M$. Suppose that V is carried into itself by parallel transport of its members with any member of C_p. Then V is said to be *holonomy invariant*. Clearly the intersection of any finite number of holonomy invariant subspaces of $T_p M$ is holonomy invariant. It then follows that if $V \subset T_p M$ is holonomy invariant and $p' \in M$ the parallel transport of V along a piecewise-smooth path c from p to p' gives rise to a subspace $V' \subset T_{p'} M$ of the same dimension as V and which is independent of the choice of c and is also holonomy invariant. The association of the subspace V' with p' at each $p' \in M$ can be shown to give rise to a distribution (in the sense of Frobenius) on M which is, in fact, integrable and is called the *holonomy invariant distribution generated by* V [10].

The infinitesimal holonomy deals locally with the curvature (and ∇). For example if $U \subset M$ is an open subset of M on which *Riem* vanishes then Φ'_p is trivial for each $p \in U$ but Φ and Φ^0 may not be since they can detect curvature elsewhere on $M \setminus U$. There is a theorem which links the values of *Riem* at each point of M with ϕ. [20]

Theorem 2.2 *(Ambrose-Singer) Let M be an n-dimensional manifold with smooth, symmetric connection ∇, associated curvature structure $\bar{R}$ and let $p \in M$. For any other $p' \in M$, any piecewise-smooth curve c from p to p' and any $X, Y, Z \in T_p M$ define a linear map f from $T_p M$ to itself by*

$$f(Z) = \tau_c^{-1}(\bar{R}(\tau_c(X), \tau_c(Y)) \tau_c(Z)). \qquad (2.43)$$

Then the set of all such linear maps for all choices of p', c, X and Y when represented in matrix form with respect to some basis of $T_p M$ spans the holonomy algebra ϕ in matrix representation when the holonomy group Φ of M is *described as a matrix Lie subgroup of $G = GL(n, \mathbb{R})$ with respect to this basis of $T_p M$.*

In other words fix p and choose p' and c and then compute all tensors of the form $R^a{}_{bcd} X'^c Y'^d$ at p' $(X', Y' \in T_{p'} M)$ and then parallel transport them to p along c and repeat this for all such p' and c to get a matrix representation the members of which span ϕ. Thus if $Riem$ is known the Ambrose-Singer theorem gives information about Φ^0 and hence about Φ if M is simply connected.

Chapter 3

Four-Dimensional Manifolds

3.1 Metrics on 4-dimensional Manifolds

In this chapter the manifolds considered will, as mentioned earlier, be assumed to be smooth, connected, second countable and Hausdorff. General references for much of this chapter are [9, 10, 13].

Let M be a manifold as above of dimension $n \geq 3$. A smooth *metric* g on M is a global, smooth, symmetric tensor field of type $(0, 2)$ on M denoted by g such that, at each $p \in M$, $g_p \equiv g(p)$ is an inner product, that is, a symmetric, non-degenerate bilinear form, on T_pM. This structure is denoted by the pair (M, g). Thus, for $p \in M$ and chart x whose domain U contains p, g may be written at p as

$$g_p = g_{ab}(dx^a)_p \otimes (dx^b)_p \qquad g_{ab} \equiv g_p((\frac{\partial}{\partial x^a})_p, (\frac{\partial}{\partial x^b})_p) = g_{ba} \qquad (3.1)$$

and this gives rise to smooth real-valued functions g_{ab} on U with $\det(g_{ab}) \neq 0$. From the theory of the Sylvester canonical form, it is noted that upon a change of coordinates from charts x to x' the matrices $g \equiv g_{ab}$ (in the chart x) and $g' \equiv g'_{ab}$ (in the chart x') are related by the tensor transformation laws which are $g' = S^T g S$ where S is the matrix $\frac{\partial x^a}{\partial x'^b}$ and so, as matrices, g and g' are congruent. Thus a basis of T_pM always exists where the components g_{ab} of g_p take the Sylvester form appropriate to the signature of g_p. There are several possibilities for the signature of g_p. If the Sylvester matrix (up to a multiplicative sign, which will be ignored here) is $(+, +, ..., +)$ $g(p)$ is said to be of *positive definite* signature and if it is $(-, +, +, ..., +)$ $g(p)$ is said to be of *Lorentz* signature (or to be *Lorentzian*). If $g(p)$ has the same signature at each $p \in M$, g is said to be of that signature. It will be seen later that since M is assumed connected g *has the same signature at each* $p \in M$. A positive definite metric g on M is called *Riemannian* by some authors but this term will not be used here since it does not seem to be universally accepted. For the purposes of this book, if $\dim M = 4$ and if M admits a metric such that, at $p \in M$, $g(p)$ has signature $(+, +, -, -)$ (again up to a multiplicative sign) then the metric is said to be of *neutral signature*.

For $u, v \in T_pM$, and if g is understood (and recalling the Einstein summation convention), the real number $g_p(u, v) = g_{ab}u^a v^b$, which is independent of the coordinate system chosen, is denoted $u \cdot v$ and called the *inner product* of u

DOI: 10.1201/ 9781003023166-3

and v. (Henceforth, ordinary symbols rather than bold symbols will be used to denote vectors in T_pM.) Then u is called *spacelike* if $\mid u \mid \equiv u \cdot u > 0$, *timelike* if $\mid u \mid < 0$ and *null* if $u \neq 0$ and $\mid u \mid = 0$. If $u \neq 0 \neq v$ and $u \cdot v = 0$, u and v are called *orthogonal*. If $\mid u \mid = \pm 1$, then u is called a *unit* vector. A collection consisting of n unit, orthogonal vectors is necessarily a basis for T_pM and is called *orthonormal*. If U and V are subspaces of T_pM and if for any $u \in U$ and any $v \in V$, $u \cdot v = 0$, U and V are called *orthogonal* subspaces. For any subspace $U \subset T_pM$, the collection of vectors $\{v \in T_pM : v \cdot u = 0, \forall u \in U\}$ is a subspace of T_pM called the *orthogonal complement* of U and denoted $U^{\perp}$. It is noted that U and V may not be "complementary" in the usual sense, in that, $\mathrm{Sp}(U \cup V)$ may not equal T_pM. It is noted here that if h and h' are inner products at p neither of which is positive definite then if they agree as to their null vectors they are proportional. A simple proof proceeds by choosing one of them in its Sylvester form.

At each $p \in U$ and for each chart x whose domain contains p one may construct the matrix inverse to g_{ab} denoted by g^{ab} and from this the bilinear form $g^{ab}(\frac{\partial}{\partial x^a})_p \otimes (\frac{\partial}{\partial x^b})_p$ on $\overset{*}{T}_pM + \overset{*}{T}_pM$ which is easily checked to be a type $(2,0)$ tensor at p and to give rise to a smooth tensor field on M denoted by g^{-1}. Sometimes (3.1) is referred to as the *covariant* form for g and this latter one the *contravariant* form for g. At any $p \in M$ and in any chart, $g_{ac}g^{cb} = \delta_a^b$.

Thus g gives rise to a vector space isomorphism $f_{g_p} : T_pM \to \overset{*}{T}_pM$ according to $v \to f_{g_p}(v)$ where $f_{g_p}(v)(u) = g_p(v, u)$ for $u, v \in T_pM$. Thus if in some chart x one has $v = v^a(\frac{\partial}{\partial x^a})_p$, then it follows that $f_{g_p}(v) = (g_{ab}v^b)(dx^a)_p$. This leads to the standard definition (for a given g) that $f_{g_p}(v) = v_a(dx^a)_p$ where $v_a \equiv g_{ab}v^b$. Thus g_p is said to *lower indices* at each $p \in M$. It is straightforward to show that one may use a similar map from the contravariant form for g to raise indices and thus, at each $p \in M$, an isomorphism $\overset{*}{T}_pM \to T_pM$ and which is the inverse operation to raising indices since $g_{ac}g^{cb} = \delta_a^b$. Thus the existence of the metric g_p gives a "canonical isomorphism" between T_pM and $\overset{*}{T}_pM$, $v^a \to v_a$, and this will always be understood. It follows that these procedures allow one to convert a smooth vector field defined on some open subset U of M into a smooth covector field on U. In addition, one may raise and lower indices of arbitrary tensors at $p \in M$ or on M. Thus a type $(1,3)$ tensor at $p \in M$ with chart components $T^a{}_{bcd}$ may be turned into a type $(0,4)$ tensor at $p \in M$ with components $T_{abcd} \equiv g_{ae}T^e{}_{bcd}$ and similarly for smooth tensor fields. The use of the same symbol (in this case T) will be understood throughout. It follows from the above that an object δ at $p \in M$ with components δ_a^b in every coordinate system is a type $(1,1)$ tensor at p (the Kronecker symbol or Kronecker delta).

Consider the pair (M, g) and let M' be a submanifold of M with natural inclusion map $i : M' \to M$. Then the pullback i^*g is a type $(0,2)$ symmetric tensor on M' but may not be a metric on M' since it may fail to be non-degenerate on M'. However, if g is positive definite i^*g is easily seen to be always non-degenerate (in fact, positive definite) on M' but this is not necessarily the case if g is not of positive definite signature since then i^*g, if non-degenerate, may take any of the Sylvester forms permitted by $\dim M'$. If i^*g is a metric on M' it is called the *induced metric* on M' (from g on M).

If (M, g) and (M', g') are two manifold-metric pairs, as above, and $\widetilde{M} \equiv M \times M'$ is the product manifold with natural (smooth) projections $i : \widetilde{M} \to M$ and $j : \widetilde{M} \to M'$ then $\widetilde{M}$ admits a metric $\tilde{g} \equiv g \times g'$ defined in an obvious way (see section 2.3) by $g \times g' \equiv i^*g + j^*g'$ and called the *product* of g and g'. Then $(\widetilde{M}, g \times g')$ is called the *metric product* of (M, g) and (M', g').

In a chart x on M with chart domain U one may always construct a *local* smooth metric of any signature by choosing an appropriate (constant) non-singular matrix at each point of U to represent the metric components on U in the chart x. However, the existence of a smooth global metric on M is a little different and requires further restrictions on M which can now be discussed. Since the manifold M has been assumed to be connected, Hausdorff and second countable, it follows that it is paracompact (for a definition see [9]). It can be shown in this case that a useful collection of (local) functions called a partition of unity exists on M and from this that a global, *positive definite metric on M necessarily exists*. The converse is also true since a global, smooth, positive definite metric on a smooth, connected, Hausdorff manifold M leads naturally to a topological metric (distance function) on M whose (natural metric) topology equals the manifold topology and which, from a well-known theorem of Stone, is paracompact and from which, under the present assumptions, second countability follows. However, such a manifold need not admit a Lorentz metric but does so if and only if M admits a 1-dimensional distribution [27] (see also, for example, [15, 13]). The usual construction of such a metric on M relies on its admitting a positive definite metric. For the restriction on a manifold needed for it to admit a metric of neutral signature see [28].

3.2 The Connection, the Curvature and Associated Tensors

Let M be a manifold and g a metric of arbitrary signature on M and suppose ∇ is a symmetric connection on M. Of course, g and ∇ may be prescribed independently but suppose one links them by the following compatibility condition. Let $p \in M$, let c be any smooth path in M passing through p and let v be any member of T_pM. Then insist that the parallel transport $v(t)$ of v along c with respect to ∇ is such that the function $g(v(t), v(t)) = |v(t)|$ is constant for all such p, v and c. Thus, in components, $g_{ab}v^a v^b$ is constant along c and so, since $v(t)$ undergoes parallel transport along c and if $T(t)$ is the tangent vector to c at the point with parameter t, one has $\nabla_T v = 0$ or, in components, $v^a{}_{;b} T^b = 0$ on c. Thus the constancy of $g_{ab} v^a v^b$ along c gives $\nabla_T (g_{ab} v^a v^b) = 0$ for all such p, v and c and which becomes $\nabla_T g = 0$ for any such T. Hence $\nabla g = 0$ on M, or in components, $g_{ab;c} = 0$ (with a semi-colon denoting a ∇-covariant derivative) in any coordinate domain on M.

It then follows from $\nabla g = 0$ (chapter 2) that, in any coordinate domain,

$$\frac{\partial g_{ab}}{\partial x^c} - \Gamma^d_{ca} g_{db} - \Gamma^d_{cb} g_{ad} = 0. \tag{3.2}$$

On rewriting (3.2) after the index permuting $a \to b \to c \to a$ and again after another such permuting and subtracting these last two equations from (3.2) one finds

$$\Gamma^a_{bc} = \frac{1}{2} g^{ad} \Big(\frac{\partial g_{db}}{\partial x^c} + \frac{\partial g_{dc}}{\partial x^b} - \frac{\partial g_{bc}}{\partial x^d} \Big). \tag{3.3}$$

Then $(3.3) \Rightarrow (3.2)$ and so, given g, such a symmetric connection necessarily exists. These connection coefficients Γ^a_{bc} arising from g are usually referred to as *Christoffel symbols* and they (and hence the *symmetric* connection ∇) are uniquely determined by g. The symmetric connection ∇ defined here is called the *Levi-Civita* connection associated with g. The above compatibility condition $\nabla g = 0$ is usually expressed by saying that ∇ is a *metric connection* (*compatible* with g) (and it is remarked that, in general, a connection may not be a metric connection for *any* metric). It is also easily checked that for the tensors δ and g^{-1} defined earlier $\nabla \delta = \nabla g^{-1} = 0$. The connection ∇ preserves inner products along any such path in that if $u, v \in T_p M$ undergo parallel transport along c, giving vectors $u(t)$ and $v(t)$ in terms of the parameter t on c, then $u(t) \cdot v(t)$ is constant along c as follows from the fact that $|u(t) + v(t)|$ is constant along c. It is easily seen that the establishment of the connection ∇ on M compatible with g shows that if M is connected (and hence path-connected) the signature of g is constant on M. This follows since the signature at $p \in M$ can be characterised by an appropriately chosen orthonormal basis at p reflecting the signature of $g(p)$ and then parallel transporting it to any other point of M.

As shown earlier, ∇ gives rise to a type $(1,3)$ curvature tensor *Riem* on M with components $R^a{}_{bcd}$. This tensor requires only a connection for its existence. However, because of the existence of the metric g, one has the (smooth) type $(0,4)$ curvature tensor with components $R_{abcd} = g_{ae} R^e{}_{bcd}$ and which has the important index symmetry conditions obtainable from chapter 2,

$$R_{abcd} = -R_{abdc} = -R_{bacd}, \qquad R_{abcd} = R_{cdab},$$
$$R_{abcd} + R_{acdb} + R_{adbc} = 0. \tag{3.4}$$

One can then contract to get the important, second order, type $(0,2)$, smooth, symmetric *Ricci tensor* denoted by *Ricc*, with components R_{ab}, and the associated smooth *Ricci scalar* denoted by R, and given by

$$R_{ab} \equiv g^{cd} R_{cadb} = R^c{}_{acb} = R_{ba}, \qquad R \equiv R_{ab} g^{ab}. \tag{3.5}$$

The symmetry of R_{ab} in the indices a and b follows easily from its definition and the second equation in (3.4). One also has the (differential) *Bianchi identity* satisfied by *Riem* with a semi-colon denoting a covariant derivative as in chapter 2,

$$R^a{}_{bcd;e} + R^a{}_{bde;c} + R^a{}_{bec;d} = 0, \tag{3.6}$$

which after a contraction with $g^{ac}g^{bd}$ gives $(R^a{}_b - \frac{R}{2}\delta^a_b)_{;a} = 0$ (the twice contracted Bianchi identity). If (M, g) is such that at each $p \in M$ *Ricc* is proportional to g, (M, g) is called an *Einstein space*. Then (3.5) shows that the proportionality function is $\frac{R}{n}$ and so, in components, $R_{ab} = \frac{R}{n}g_{ab}$ and the twice contracted Bianchi identity then shows that, since $n \neq 2$, R is constant on M. If $R \neq 0$, (M, g) is called a (proper) *Einstein space* and if $R = 0$, *Ricc* $= 0$ and (M, g) is called *Ricci flat*. (In the case of Lorentz signature, (M, g) is called *a vacuum* if *Ricc* $= 0$ for reasons related to its use in Einstein's general relativity theory.) It is also useful to define the smooth, symmetric, *tracefree Ricci tensor* $\widetilde{Ricc}$ with components $\widetilde{R_{ab}} \equiv R_{ab} - \frac{R}{n}g_{ab}$ ($\widetilde{R}^c{}_c = 0$). Thus $\widetilde{Ricc} = 0$ on M if and only the Einstein space condition holds on M. If (M, g) is flat, so that *Riem* $\equiv 0$ on M, it can be shown that, given $p \in M$, there exists a chart x with domain U containing p such that the metric g has constant tensor components g_{ab} on U which may be selected as the appropriate Sylvester matrix for the signature of g. A metric whose Levi-Civita connection leads to an identically zero tensor *Riem* is sometimes referred to as a *flat* metric and ∇ is then a flat connection.

If g and g' are two (global) smooth metrics on M satisfying $g' = \phi g$ for some smooth $\phi : M \to M$, g and g' are said to be *conformally related* with *conformal factor* ϕ. For this situation there is an important, smooth, type $(1, 3)$ tensor, the *Weyl conformal tensor*, denoted by C and with components $C^a{}_{bcd}$ in any coordinate domain given by [84, 30]

$$C^a{}_{bcd} = R^a{}_{bcd} + \frac{1}{n-2}(\delta^a{}_d R_{bc} - \delta^a{}_c R_{bd} + g_{bc}R^a{}_d - g_{bd}R^a{}_c)$$
$$+ \frac{R}{(n-1)(n-2)}(\delta^a{}_c g_{bd} - \delta^a{}_d g_{bc}). \tag{3.7}$$

It is noted that the Weyl conformal tensor requires a metric for its existence. It was introduced by Weyl for the reason that *if g and g' are two conformally related metrics on M their Weyl conformal tensors C and C' are equal.* This tensor has the important property of being tracefree, in the sense that $C^c{}_{acb} = 0$. The Weyl tensor can be written as a smooth, type $(0, 4)$ tensor, with components $C_{abcd} = g_{ae}C^e{}_{bcd}$. This latter tensor is very useful but it obviously does not have the above "conformally invariant" property enjoyed by C. One can now write down a useful expression which tidies up (3.7) by introducing a smooth type $(0, 4)$ tensor E on M,

$$R_{abcd} = C_{abcd} + E_{abcd} + \frac{R}{n(n-1)}(g_{ac}g_{db} - g_{ad}g_{cb}), \tag{3.8}$$

where

$$E_{abcd} = \frac{1}{n-2}(\widetilde{R}_{ac}g_{db} - \widetilde{R}_{ad}g_{cb} + \widetilde{R}_{bd}g_{ac} - \widetilde{R}_{bc}g_{ad}). \tag{3.9}$$

The tensor E, referred to here simply as the E *tensor*, has the index symmetry properties given by

$$E_{abcd} = -E_{bacd} = -E_{abdc}, \qquad E_{abcd} = E_{cdab},$$
$$E_{abcd} + E_{acdb} + E_{adbc} = 0, \qquad E^c{}_{acb} = \widetilde{R}_{ab} \qquad E^{ab}{}_{ab} = 0. \tag{3.10}$$

Thus for $p \in M$, $E(p) = 0$ if and only if $\widetilde{Ricc}(p) = 0$ if and only if the Einstein space condition holds at p. The Weyl conformal type $(0,4)$ tensor has the properties

$$C_{abcd} = -C_{bacd} = -C_{abdc}, \qquad C_{abcd} = C_{cdab},$$
$$C_{abcd} + C_{acdb} + C_{adbc} = 0, \qquad C^c{}_{acb} = 0. \tag{3.11}$$

In the event that $C \equiv 0$ on M, (M, g) is called *conformally flat* and it can then be shown that each $p \in M$ admits an open neighbourhood U such that the metric g, restricted to U, is conformally related to a flat metric on U. Conversely if these latter conditions hold, (M, g) is conformally flat.

3.3 Algebraic Remarks, Bivectors and Duals

For the rest of this chapter, it is assumed that, in addition to the other properties forced upon M, one now has $\dim M = 4$. There are thus three possibilities for the signature of g on M (up to multiplicative signs, as before); $(+, +, +, +)$ (positive definite signature), $(-, +, +, +)$ (Lorentz signature) and $(+, +, -, -)$ (neutral signature). A few remarks can now be made regarding tensor classification on such manifolds. Before making them a few important comments will be given and which involve the identification of tensors of different types but which are related by the metric. The concept of raising and lowering indices, using the metric, gives isomorphisms between tensor spaces and as long as M admits a metric and this metric is specified the identification of these spaces will be assumed. For example, if W is a tensor of type $(0, 2)$ with components W_{ab}, it uniquely determines a tensor of type $(1, 1)$ with components $W^a{}_b \equiv g^{ac}W_{cb}$, a type $(1, 1)$ tensor with components $W_a{}^b \equiv g^{bc}W_{ac}$ and a type $(2, 0)$ tensor with components $W^{ab} \equiv g^{ac}g^{bd}W_{cd}$. These tensors (taking care with the position of the indices) will sometimes be used interchangeably. This is usually assumed but there are places in this book where care is needed because of the potential existence of another metric on M. It is remarked here that, on occasions, a contraction will be needed between *complex* tensors. This will be accomplished naively by multiplying out the products in the contraction. Thus, for example, if $p, q, r, s \in T_pM$ and $p + iq$ and $r + is$ are members of the complexification of T_pM then, with an abuse of notation, $g(p+iq, r+is) = g(p, r) - g(q, s) + i(g(p, s) + g(q, r))$. It is convenient to keep the inner product notation and to write this as $(p + iq) \cdot (r + is)$ (and similarly to write $(p+iq) \cdot (p+iq)$ as $|p+iq|$). It is also convenient, if $|p+iq| = 0$ but $p + iq \neq 0$, to refer to $p + iq$ as "null".

Let (M, g) be as before with g of any signature and let S be a *symmetric* type $(0, 2)$ tensor at $p \in M$ so that in any chart x whose domain contains p, S has components $S_{ab} = S_{ba}$. So S may be regarded as a linear map on T_pM given by $v^a \to S^a{}_b v^b$ for $v \in T_pM$ (and recalling the remark above, $S^a{}_b \equiv g^{ac}S_{cb}$). This will be referred to as the linear map on T_pM *associated* with S. Then the classification theory discussed in chapter 1 may be applied to $S(p)$ (but noting that the index contracted in this expression for eigenvectors has been changed from the first, as in chapter 1, to the second index for convenience. Since S is symmetric this makes no difference). Thus a (real or complex) vector w is called an *eigenvector* of S and $\lambda \in \mathbb{C}$ its corresponding *eigenvalue (with respect to g)* if one has $S^a{}_b w^b = \lambda w^a = \lambda \delta^a_b w^b$. This is equivalent to $S_{ab} w^b = \lambda g_{ab} w^b$. It is important to note that in the first of these $S^a{}_b$ is, in general, no longer symmetric whilst in the second, although S_{ab} is symmetric, the signature of g_p must be taken into account (cf the work in section 1.5). The subspace $Sp(w)$ spanned by w is the *eigendirection* determined by w. Also suppose w, w' are independent (real or complex) eigenvectors with respective eigenvalues $\mu, \nu \in \mathbb{C}$, $S_{ab} w^b = \mu g_{ab} w^b$ and $S_{ab} w'^b = \nu g_{ab} w'^b$. Then using the symmetry of S one finds $(\mu - \nu)(w \cdot w') = 0$ and so if $\mu \neq \nu$, then $w \cdot w' = 0$. Further, if $\mu = \nu$ one may always choose w and w' within $Sp(w, w')$ to be orthogonal. To see this note that if all members of $Sp(w, w')$ are null then $w + w'$ is null and hence $w \cdot w' = 0$. Otherwise, some such w satisfies $w \cdot w \neq 0$ and if $w \cdot w' \neq 0$ there exists $\lambda \in \mathbb{C}$ such that w and $w + \lambda w'$ are orthogonal eigenvectors. It is also noted here that if $U \subset T_pM$ is a real invariant subspace for (the map associated with) S, then its orthogonal complement $U^\perp$ is also invariant. To see this note that, since U is invariant, for any $u \in U$ and any $r \in U^\perp$, $S_{ab} u^b r^a = 0$. Hence $S^a{}_b r^b$ lies in $U^\perp$ and this completes the proof. In fact, the map associated with S must admit an invariant 2-space. To see this note that f must admit a (real or complex) eigenvector. If f has a complex (non-real) eigenvector its real and imaginary parts span an invariant 2-space. If f has all eigenvalues real then either there are at least two independent ones, whose span is then invariant or there is only one with associated elementary divisor non-simple. In this latter case the first two members of a Jordan basis span an invariant 2-space (chapter 1). It is stated here that the term "complex" as applied to an eigenvector means that it is not real and has a non-real eigenvalue. This avoids the situation when, say, r and s are real eigenvectors for some tensor each with real eigenvalue a then $r \pm is$ are trivially complex eigenvectors with real eigenvalue a. Later in this chapter this convention will also be adopted in dealing with "complex" eigenbivectors.

Again with g of any signature let F be a type $(0, 2)$ skew-symmetric tensor at $p \in M$ so that in any chart domain containing p the components of F satisfy $F_{ab} = -F_{ba}$. Henceforth such skew-symmetric tensors, sometimes called 2-forms, will be referred to here as *bivectors* (and, recalling the above remarks, the liberty will be taken on occasion of writing such a bivector as the tensor of type $(1, 1)$ with components $F^a{}_b$). The 6-dimensional real vector space of all such bivectors at p is denoted $\Lambda_p M$. Since F is skew-symmetric, its rank, as a matrix, is an even integer as is known from elementary algebra, that is, 2 or 4. If F has rank 2 it is called *simple* and in this case one may find two independent members $k, k' \in T_p M$ such that $F_{ab} k^b = F_{ab} k'^b = 0$. Such a vector k (or k') will be said to *annihilate* F. One can then check that there exists $p, q \in T_p M$ such that $F^{ab} = p^a q^b - q^a p^b$. In fact, one may, in addition, choose p and q orthogonal by an argument similar to one given above. Of course, p and q are not uniquely determined by F but the 2-dimensional subspace (referred to as a 2-*space*) of $T_p M$ spanned by p and q is uniquely determined and called the *blade* of F. Sometimes one writes, somewhat informally and where no confusion can arise, $F = p \wedge q$ and, indeed, the symbol $p \wedge q$ will also sometimes be used to denote the associated blade. For purposes of calculation, $p \wedge q = (p^a q^b - q^a p^b)$. [Sometimes the blade of F and its orthogonal complement are referred to as the *canonical blades* of F but the *blade* of F will always mean that which is given above.] If F has rank 4 it is called *non-simple* (and this rank classification is independent of the tensor type chosen for F since the matrix representing g is non-singular). Again F may be regarded as a linear map on $T_p M$ given by $v^a \to F^a{}_b v^b$. [Once again the contracting index has been changed (cf chapter 1) from the first to the second, as was done above for the tensor S. In this case a change of sign arises, but no other problems, occur.] Again one calls w a (real or complex) *eigenvector* of F with corresponding eigenvalue $\lambda \in \mathbb{C}$ if $F^a{}_b w^b = \lambda w^a = \lambda \delta^a_b w^b$ or, equivalently, $F_{ab} w^b = \lambda g_{ab} w^b$. Two things are noted here from the skew-symmetry of F; if w is an eigenvector of F with eigenvalue λ a contraction with w easily shows that $\lambda(w \cdot w) = 0$ and so if w is not null, $\lambda = 0$, and if $w, w' \in T_p M$ are eigenvectors of F with eigenvalues μ and ν then $(\mu + \nu)(w \cdot w') = 0$ and so either $\mu = -\nu$ or $w \cdot w' = 0$. It is remarked here that for a real, skew-symmetric tensor at p the real and imaginary parts of a complex (non-real) eigenvector span a (real) invariant 2-space at p. Further, a similar argument to one given above in the symmetric case shows that an invariant 2-space always exists and that if U is an invariant subspace for (the map associated with) F then so also is $U^\perp$.

If $F \in \Lambda_p M$ its *dual* is the bivector $\overset{*}{F} \in \Lambda_p M$ defined in components by $\overset{*}{F}_{ab} = \frac{1}{2}\epsilon_{abcd} F^{cd} = \frac{1}{2}\sqrt{|(\det g)|}\, \delta_{abcd} F^{cd}$ where δ_{abcd} is the usual alternating symbol and $\epsilon_{abcd} = \sqrt{|(\det g)|}\, \delta_{abcd}$. ($\delta_{abcd} = -\delta_{bacd} = -\delta_{abdc}$ and $\delta_{abcd} = \delta_{cdab}$ and similarly for ϵ_{abcd}.) Thus $*$ is a linear map on $\Lambda_p M$. [It is remarked here that $\overset{*}{F}$ only behaves as a tensor under a coordinate transformation with positive Jacobian. This will be seen to cause no essential problems in what is to follow.] It can then be shown from the properties of $*$ (see, for example,

[33, 14]) that the dual of $\overset{*}{F}$ satisfies $\overset{**}{F} = \pm F$ with the plus sign applying for signatures $(+, +, +, +)$ and $(+, +, -, -)$ (that is, $\det g > 0$) and the minus sign for $(-, +, +, +)$ ($\det g < 0$). Thus the linear map $*$ has eigenvalues ± 1 for positive definite and neutral signatures and $\pm i$ for Lorentz signature. The following general relations hold for all signatures.

Lemma 3.1 *The following conditions are equivalent for a non-zero bivector F at $p \in M$.*

(i) F *is simple,*

(ii) *There exists $k \in T_pM$ ($k \neq 0$) satisfying $F_{ab}k^b = 0$,*

(iii) $\overset{*}{F}$ *is simple,*

(iv) $\overset{*}{F}_{ab}F^{bc} = 0$,

(v) $\overset{*}{F}_{ab}F^{ab} = 0$,

(vi) $F_{ab}k_c + F_{bc}k_a + F_{ca}k_b = 0$ *for some non-zero $k \in T_pM$. (k necessarily lies in the blade of F),*

(vii) $F_{ab}F_{cd} + F_{ac}F_{db} + F_{ad}F_{bc} = 0$.

Proof By definition, (i) and (ii) are equivalent and then (i) and (iii) are equivalent by a direct substitution of the expression given earlier for a simple F into the expression for $\overset{*}{F}$ and use of the properties of the alternating symbol. Thus (i), (ii) and (iii) are equivalent (and it is easily seen that, for simple F, the blades of F and $\overset{*}{F}$ are orthogonal with the annihilators of F spanning the blade of $\overset{*}{F}$ and vice versa). A simple calculation then shows that if F is simple and if k is a non-zero member of the blade of F then (vi) holds and that if (vi) holds a contraction with any $k' \in T_pM$ satisfying $k \cdot k' \neq 0$ shows that F is simple and that k lies in the blade of F. Thus (i), (ii), (iii) and (vi) are equivalent. The orthogonality of the blades of a (simple) F and $\overset{*}{F}$ then show that $(i) \Rightarrow (iv)$ and if (iv) holds, and since $F \neq 0$ means there exists $k \in T_pM$ such that $F_{ab}k^b \neq 0$, a contraction of (iv) with k_c shows that $\overset{*}{F}$ is simple. A contraction of (vii) with this vector k then reveals (vi) and so F is simple and if F is simple (vii) is easily shown to hold. Thus (i), (ii), (iii), (iv), (vi) and (vii) are equivalent. Finally if F is simple, (v) easily follows from (iv) and if (v) holds then that condition is equivalent to $\epsilon_{abcd}F^{ab}F^{cd} = 0$ which, with some manipulation of the ϵ symbol, is equivalent to (vii). □

The following result can now be proved.

Lemma 3.2 *Let V be a subspace of Λ_pM such that all members of V are simple. Then $\dim V \leqslant 3$ and all such subspaces V may be easily found.*

Proof The proof that dim$V$ cannot be greater than 3 involves assuming that dim$V \geq 4$ and then constructing a 3-dimensional subspace U of $\Lambda_p M$ all of whose non-trivial members are non-simple [34]. Then using the formula $\dim(U \cap V) = \dim U + \dim V - \dim(U + V)$ gives the contradiction that $\dim(U \cap V) \geq 1$ and so some member of U is non-simple. The construction of U depends on the signature of g and will be given in the appropriate place. It is noted first that if V has the properties in the statement of the lemma, then so does the 3-dimensional subspace $\overset{*}{V} \equiv \{\overset{*}{F} : F \in V\}$. Suppose dim$V = 1$. Then V is spanned by a simple bivector. If dim$V = 2$ let V be spanned by independent $F, G \in \Lambda_p M$. Then F, G and all linear combinations $F + \lambda G$ are simple. For this last bivector one has, from lemma 3.1, $(F_{ab} + \lambda G_{ab})(\overset{*}{F}{}^b{}_c + \lambda \overset{*}{G}{}^b{}_c) = 0$ and, using the fact that F and G are simple, one finds $G_{ab}\overset{*}{F}{}^b{}_c + F_{ab}\overset{*}{G}{}^b{}_c = 0$. Now there exists $0 \neq k \in T_p M$ such that $F_{ab}k^b = 0$ and so on contracting the previous equation with k^a one finds $(G_{ab}k^a)\overset{*}{F}{}^b{}_c = 0$. If $G_{ab}k^a = 0$, k lies in the intersection of the blades of $\overset{*}{F}$ and $\overset{*}{G}$ and hence the blades of F and G have a non-zero common member. If $G_{ab}k^a \equiv k'_b$ is not zero then k' annihilates $\overset{*}{F}$ and hence lies in the blade of F (as well as that of G). Thus if dim$V = 2$ then V is spanned by simple bivectors whose blades intersect in a 1-dimensional subspace of $T_p M$. Then, of course, the blades of $\overset{*}{F}$ and $\overset{*}{G}$ also intersect in a 1-dimensional subspace of $T_p M$.

Now suppose dim$V = 3$ and let V be spanned by simple independent bivectors F, G and H. Consider the subspaces of V spanned by the pairs $\{F, G\}$, $\{F, H\}$ and $\{G, H\}$ and which satisfy the conditions of the previous part of the lemma. These pairs supply, respectively, non-zero $k_1, k_2, k_3 \in T_p M$ satisfying $F_{ab}k_1^b = G_{ab}k_1^b = 0$ and similarly for the other pairs. Consider the span Sp(k_1, k_2, k_3). If this is 1-dimensional there exists $v \in T_p M$ which annihilates each of F, G and H and hence lies in each of the blades of $\overset{*}{F}$, $\overset{*}{G}$ and $\overset{*}{H}$. If this span is 2-dimensional say with k_1 and k_2 independent and k_3 a linear combination of them, then k_1 and k_2 span the blade of $\overset{*}{F}$ and $k_3 = \alpha k_1 + \beta k_2$ $(\alpha, \beta \in \mathbb{R})$. Then whatever the choices for α and β one sees that $\overset{*}{F}$, $\overset{*}{G}$ and $\overset{*}{H}$ are not independent and hence nor are F, G and H and a contradiction follows. If the span of (k_1, k_2, k_3) is 3-dimensional then k_1, k_2 and k_3 are, respectively, in the intersections of the pairs of blades of $\{\overset{*}{F}, \overset{*}{G}\}$, $\{\overset{*}{F}, \overset{*}{H}\}$ and $\{\overset{*}{G}, \overset{*}{H}\}$. Thus $\overset{*}{F}$, $\overset{*}{G}$ and $\overset{*}{H}$ are proportional to $k_1 \wedge k_2$, $k_1 \wedge k_3$ and $k_2 \wedge k_3$, respectively, and so any $0 \neq k \in T_p M$ orthogonal to k_1, k_2 and k_3 is unique up to a scaling and is common to the blades of F, G and H. Thus if dim $V = 3$, $V = Sp(F, G, H)$, either the blades of F, G, H intersect in a 1-dimensional subspace of $T_p M$ or the blades of $\overset{*}{F}, \overset{*}{G}, \overset{*}{H}$ intersect in a 1-dimensional subspace of $T_p M$. $\square$

From the metric g on M one may define a *(bivector) metric* P on $\Lambda_p M$, where for $F, G \in \Lambda_p M$, $P(F, G) \equiv P_{abcd} F^{ab} G^{cd}$ where $P_{abcd} = \frac{1}{2}(g_{ac}g_{bd} - g_{ad}g_{bc})(p)$ and so $P_{abcd} = -P_{bacd} = -P_{abdc}$, $P_{abcd} = P_{cdab}$, $P_{abcd} + P_{acdb} + P_{adbc} = 0$ and $P^c{}_{acb} = \frac{3}{2}g_{ab}$. For any bivector F one may use P to raise and lower skew-symmetric index pairs since $F_{ab} = P_{abcd}F^{cd}$ and so $P(F, G) = F_{ab}G^{ab}$. Usually one writes $P(F, G)$ as $F \cdot G$ and $F \cdot F$ is denoted $|F|$. It is easily checked that if g has positive definite signature P also has positive definite signature, if g has Lorentz signature P has signature $(+, +, +, -, -, -)$ and if g has neutral signature P has signature $(+, +, -, -, -, -)$. It is also useful at this point to introduce the notation for the complete symmetrisation and complete skew-symmetrisation of tensor indices (or a collection of tensor indices) by enclosing them inside round or square brackets, respectively. [This notation will not be excessively used: writing out the full expression, whilst longer, is often more easily visualised.] Thus, for example, if T is a type $(0, 2)$ tensor and W is a type $(0, 3)$ tensor with respective components T_{ab} and W_{abc} one defines

$$T_{(ab)} = \frac{1}{2}(T_{ab} + T_{ba}), \qquad T_{[ab]} = \frac{1}{2}(T_{ab} - T_{ba}) \tag{3.12}$$

and

$$W_{(abc)} = \frac{1}{6}(W_{abc} + W_{bca} + W_{cab} + W_{bac} + W_{cba} + W_{acb}), \tag{3.13}$$

$$W_{[abc]} = \frac{1}{6}(W_{abc} + W_{bca} + W_{cab} - W_{bac} - W_{cba} - W_{acb}). \tag{3.14}$$

Thus for the curvature tensor, the third symmetry in (3.4) may be written $R_{a[bcd]} = 0$, for the tensor E in (3.10) $E_{a[bcd]} = 0$, for the Weyl tensor in (3.11) $C^a{}_{[bcd]} = 0$ and for the bivector metric $P_{a[bcd]} = 0$ whilst conditions (vi) and (vii) in lemma 3.1 may be written $F_{[ab}k_{c]} = 0$ and $F_{a[b}F_{cd]} = 0$. With this notation one may establish the following result which will be useful later in decompositions of the curvature, Weyl and E tensors.

Lemma 3.3 (i) *Let F and G be simple bivectors at $p \in M$. Then*

$$F_{a[b}G_{cd]} + G_{a[b}F_{cd]} = 0 \tag{3.15}$$

if and only if the blades of F and G intersect non-trivially.

(ii) *Let $F = P + Q$ and $G = P - Q$ be bivectors at $p \in M$ with P and Q simple. Then (3.15) holds for F and G.*

Proof For (i) if F and G are proportional the result follows trivially from lemma 3.1. So suppose that F and G are (simple and) independent and satisfy (3.15) and let $k \in T_p M$ be an annihilator of F with $G_{ab}k^b \equiv k'_a \neq 0$, so that k' is in the blade of G. Then a contraction of (3.15) with k^a and use of lemma 3.1(vi) shows that k' lies in the blades of F (and G). Conversely, if the blades of F and G intersect non-trivially then $H = F + G$ is simple and a substitution

into lemma 3.1(vii) and use of the fact that F and G are simple reveals (3.15). For part (ii), substitute $F = P + Q$ and $G = P - Q$ into the left hand side of (3.15) and after some obvious cancellations and by using the fact that P and Q are simple (lemma 3.1) one achieves zero. [It is remarked here that (3.15) could hold with F simple and G not. For example and for a basis p, q, r, s at p, $F = p \wedge s$ and $G = p \wedge q + r \wedge s$, as is clear from the first part of this lemma]
□

The 6-dimensional real vector space $\Lambda_p M$ of bivectors at p can be given the structure of the 6-dimensional manifold $\mathbb{R}^6$. Consider the 6-dimensional open (hence regular) submanifold $\Lambda'_p M \subset \Lambda_p M$ of all non-zero bivectors and let $\sim$ be the equivalence relation on $\Lambda'_p M$ given, for $F, G \in \Lambda'_p M$, by $F \sim G$ if and only if $F = \nu G$ for $0 \neq \nu \in \mathbb{R}$ (see, e.g., [31]). Then $\Lambda'_p M / \sim$ is a quotient manifold of $\Lambda'_p M$ diffeomorphic to the projective space $P\mathbb{R}^5$ with associated smooth submersion $\mu : \Lambda'_p M \to \Lambda'_p M / \sim$ and whose members are sometimes referred to as *projective bivectors*. The advantage of this construction arises from the fact that simple bivectors at $p \in M$ determine their blades uniquely but a 2-space in $T_p M$ only determines an equivalence class (in the above sense) of simple bivectors. To introduce the set of simple bivectors consider the smooth map $f : \Lambda'_p M \to \mathbb{R}$ given by $F \to F_{ab} \overset{*}{F}{}^{ab} = \frac{1}{2} \epsilon_{abcd} F^{ab} F^{cd}$. Choose a basis x, y, z, t for $T_p M$ (with the basis members being mutually orthogonal unit vectors so that it applies to any of the three signatures) and then a (bivector) basis $x \wedge y, x \wedge z, y \wedge z, x \wedge t, y \wedge t, z \wedge t$ of $\Lambda_p M$ so that the first three members of this latter basis are, up to signs, the duals of the last three. Then it follows that, for $0 \neq F \in \Lambda_p M$, $f(F)$ is quadratic in the components of F in the bivector basis. The simple members of $\Lambda_p M$ constitute the set $f^{-1}\{0\}$ (lemma 3.1(v)) and using the above bases, f may be checked to have rank equal to 1 on $f^{-1}\{0\}$ whatever the signature of g. Thus the set of all simple bivectors at p constitute a 5−dimensional regular submanifold of $\Lambda'_p M$ and hence of $\Lambda_p M$ (chapter 2). Now $\Lambda_p M$ is connected, Hausdorff and second countable and hence admits a positive definite metric, say γ. So consider the smooth map $h : \Lambda'_p M \to \mathbb{R}$ given by $F \to \frac{f(F)}{\gamma(F,F)}$. This map respects $\sim$ and so leads to a smooth map $h' : \Lambda'_p M / \sim \to \mathbb{R}$ where $h = h' \circ \mu$ (chapter 2) and where $h'^{-1}\{0\}$ is the set of *projective simple bivectors*. Now since f has rank 1 on the set of all simple bivectors at p, $f^{-1}\{0\}$, it can be checked that h has rank 1 on the set $h'^{-1}\{0\}$ of all projective simple bivectors. It follows that, since $h = h' \circ \mu$ with μ a submersion, h' also has rank 1 on the set $h'^{-1}\{0\}$ of all projective simple bivectors and hence that this latter set can be given the structure of a 4-dimensional regular submanifold of the manifold of all projective bivectors (section 2.7). Useful information for rank calculations for such maps can be found in [9].

To complete this argument consider the set of all 2-spaces of $T_p M$. This set is in bijective correspondence with the manifold of projective simple bivectors above. Then such a 2-space U is determined by 2 independent vectors at p thought of as column vectors of a 4×2 matrix of rank 2 in some basis of $T_p M$.

Suppose that the first two rows of this matrix are independent. Then U determines exactly one such matrix whose columns are $(1, 0, x, y)$ and $(0, 1, a, b)$. So the subset of the set of 2-spaces to which this applies is in bijective correspondence with $\mathbb{R}^4$ by associating each member with (x, y, a, b) and one is lead to a chart for this set of 2-spaces at p. This can then be extended to give a smooth 4-dimensional manifold structure to the set of all 2-spaces of $T_p M$ called the *Grassmann manifold* of all 2-spaces of $T_p M$ and is denoted by $G(2, \mathbb{R}^4)$ (cf section 2.7). Any two independent members of such a 2-space with components say $X^a = (1, 0, x, y)$ and $Y^a = (0, 1, a, b)$ then give rise to the simple bivector $F^{ab} = X^a Y^b - Y^a X^b$ with coordinates (in $\mathbb{R}^6$) in some pre-chosen order $(1, a, b, -x, -y, xb - ya)$ and so the coordinates of the corresponding projective bivector (in $P\mathbb{R}^5$) may be given by $(a, b, -x, -y, xb - ya)$. Since $G(2, \mathbb{R}^4)$ and the set of projective simple bivectors at p are in a bijective correspondence, with the latter a regular submanifold of the manifold of projective bivectors $P\mathbb{R}^5$ at p, one may identify $G(2, \mathbb{R}^4)$ as a subset of $P\mathbb{R}^5$ and then the map $(x, y, a, b) \rightarrow (a, b, -x, -y, xb - ya)$ is a smooth coordinate representative of the inclusion map $G(2, \mathbb{R}^4)$ into $P\mathbb{R}^5$. This map is of rank 4 and shows that $G(2, \mathbb{R}^4)$ is a 4-dimensional submanifold of $P\mathbb{R}^5$. Thus the set $G(2, \mathbb{R}^4)$ and the set of projective simple bivectors can be identified and each has the structure of a 4-dimensional submanifold of $P\mathbb{R}^5$ with the latter regular. It follows (section 2.7) that these manifolds are diffeomorphic. Thus one may regard projective simple bivectors as the Grassmann manifold $G(2, \mathbb{R}^4)$. It is remarked here that $G(2, \mathbb{R}^4)$ is Hausdorff, compact, second countable and connected [9] and is a closed subset of $P\mathbb{R}^5$. The work in the last two paragraphs made no use of a metric on M and hence applies to each of the three signatures to be considered for M.

The 6-dimensional vector space $\Lambda_p M$ of bivectors at p can be given the structure of a Lie algebra in the following way (and which involves, as explained earlier, the identification of F_{ab} with $F^a{}_b \equiv g^{ac} F_{cb}$ or with $F_a{}^b \equiv g^{cb} F_{ac}$). For $F, G \in \Lambda_p M$ define in one (and hence any) coordinate system the product $[F, G] \equiv F_{ac} G^c{}_b - G_{ac} F^c{}_b$ which is clearly skew-symmetric in the indices a and b. This product, sometimes referred to as the *commutator* of F and G (in matrix language it is $FG - GF$), can be checked to satisfy the conditions required of a Lie algebra given in chapter 1. The symbol $\Lambda_p M$ will still be used to denote this Lie algebra. Its subalgebras will turn out to be important.

One can extend the dual operator to the type $(0, 4)$ tensors E and C but noting that because of the existence of two pairs of skew symmetric indices one has two duals, right and left, indicated by the positional placing of the duality operator symbol, for each of these tensors and which are given by (see, e.g. [14])

$$^*C_{abcd} = \frac{1}{2}\sqrt{(det|g|)}\delta_{abef}C^{ef}{}_{cd}, \qquad C^*_{abcd} = \frac{1}{2}\sqrt{(det|g|)}C_{ab}{}^{ef}\delta_{cdef} \quad (3.16)$$

and with analogous expressions for the tensor E. Thus the type $(0,4)$ tensors C^*, $^*C$, E^* and $^*E$ are skew-symmetric in the first and also in the second pair of indices. One may then apply a useful formulae, known as the Lanczos identity (see, e.g.[14], page 51, but care should be taken since the definition of the tensor E there differs from that used here), to get, in an obvious notation,

$$^*C = C^*, \qquad ^*C^* = \epsilon C, \qquad ^*E = -E^*, \qquad ^*E^* = -\epsilon E, \qquad (3.17)$$

where $\epsilon = 1$ for positive definite and neutral signatures and $\epsilon = -1$ for Lorentz signature. It is easily checked that the first two equations in (3.17) are equivalent, as are the last two (cf the expression $\overset{**}{F} = \epsilon F$ for these signatures given earlier this section). It should be noted that some of the above relations are independent of signature and so do not depend on ϵ. It is also noted that from the first of (3.17) that

$$^*C_{abcd} = C^*_{abcd} = \frac{1}{2} C_{abmn} \epsilon^{mn}{}_{cd} = \frac{1}{2} C_{mnab} \epsilon^{mn}{}_{cd} = {}^* C_{cdab}. \qquad (3.18)$$

Thus the type $(0,4)$ tensor $^*C$ (and hence the type $(0,4)$ tensor C^*) are symmetric with respect to the interchange of the first two indices with the last two indices. This result is false for the tensor E where a change of sign is involved, $^*E_{abcd} = -{}^*E_{cdab}$. It can also be checked that the condition $C_{a[bcd]} = 0$ (see (3.11)) is equivalent to $^*C^c{}_{acb} = 0$ and hence to $C^{*c}{}_{acb} = 0$ and that $C^c{}_{acb} = 0$ (see (3.11)) is equivalent to $^*C_{a[bcd]} = 0$ and thus to $C^*_{a[bcd]} = 0$. Also $^*E^c{}_{acb} = E^{*c}{}_{acb} = 0$ and (see (3.10)) $E^c{}_{acb} = \widetilde{R_{ab}}$. It is also remarked for future use that the bivector metric P satisfies the following conditions (with the definition of ϵ and shorthand notation as above) [14]

$$^*P^* = \epsilon P, \qquad ^*P = P^*. \qquad (3.19)$$

Equations (3.17) and (3.19) allow the computation of the duals $^*Riem$, $Riem^*$ and $^*Riem^*$ for each signature.

Lemma 3.4 *For any bivectors F, G and H,*

(i) $F \cdot [G, H] = G \cdot [H, F] = H \cdot [F, G]$,

(ii) $F \cdot [F, G] = G \cdot [F, G] = 0$,

(iii) $F \cdot \overset{*}{G} = \overset{*}{F} \cdot G$, $\qquad \overset{*}{F} \cdot \overset{*}{G} = \epsilon F \cdot G \ (\Rightarrow |\overset{*}{F}| = \epsilon |F|)$.

Proof For (i) one computes directly, in indices, as follows. $F \cdot [G, H] = F^b{}_a(G^a{}_c H^c{}_b - H^a{}_c G^c{}_b) = G^a{}_c(F^b{}_a H^c{}_b - F^c{}_b H^b{}_a) = G^a{}_c(H^c{}_b F^b{}_a - F^c{}_b H^b{}_a) = G \cdot [H, F]$, etc. For part (ii) one simply puts $F = H$ into the result of part (i) [67]. The proof of the first result in part (iii) consists simply of transferring the dual across. The final part then follows. $\qquad \square$

3.4   The Positive Definite Case and Tensor Classification

Now suppose that g has signature $(+, +, +, +)$ on M. If x, y, z, w is an orthonormal basis at $p \in M$, one has the completeness relation relating this basis with the metric at p given by $g_{ab} = x_a x_b + y_a y_b + z_a z_b + w_a w_b$ at p. Conversely this last relation for $x, y, z, w \in \overset{*}{T}_p M$ is easily checked to imply that x, y, z, w give rise to an orthonormal basis at p. To see this suppose the last relation holds at p for $x, y, z, w \in \overset{*}{T}_p M$ which are then clearly linearly independent. A contraction with x^b gives $x_a = (x \cdot x) x_a + (x \cdot y) y_a + (x \cdot z) z_a + (x \cdot w) w_a$ and so $x \cdot x = 1$, $x \cdot y = x \cdot z = x \cdot w = 0$. Similar contractions with y^b, z^b and w^b complete the proof. If S is a symmetric tensor at $p \in M$ and if one chooses a chart x whose domain contains p and in which the components g_{ab} of g at p take the Sylvester form $\delta_{ab} \equiv \text{diag}(1, 1, 1, 1)$, as one always can, the eigenvector-eigenvalue problem for S discussed at the beginning of the last section gives for an eigenvector v and corresponding eigenvalue λ of S at p, $S_{ab} v^b = \lambda \delta_{ab} v^b$. It follows from section 1.5 that, since S is symmetric, it is diagonalisable over $\mathbb{R}$ and of Segre type $\{1111\}$ or one of its degeneracies. This classification thus applies to the tensors $Ricc$ and $\widetilde{Ricc}$ (and their Segre types are necessarily the same, including degeneracies, from the definition of $\widetilde{Ricc}$). It is remarked (and easily checked) that in this case any invariant 2-space of S contains two independent (real) eigenvectors which may be chosen to be orthogonal.

If F is a simple bivector at $p \in M$ one may always write $F_{ab} = \alpha(u_a v_b - v_a u_b)$ for real associated $u, v \in \overset{*}{T}_p M$ with $| u | = | v | = 1$ and $\alpha \in \mathbb{R}$. One may always choose u and v to be orthogonal and then $F^a{}_b v^b = \alpha u^a$, $F^a{}_b u^b = -\alpha v^a$ and if $r, s \in T_p M$ with the 2-space $r \wedge s$ orthogonal to the 2-space $u \wedge v$ (and choosing r and s unit and orthogonal), $F^a{}_b r^b = F^a{}_b s^b = 0$. Thus u, v, r, s constitute an orthonormal tetrad and $F^a{}_b (u^b \pm iv^b) = \pm i\alpha(u^a \pm iv^a)$ so that F is diagonalisable over $\mathbb{C}$ with eigenvalues $i\alpha, -i\alpha, 0, 0$ and this is designated by the Segre symbol $\{z\bar{z}11\}$ with the symbol "$z\bar{z}$" denoting a complex conjugate pair of (non-real) eigenvalues. In the above language, $\overset{*}{F}$ is proportional to $r \wedge s$ and its blade is the zero-eigenspace of F. Thus in this case F uniquely determines the 2-spaces $p \wedge q$ and $r \wedge s$. [It is noted here that in the positive definite case a subspace of $T_p M$ together with its orthogonal complement, as defined earlier, are orthogonal complements in the usual sense of the word, that is, their span is $T_p M$.]

Now suppose that F is a non-simple bivector at p. Then the non-simple condition shows that all its eigenvalues are non-zero. Results in section 3.3 above show first that, since g is positive definite, any *real* eigenvector of F (which cannot be null) must have a zero eigenvalue and this contradiction reveals that all eigenvalues are (non-zero and) complex. Second if k and $\bar{k}$

are a complex conjugate pair of (complex) eigenvectors, $F_{ab}k^b = \lambda k_a$ and $F_{ab}\bar{k}^b = \bar{\lambda}\bar{k}_a$, $\lambda \in \mathbb{C}$ and $k = x + iy$ for $x, y \in T_pM$ then $\lambda + \bar{\lambda} = 0$ or $k \cdot \bar{k} = 0$. The latter gives the contradiction $|x| + |y| = 0$ and so $\lambda = i\alpha$ for $0 \neq \alpha \in \mathbb{R}$. Then, since $\alpha \neq 0$, the condition that $x \pm iy$ are null, $|x \pm iy| = 0$, shows that $|x| = |y|$ and $x \cdot y = 0$ and so x and y are orthogonal. So $x \wedge y$ is a (real) invariant 2-space for F and its orthogonal complement, which is also invariant, is easily seen to yield another conjugate pair of eigenvectors $z \pm iw$ with eigenvalues $\pm i\beta$ and with $z, w \in T_pM$ and $0 \neq \beta \in \mathbb{R}$. If $\alpha \neq \pm\beta$ the Segre type of F is $\{z\bar{z}w\bar{w}\}$ and one has $(x \pm iy) \cdot (z \pm iw) = 0$, that is, $x \cdot z = x \cdot w = y \cdot z = y \cdot w = 0$. Thus in this case the eigenvectors are each determined up to a complex scaling and so the 2-spaces $x \wedge y$ and $z \wedge w$ are *uniquely determined* by F and orthogonal and are referred to as the *canonical blades of F*. The vectors x, y, z, w, after scalings, form an orthonormal tetrad at p and

$$F_{ab} = \alpha(x_a y_b - y_a x_b) + \beta(z_a w_b - w_a z_b). \tag{3.20}$$

Now suppose F is non-simple but the eigenvalues above are equal, say, $\alpha = \beta$ (the case $\alpha = -\beta$ is similar). The Segre type of F is then written $\{(zz)(\bar{z}\bar{z})\}$ and the eigenvectors of F may be taken as $x + iy$ and $z + iw$ (eigenvalue $i\alpha$) and $x - iy$ and $z - iw$ (eigenvalue $-i\alpha$). However, the 2-spaces $x \wedge y$ and $z \wedge w$ are not now uniquely determined by F. To see this let x', y', z', w' be another orthonormal basis at p, let $F' = \alpha(x' \wedge y' + z' \wedge w')$ and suppose $F = F'$. Then $x' + iy'$ is an eigenvector of $F' = F$ and thus lies in the $i\alpha-$ eigenspace of F spanned by $x + iy$ and $z + iw$. Assuming that F' is not obtained from F by the trivial switch $x' \wedge y' = z \wedge w$, one can write

$$x' + iy' = K(x+iy) + (c+id)(z+iw), \quad z' + iw' = K'(z+iw) + (m+in)(x+iy) \tag{3.21}$$

where $K, K', c, d, m, n \in \mathbb{R}$ with $K > 0 < K'$. Here, complex scalings have been used to make K and K' real and positive and correspond to ignoring rotations of x, y in $x \wedge y$ and similarly z, w in $z \wedge w$ (which would not change the 2-spaces $x \wedge y$ and $z \wedge w$). Applying the usual orthonormality condition on the basis x', y', z', w' then shows that F may be written either as proportional to $\alpha(x \wedge y + z \wedge w)$ or to $\alpha(x' \wedge y' + z' \wedge w')$ provided that (after normalising)

$$x' = K(x + cz - dw), \quad y' = K(y + dz + cw), \quad z' = K(z - cx - dy),$$
$$w' = K(w - cy + dx), \quad (K = (1 + c^2 + d^2)^{-\frac{1}{2}}). \tag{3.22}$$

In this case any such pair of blades $x' \wedge y'$ and $z' \wedge w'$ (and including the pair $x \wedge y$ and $z \wedge w$) will be referred to as a *canonical blade pair* for F.

Thus, algebraically, a bivector at p may be either simple, non-simple of Segre type $\{z\bar{z}w\bar{w}\}$ with no degeneracies or non-simple and of Segre type $\{(zz)(\bar{z}\bar{z})\}$. These remarks essentially complete the classification of bivectors for positive definite signature.

This argument regarding the ambiguity in the above tetrad for F may be viewed in another important way (see, e.g. [32, 34]). Define two 3-dimensional subspaces $\overset{+}{S}_p$ and $\overset{-}{S}_p$ of $\Lambda_p M$ by

$$\overset{+}{S}_p = \{F \in \Lambda_p M : \overset{*}{F} = F\}, \qquad \overset{-}{S}_p = \{F \in \Lambda_p M : \overset{*}{F} = -F\}. \tag{3.23}$$

The subspaces $\overset{+}{S}_p$ and $\overset{-}{S}_p$ are sometimes referred to as the subspaces of *self dual* and *anti-self dual* bivectors at p, respectively. Then clearly $\overset{+}{S}_p \cap \overset{-}{S}_p = \{0\}$ and since any $F \in \Lambda_p M$ may be written as $\frac{1}{2}(F + \overset{*}{F}) + \frac{1}{2}(F - \overset{*}{F})$, that is, as the sum of a member of $\overset{+}{S}_p$ and a member of $\overset{-}{S}_p$ and clearly in a unique way, one sees that $\Lambda_p M$ is a direct sum

$$\Lambda_p M = \overset{+}{S}_p + \overset{-}{S}_p. \tag{3.24}$$

Thus a bivector G is in $\overset{+}{S}_p$ (respectively, in $\overset{-}{S}_p$) if and only if $G = F + \overset{*}{F}$ (respectively, $G = F - \overset{*}{F}$) for some bivector F. By viewing the dual operation as a linear map on $\Lambda_p M$ it follows, since $\epsilon = 1$ for this signature, that $\overset{**}{F} = F$, and so this map has only the eigenvalues ± 1 and the decomposition (3.23) shows that the corresponding eigenspaces are $\overset{+}{S}_p$ and $\overset{-}{S}_p$, respectively. It is clear that if F is simple it is independent (in $\Lambda_p M$) of $\overset{*}{F}$. As a consequence of this and for any $F \in \Lambda_p M$, F and $\overset{*}{F}$ are independent if and only if either F is simple, or F is non-simple and $F \notin \widetilde{S}_p$ where $\widetilde{S}_p = \overset{+}{S}_p \cup \overset{-}{S}_p$. It follows that all non-trivial members of $\widetilde{S}_p$ are non-simple [and $\overset{+}{S}_p$ or $\overset{-}{S}_p$ give convenient examples required for the completion of the proof of lemma 3.2 for this signature (private communication from Z Wang)]. In fact if one chooses an orthonormal tetrad x, y, z, w at p and arranges its orientation so that $(x \wedge y)^* = z \wedge w$, $(x \wedge z)^* = w \wedge y$ and $(x \wedge w)^* = y \wedge z$ then $\overset{+}{S}_p$ has a basis consisting of $(x \wedge y + z \wedge w), (x \wedge z + w \wedge y)$ and $(x \wedge w + y \wedge z)$ whilst a basis for $\overset{-}{S}_p$ is $(x \wedge y - z \wedge w), (x \wedge z - w \wedge y)$ and $(x \wedge w - y \wedge z)$. In fact, the non-simple bivector (3.20) is in $\widetilde{S}_p$ if and only if $\alpha = \pm \beta$. The following results now hold.

Lemma 3.5 (i) *If $F \in \overset{+}{S}_p$ and $G \in \overset{-}{S}_p$ then $F \cdot G = 0$ and $[F, G] = 0$. For $Q, R \in \Lambda_p M$, $[Q, \overset{*}{R}] = [\overset{*}{Q}, R] = [Q, R]^*$ and $[\overset{*}{Q}, \overset{*}{R}] = [Q, R]$.*

(ii) *If A, B, C denote, respectively, the basis members above for $\overset{+}{S}_p$ then $|A| = |B| = |C| = 4$ and $A \cdot B = A \cdot C = B \cdot C = 0$ and using the commutator operator, $[A, B] = -2C$, $[A, C] = 2B$ and $[B, C] = -2A$. Writing $2A' = -A$, $2B' = -B$ and $2C' = -C$ one gets the more usual form*

$[A', B'] = C'$, $[B', C'] = A'$ and $[C', A'] = B'$ for the algebra product and with A', B', C' an orthonormal basis for $\overset{+}{S}_p$. Thus $\overset{+}{S}_p$ is a Lie subalgebra of $\Lambda_p M$ and, in fact, is isomorphic to the Lie algebra $o(3)$ (see, e.g. [83]). Similar comments apply to the subspace $\overset{-}{S}_p$. Thus $\overset{+}{S}_p$ and $\overset{-}{S}_p$ are isomorphic. Neither has a 2-dimensional subalgebra. Thus if $A, B \in \overset{+}{S}_p$ are independent, $[A, B] \neq 0$, and similarly for $\overset{-}{S}_p$.

(iii) Suppose $0 \neq A \in \Lambda_p M \setminus \widetilde{S}_p$. Then for $X \in \Lambda_p M$, $[X, A] = 0 \Rightarrow X$ is a linear combination of A and $\overset{*}{A}$. If $A \in \overset{+}{S}_p$ and $[X, A] = 0$, then $X = \lambda A + Q$ where $\lambda \in \mathbb{R}$ and $Q \in \overset{-}{S}_p$ and similarly for $A \in \overset{-}{S}_p$.

(iv) Let $F \in \overset{+}{S}_p$ and $G \in \overset{-}{S}_p$ with $|F| = \alpha(> 0)$ and $|G| = \beta(> 0)$. Then there exists a unique pair of simple bivectors in the span $Sp(F, G)$ and which are duals of each other and given by $H \equiv F + \kappa G$ and $\overset{*}{H} \equiv F - \kappa G$ where $\kappa = (\frac{\alpha}{\beta})^{\frac{1}{2}}$. Thus $2F = H + \overset{*}{H}$ and $2\kappa G = H - \overset{*}{H}$. It thus follows from lemma 3.3(ii) that one has $F_{a[b}G_{cd]} + G_{a[b}F_{cd]} = 0$.

(v) If $F, G \in \overset{+}{S}_p$ are non-zero and have a common canonical blade amongst their collective canonical blade pairs then they have a common canonical blade pair and are proportional. Similar comments apply to $\overset{-}{S}_p$.

(vi) If $F, G, H \in \overset{+}{S}_p$, or $F, G, H \in \overset{-}{S}_p$ the relations $|[F, G]| = |F||G| - (F \cdot G)^2$ and $[F, [G, H]] = (F \cdot H)G - (F \cdot G)H$ hold.

(vii) If f is a Lie algebra isomorphism from any of $\overset{+}{S}_p$ and $\overset{-}{S}_p$ to any of $\overset{+}{S}_p$ and $\overset{-}{S}_p$ then for $F \in \overset{+}{S}_p$ (or $\overset{-}{S}_p$), $|F| = |f(F)|$.

Proof The first result in part (i) follows from lemma 3.4(iii) whilst the second result in part (i) follows from a direct calculation from the above bases for $\overset{+}{S}_p$ and $\overset{-}{S}_p$. For the rest of part (i) one writes $Q = F + G$ with $F \in \overset{+}{S}_p$ and $G \in \overset{-}{S}_p$ and similarly for R and uses the previous result. The fact that $\overset{+}{S}_p$ and $\overset{-}{S}_p$ are subalgebras is important here.

The first of part (ii) requires a calculation from the above bases for $\overset{+}{S}_p$ and $\overset{-}{S}_p$. For the last part of (ii), let G, H span a 2-dimensional subalgebra of $\overset{+}{S}_p$. Then it is clear that, by taking judicious linear combinations one can arrange that, in terms of the basis A, B, C, $G = A + aB$, $H = C$, or that $G = A + bC$, $H = B + cC$ for $a, b, c \in \mathbb{R}$. Then setting $[G, H] = \mu G + \nu H$ $(\mu, \nu \in \mathbb{R})$ no solutions for the real pair (μ, ν) are possible. Similar comments apply to $\overset{-}{S}_p$. The final statement in (ii) is clear.

For part (iii) write $X = \overset{+}{X} + \overset{-}{X}$ and $A = \overset{+}{A} + \overset{-}{A}$ for $\overset{+}{X}, \overset{+}{A} \in \overset{+}{S}_p$ and $\overset{-}{X}, \overset{-}{A} \in \overset{-}{S}_p$. Then one has $[\overset{+}{X} + \overset{-}{X}, \overset{+}{A} + \overset{-}{A}] = 0$ and so, since $\overset{+}{S}_p$ and $\overset{-}{S}_p$ are subalgebras with only trivial intersection, parts (i) and (ii) give $[\overset{+}{X}, \overset{+}{A}] = [\overset{-}{X}, \overset{-}{A}] = 0$. Thus if $A \in \Lambda_p M \setminus \widetilde{\overset{+}{S}_p}$, $\overset{+}{A}$ and $\overset{-}{A}$ are each non-zero and since neither $\overset{+}{S}_p$ nor $\overset{-}{S}_p$ has a 2-dimensional subalgebra, $\overset{+}{X}$ and $\overset{-}{X}$ are multiples, possible zero, respectively, of $\overset{+}{A}$ and $\overset{-}{A}$ and the result follows. If $A \in \overset{-}{S}_p$ so that $\overset{-}{A} = 0$, one similarly gets $[\overset{+}{X}, A] = 0$ and $\overset{+}{X}$ is a multiple of A (possibly zero) and $\bar{X}$ is unrestricted). The result follows.

For part (iv) let $F \in \overset{+}{S}_p$ and $G \in \overset{-}{S}_p$ with $|F| = \alpha > 0$ and $|G| = \beta > 0$ and consider $F + \kappa G$ for $\kappa \in \mathbb{R}$. This bivector is simple if and only if $\kappa = \pm(\frac{\alpha}{\beta})^{\frac{1}{2}}$ (lemma 3.1(v)) and so F and G fix a unique pair of simple bivectors in their span in $\Lambda_p M$ given by $H \equiv F + \kappa G$ and $\overset{*}{H} \equiv F - \kappa G$ with orthogonal blades, say, $x \wedge y$ and $z \wedge w$ which are a pair of canonical blades for F and G since $2F = H + \overset{*}{H}$ and $2\kappa G = H - \overset{*}{H}$. [It is remarked that if one fixes F and varies G one achieves the blade pair ambiguity mentioned earlier for $F \in \overset{+}{S}_p$.] The rest of part (iv) is clear.

For part (v) if F and G have a common canonical blade represented by (some multiple) of a simple bivector A, one may write $F = \lambda A + B$ and $G = \mu A + C$ where B and C are simple bivectors with $\overset{*}{B} = \lambda A$ and $\overset{*}{C} = \mu A$ ($\lambda, \mu \in \mathbb{R}$ with each non-zero). It follows that $B = \frac{\lambda}{\mu} C$ and so $F = \frac{\lambda}{\mu} G$. [Curiously, if $F \in \overset{+}{S}_p$ with $F = A + B$ for simple bivectors A and B it does not follow that $\overset{*}{A} = B$. For example, in the above basis for $T_p M$, $F = x \wedge y + z \wedge w \in \overset{+}{S}_p$ and so $F = (x \wedge y + x \wedge w) + (z \wedge w - x \wedge w)$. Then $x \wedge y + x \wedge w$ and $z \wedge w - x \wedge w$ are non-zero and simple but not a dual pair. This example was suggested to the author by Z. Wang.]

Part (vi) is readily checked to be the case by expanding out F, G and H in terms of the basis A', B', C' given in part (ii) and computing.

For part (vii) consider the case $f : \overset{+}{S}_p \to \overset{-}{S}_p$ (the others are similar) and choose the orthonormal basis A', B', C' above for $\overset{+}{S}_p$. Then $A'' \equiv f(A')$, $B'' \equiv f(B')$ and $C'' \equiv f(C')$ give a basis for $\overset{-}{S}_p$. Now since f is a Lie algebra isomorphism $[A'', B''] = f([A', B']) = f(C') = C''$, etc and so, from lemma 3.4(ii), A'', B'' and C'' are mutually orthogonal. Then the first part of (vi) above shows that $|C''| = |[A'', B'']| = |A''||B''|$ and similarly $|A''| = |[B'', C'']| = |B''||C''|$ and $|B''| = |[C'', A'']| = |C''||A''|$ from which it follows from the positive definite nature of the bivector metric P that $|A''| = |B''| = |C''| = 1$. The result now follows by linearity. $\qquad \square$

Thus $\overset{+}{S}_p$ and $\overset{-}{S}_p$ are each isomorphic to $o(3)$ and (3.24) and lemma 3.5(*i*) show that $\Lambda_p M$ together with the commutator product is the *product Lie algebra* $o(3) + o(3)$. Parts (*vi*) and (*vii*) were suggested to the author by Z Wang [67].

Now let F be a bivector at p, regarded now as a type $(1,1)$ tensor with components $F^a{}_b$ and let h be a symmetric type $(0,2)$ tensor at p. Then consider the type $(0,2)$ tensor hF at p with components $h_{ac}F^c{}_b$ [13]. If one equates h with the metric g, the tensor hF is skew-symmetric with components F_{ab} but, otherwise, this is not necessarily true. But suppose h is such that hF is skew-symmetric [13, 34] and cf [85].

Lemma 3.6 *Suppose that h is a non-zero, symmetric tensor and F a bivector at $p \in M$ such that the above tensor hF is skew-symmetric, that is,*

$$h_{ac}F^c{}_b + h_{bc}F^c{}_a = 0. \tag{3.25}$$

(*i*) *If U is a (real) eigenspace of F it is an invariant subspace for h. Thus if U is 1-dimensional, it gives an eigendirection for h.*

(*ii*) *If F is simple, its blade is an eigenspace of h with respect to g, that is, if $u, v \in T_p M$ span the blade of F, $h_{ab}u^b = \lambda g_{ab}u^b = \lambda u_a$ and similarly $h_{ab}v^b = \lambda v_a$, $\lambda \in \mathbb{R}$.*

(*iii*) *If F is not simple and not a member of $\widetilde{S}_p$ the (unique pair of) canonical blades of F are each eigenspaces of h.*

(*iv*) *If $F \in \widetilde{S}_p$ then any solution of (3.25) for h, excluding multiples of g, has a pair of 2-dimensional eigenspaces which coincide with a canonical pair of blades for F. Each representation (and canonical pair of blades) for F, as in (3.20) and (3.22), gives rise to a distinct solution for h whose eigenspaces are these canonical blades.*

(*v*) *If F and G satisfy (3.25), then $[F, G]$ also satisfies (3.25). Thus for a fixed h, the solutions F of (3.25) form a subalgebra of $\Lambda_p M$.*

Proof For part (*i*) suppose k is an eigenvector of F with eigenvalue α. Then a contraction of (3.25) with k^b reveals that $h_{ab}k^b$ is an eigenvector of F with eigenvalue α and hence lies in the α-eigenspace of F. Thus any eigenspace of F is an invariant space of h.

For part (*ii*) the blade of $\overset{*}{F}$ is the 0-eigenspace of F and hence is invariant for h by part (*i*). Thus the blade of F, being orthogonal to that of $\overset{*}{F}$ is invariant for h and hence h admits two independent eigenvectors in this 2-space since the latter has a positive definite induced metric. Writing $F^{ab} = p^a q^b - q^a p^b$ for $p, q \in T_p M$ chosen so that $p \cdot p = q \cdot q = 1$ and $p \cdot q = 0$ contractions of (3.25) with $p^a p^b$ and $q^a p^b$ show that $h_{ab}p^a q^b = 0$ and $h_{ab}p^a p^b = h_{ab}q^a q^b$ and then the fact that the blade of F is invariant for h shows that p and q are eigenvectors of h with the same eigenvalue.

For part (*iii*) one writes out $F = (p^a q^b - q^a p^b) + \mu(r^a s^b - s^a r^b)$ ($\mu \neq \pm 1$) for some orthonormal tetrad $p, q, r, s \in T_p M$. Since $F \notin \widetilde{S}_p$ the 2-spaces $p \wedge q$

and $r \wedge s$ are uniquely determined by F, and $p \pm iq$ and $r \pm is$ are eigenvectors of F with, collectively, four distinct (complex) eigenvalues. Thus $p \pm iq$ and $r \pm is$ are eigenvectors for h and hence $p \wedge q$ and $r \wedge s$ are invariant for h. So each of p, q, r, s may be chosen as eigenvectors of h. A back substitution into (3.25) then reveals that $p \wedge q$ and $r \wedge s$ are eigenspaces for h.

For part (iv), and with $F \in \widetilde{S_p}$ fixed, note that $h \neq 0$ is symmetric and the metric g is positive definite. Thus h is diagonalisable over $\mathbb{R}$. Let $x \in T_pM$ be an eigenvector for h, $h_{ab}x^b = \lambda g_{ab}x^b = \lambda x_a$ with $\lambda \in \mathbb{R}$. Then a contraction of (3.25) with x^a shows that $y^a \equiv F^a{}_b x^b$ (which is non-zero since F is non-simple) is also an eigenvector of h with the same eigenvalue λ and is independent of, and orthogonal to, x. Since h is diagonalisable, one can choose another eigenvector $z \in T_pM$ for h, $h_{ab}z^b = \nu z_a$, $(\nu \in \mathbb{R})$ in the 2-space orthogonal to $x \wedge y$ and then another, $w \in T_pM$ follows, as above, given by $w^a = F^a{}_b z^b \neq 0$ with $h_{ab}w^b = \nu w_a$. By choice, $x \cdot y = x \cdot z = y \cdot z = z \cdot w = 0$ and a contraction of (3.25) with $x^a z^b$ shows that $x \cdot w = 0$. To see that x, y, z, w are independent let $ax + by + cz + dw = 0$ $(a, b, c, d \in \mathbb{R})$ and contract, respectively, with x and z to get $a = c = 0$ and so $by + dw = 0$. But this last equation is $F_{ab}(bx^b + dz^b) = 0$ from which $b = d = 0$ follows since F is non-simple. Thus x, y, z, w are independent members of T_pM. Now, if $\lambda = \nu$, h is a multiple of the metric $g(p)$ and this is clearly a solution of (3.25). If $\lambda \neq \nu$ then the eigenspaces $x \wedge y$ and $z \wedge w$ are orthogonal and so, after scalings, x, y, z, w form an orthonormal basis at p. Next write F as a linear combination of the basis bivectors $x \wedge y, x \wedge z, ..., z \wedge w$ and use the definitions of y and w above to reduce F to a linear combination of $x \wedge y, z \wedge w$ and $y \wedge w$. Finally the condition that $F \in \widetilde{S_p}$ shows that F is a linear combination of $x \wedge y$ and $z \wedge w$ only and, in fact, a multiple of $x \wedge y \pm z \wedge w$ (depending on whether $F \in \overset{+}{S_p}$ or $\overset{-}{S_p}$). Further, $h_{ab} = \lambda(x_a x_b + y_a y_b) + \nu(z_a z_b + w_a w_b)$, $(\lambda \neq \nu)$. Thus the 2-spaces $x \wedge y$ and $z \wedge w$ are an orthogonal pair of eigenspaces for h and a canonical blade pair for F. Of course each representation of $F \in \widetilde{S_p}$ and associated canonical blade pair satisfies (3.25) for h with these blades as eigenspaces for h.

Part (v) is easy to prove directly from the definition of the commutator by rearranging indices. Thus one writes $h_{ac}(F^c{}_d G^d{}_b - G^c{}_d F^d{}_b) + h_{bc}(F^c{}_d G^d{}_a - G^c{}_d F^d{}_a)$, expands this out using (3.25) and then shows that the first and fourth terms cancel as do the second and third. $\qquad\square$

3.5 The Curvature and Weyl Conformal Tensors

For the manifold (M, g) with g a positive definite metric on M one has a bivector metric P on the 6-dimensional vector space of bivectors Λ_pM of (positive definite) signature denoted, as above for $F, G \in \Lambda_pM$, by

$P(F, G) = F \cdot G$. One can now define the linear *curvature map* f from $\Lambda_p M$ to the vector space of all type $(1,1)$ tensors at p in terms of the curvature tensor *Riem* by [13]

$$f : F^{ab} \rightarrow R^a{}_{bcd} F^{cd}. \tag{3.26}$$

The *range space* of f is denoted by $rgf(p)$ and is a subspace of the vector space of all type $(1,1)$ tensors at p consisting precisely of tensors of the form $R^a{}_{bcd} G^{cd}$ for some $G \in \Lambda_p M$. Of course, if the metric g is specified, this map may (and will), with an abuse of notation and index positioning, be regarded as the linear map on $\Lambda_p M$ given by $F^{ab} \rightarrow R^{ab}{}_{cd} F^{cd}$. In either case the map f has the same rank and this rank, written $\text{rank} f_p$, is called the *curvature rank* of (M, g) (or of *Riem*) at p [13]. Of course the associated type $(0, 4)$ curvature tensor may be written as a sum of symmetrised product of members of $rgf(p)$. The curvature map, considered as the map $F^{ab} \rightarrow R^{ab}{}_{cd} F^{cd}$, is self-adjoint with respect to the bivector metric P because of the symmetries of *Riem* and so the eigenvalues of this version of the curvature map f are all real and diagonalisability over $\mathbb{R}$ follows. In this form one is essentially treating the tensor with components R_{abcd} (or $R^{ab}{}_{cd}$) as a 6×6 matrix R_{AB} (or $R^A{}_B$) where the "block" index A represents the (skew-symmetric) index pair ab and B the skew-symmetric pair cd. Block indices are raised and lowered using the bivector metric P, written in the obvious form P_{AB}, and P_{AB} and R_{AB} are symmetric.

Now let $\overline{rgf(p)}$ denote the smallest subalgebra of $\Lambda_p M$ containing $rgf(p)$, that is, the intersection of all the subalgebras of $\Lambda_p M$ containing $rgf(p)$ [35]. It is convenient to classify the map f at p into five mutually disjoint and exhaustive *curvature classes* A, B, C, D and O with O meaning that f is the zero map, that is, the curvature tensor vanishes at p, $Riem(p) = 0$. These curvature classes are [13, 55, 85];

Class D. This arises when $\dim rgf(p) = 1$ with $\overline{rgf(p)}$ being spanned by a (necessarily) simple bivector. In this case $rgf(p) = \overline{rgf(p)}$.

Class C. This arises when there exists a *unique* (up to scaling) $0 \neq k \in T_p M$ which annihilates every member of $rgf(p)$. Thus $\dim rgf(p)$ equals 2 or 3 (cf lemma 3.2).

Class B. This arises when $\overline{rgf(p)} = Sp(P, Q)$ where P and Q are independent members of $\Lambda_p M$ with *no* common annihilator and with $[P, Q] = 0$. Thus $rgf(p) = \overline{rgf(p)}$ and has dimension 2.

Class A. This arises when $\overline{rgf(p)}$ is not of class B, C, D or O and then $\dim rgf(p) \geq 2$.

As described here this classification into curvature classes is independent of signature. It will be developed in the sequel for each signature separately.

It is remarked that, for class D, the curvature tensor must take the form $R_{abcd} = F_{ab} F_{cd}$ at p where F spans $rgf(p)$ and then it follows from (3.4) that $F_{a[b} F_{cd]} = 0$ and then from lemma 3.1(vii) that F is necessarily simple. Also for class B one may write, for the positive definite case, $P = \overset{+}{P} + \overset{-}{P}$ with $\overset{+}{P} \in \overset{+}{S}_p$ and $\overset{-}{P} \in \overset{-}{S}_p$ and similarly for Q. Then since $\overset{+}{S}_p$ and $\overset{-}{S}_p$ are Lie algebras $[P, Q] = 0 \Rightarrow [\overset{+}{P}, \overset{+}{Q}] + [\overset{-}{P}, \overset{-}{Q}] = 0 \Rightarrow [\overset{+}{P}, \overset{+}{Q}] = [\overset{-}{P}, \overset{-}{Q}] = 0$. Thus, since $o(3)$ contains no 2-dimensional subalgebras, lemma 3.5(iii) shows that $\overset{+}{Q}$ and $\overset{+}{P}$ are proportional as are $\overset{-}{P}$ and $\overset{-}{Q}$. If $\overset{+}{P} \neq 0 \neq \overset{-}{P}$, then $rgf(p) = Sp(\overset{+}{P}, \overset{-}{P})$ and similarly if $\overset{+}{Q} \neq 0 \neq \overset{-}{Q}$, $rgf(p) = Sp(\overset{+}{Q}, \overset{-}{Q})$. Otherwise, since P and Q are independent, one must have $\overset{+}{P} = \overset{-}{Q} = 0$ leading to $rgf(p) = Sp(\overset{+}{Q}, \overset{-}{P})$, or $\overset{-}{P} = \overset{+}{Q} = 0$ in which case $rgf(p) = Sp(\overset{+}{P}, \overset{-}{Q})$. Thus for class B one may span $rgf(p)$ with two members, one from each of $\overset{+}{S}_p$ and $\overset{-}{S}_p$. Then lemma 3.5(iv) shows that the bivectors whose span is $rgf(p)$ *may be chosen to be a dual pair of simple bivectors which necessarily have no common annihilator.*

It is easy to check that this decomposition into curvature classes is mutually disjoint and exhaustive. To see this note that if $\dim rgf(p) = 0$ (respectively 1) then the curvature class at p is O (respectively, D). If $rgf(p) = Sp(F, G)$ is 2-dimensional with a (unique up to scaling) common annihilator, the curvature class at p is C whilst if no such annihilator exists either $[F, G] = 0$ or $[F, G] \neq 0$, (these conditions being easily checked to be independent of the choice of F, G) and so the curvature class at p is B or A, respectively. If $\dim rgf(p) = 3$, the curvature class is C if a common annihilator exists and otherwise it is A. If $\dim rgf(p) > 3$, lemma 3.2 shows that the curvature class at p is A. It is remarked here that if the curvature class at p is A, each member of $rgf(p)$ could be simple (and then $\dim rgf(p) = 3$) as the proof of lemma 3.2 shows. If, however, the curvature class at p is A and $\dim rgf(p) = 2$, $rgf(p)$ must contain a non-simple member. These remarks are independent of signature.

If one now also uses the symbol A to denote the subset of precisely those points of M at which the curvature class is A, and similarly for B, C, D and O, any two distinct members of the set of subsets $\{A, B, C, D, O\}$ have empty intersection. Further the union of these subsets equals M and so $M = A \cup B \cup C \cup D \cup O$, is a *disjoint* decomposition of M. A final remark is that, because of the symmetries of *Riem*, the curvature tensor at p may be decomposed into symmetrised products of the (independent) bivectors in $rgf(p)$ and so it is easily seen [13, 34] that the equation

$$R^a{}_{bcd} k^d = 0 \tag{3.27}$$

has a non-trivial solution for k only at those points $p \in M$ where k annihilates each member of $rgf(p)$ that is, at all points in the subset $C \cup D \cup O$ of M and has only trivial solutions for k at those points in the subset $A \cup B$ of M. However, for later purposes, one needs to be able to do calculus on these subsets and hence a more refined decomposition is needed [13, 35]. If $0 \neq k \in T_pM$ satisfies (3.27) it will be said to *annihilate Riem* at p.

Theorem 3.1 *Let (M, g) be a 4-dimensional manifold with positive definite metric g. Then one may* disjointly *decompose M as*

$$M = int A \cup int B \cup int C \cup int D \cup int O \cup Z \qquad (3.28)$$

where the interior operator in M is used and Z is the subset of M uniquely defined by the disjointness of the decomposition and is closed with $int Z = \emptyset$. The subsets A, $A \cup B$, $A \cup B \cup C$ and $A \cup B \cup C \cup D$ are open in M and so $int A = A$. Since M is connected $Z = \emptyset$ if and only if the curvature class is the same at each $p \in M$.

Proof That M may be decomposed into this disjoint decomposition is clear and then $M \setminus Z$ is open in M and so Z is closed. To continue the proof define for $p \in M$ the subspace $U_p \subset T_p M$ to be the span of the union of all the blades of all members of $rgf(p)$ (the blades of each simple member of $rgf(p)$ together with each of the canonical blade pair(s) for non-simple members of $rgf(p)$). So for $p \in O$ (respectively, D, C, B and A) dim $U_p = 0$ (respectively, 2, 3, 4 and 4) and $\dim rgf(p) = 0$ (respectively, 1, 2 or 3, 2, ≥ 2). A rank theorem (chapter 1) and the smoothness of $Riem$ then shows that if $\dim U_p = k$ ($0 \leq k \leq 4$) there exists an open neighbourhood V of p such that $\dim U_{p'} \geq k$ for each $p' \in V$, whilst if $\dim rgf(p) = k$ there exists an open neighbourhood W of p such that $\dim rgf(p') \geq k$ for each $p' \in W$. Now suppose $p \in A$ so that $\dim rgf(p) \geq 2$ and $\dim U_p = 4$. Then there exists an open neighbourhood V' of p such that $\dim U_{p'} = 4$ for $p' \in V'$ and so $V' \subset A \cup B$. If $\dim rgf(p) \geq 3$, however, there exists an open neighbourhood W' of p such that $\dim rgf(p') \geq 3$ for $p' \in W'$. Thus $V' \cap W' \subset A$. If $\dim rgf(p) = 2$, $rgf(p)$ must contain a non-simple bivector F and another independent member F'. Thus the members of $rgf(p)$ have no common annihilator (since F has no annihilators) and since $p \notin B$, $[F, F'] \neq 0$ at p. By continuity there exists an open neighbourhood V'' of p and smooth bivectors G, G', H and H' on V'' such that $G_{ab} = R_{abcd}H^{cd}$ and $G'_{ab} = R_{abcd}H'^{cd}$ and with $G(p) = F$ and $G'(p) = F'$ and V'' may be chosen so that G is non-simple and $[G, G'] \neq 0$ and so $\dim U_{p'} = 4$ on V''. But then $G(p')$ and $G'(p')$ are in $rgf(p')$ for each $p' \in V''$ and so $V'' \cap B = \emptyset$ and hence $V'' \subset A$. It follows that A is open in M. If $p \in B$ then $rgf(p)$ contains a non-simple member which, as above, may be extended to a smooth, non-simple member on some open neighbourhood V''' of p and which is in $rgf(p')$ for each $p' \in V'''$. Thus $V''' \subset A \cup B$ and so $A \cup B$ is open in M.

Another consideration of (the) rank (of rgf) then shows that $A \cup B \cup C$ is open in M, being the set of points where $\text{rank}(rgf) \geq 2$ and similarly $A \cup B \cup C \cup D$ is open in M. Finally suppose $Z \neq \emptyset$ and let $U \subset Z$ be open. Then by the disjointness of the decomposition of M and the fact that $int A = A$ one sees that $U \cap A = \emptyset$. Now $U \cap B(= U \cap (A \cup B))$ is open, since $A \cup B$ is open, and if non-empty, contradicts $A \cap int B = \emptyset$ since then $U \cap B \subset int B$. So $U \cap B = \emptyset$. Now $U \cap C(= U \cap (A \cup B \cup C))$ is open in M and, if non-empty, contradicts $U \cap int C = \emptyset$. It follows that $U \cap C = \emptyset$. Similarly one shows that $U \cap D = U \cap O = \emptyset$. Thus $U = \emptyset$ and $int Z = \emptyset$ which completes the proof the last sentence of the theorem being clear. □

This theorem may be restated by saying that $M \setminus Z$, which is clearly non-empty, is open and dense in M and then any point in $M \setminus Z$ admits a neighbourhood on which the curvature class is constant.

Now consider the Weyl conformal tensor C described in (3.7) and (3.11). From (3.16) and (3.17) one has $\epsilon = 1$ for positive definite signature, $^*C = C^*$ and $^*C^* = \overset{**}{C}$ (cf the condition $\overset{**}{F} = F$ for a bivector for this signature). The symmetries of C show that, at p, one can write out C_{abcd} as a sum of symmetrised product of members of a basis for $\Lambda_p M$. Just as for the tensor *Riem* one can introduce a linear map f_C from $\Lambda_p M$ to the vector space of type $(1,1)$ tensors at p given by $f_C : F^{ab} \rightarrow C^a{}_{bcd} F^{cd}$ called the *Weyl map* at p [32] and whose rank is referred to as the *Weyl rank* at p. As before, since g is given, one may introduce the related map (also denoted by f_C) given by $F^{ab} \rightarrow C_{abcd} F^{cd}$ and then in an obvious shorthand way (using the identifications arising from the metric g) as $f_C : F \rightarrow CF$. Then $(f_C F)^* = (CF)^* = (^*C)F = (C^*)F = C\overset{*}{F}$ and thus f_C maps the subspaces $\overset{+}{S_p}$ and $\overset{-}{S_p}$ of $\Lambda_p M$ into themselves, that is, they are invariant subspaces of f_C. It also follows that if $F \in rg f_C(p)$ then, at p, $F = CG$ for some $G \in \Lambda_p M$ and then $\overset{*}{F} =^* CG = C^* G = C\overset{*}{G}$ which shows that $\overset{*}{F}$ is also in $rg f_C(p)$ (not necessarily independent of F). Since the bivector metric is positive definite, the "symmetric" map f_C may be diagonalised over $\mathbb{R}$ and thus trivially classified by the degeneracies of the Segre type $\{111111\}$.

One can decompose the type $(0,4)$ Weyl tensor as

$$C = \overset{+}{W} + \overset{-}{W}, \qquad \overset{+}{W} = \frac{1}{2}(C +^* C), \qquad \overset{-}{W} = \frac{1}{2}(C -^* C), \qquad (3.29)$$

where the type $(0,4)$ tensors $\overset{+}{W}$ and $\overset{-}{W}$ are the *self dual* and *anti-self dual* parts of C, respectively, and satisfy $\overset{+}{W}{}^* = {}^*\overset{+}{W} = \overset{+}{W}$ and $\overset{-}{W}{}^* = {}^*\overset{-}{W} = -\overset{-}{W}$. It also follows (see (3.11), (3.17) and (3.18)) that $\overset{+}{W}_{abcd} = \overset{+}{W}_{cdab}$, that $\overset{+}{W}_{a[bcd]} = 0$ and that $\overset{+}{W}{}^c{}_{acb} = 0$ and similarly for $\overset{-}{W}$. The tensors $\overset{+}{W}$ and $\overset{-}{W}$ give rise, in an obvious way, to maps $\overset{+}{f_C}$ and $\overset{-}{f_C}$ constructed from them as f_C was from C with $\overset{+}{f_C}$ a linear map on the 3-dimensional real vector space $\overset{+}{S_p}$ (acting trivially on $\overset{-}{S_p}$) and $\overset{-}{f_C}$ a linear map on $\overset{-}{S_p}$ (acting trivially on $\overset{+}{S_p}$) and with $f_C = \overset{+}{f_C} + \overset{-}{f_C}$ and then $\overset{+}{W}F = \frac{1}{2}(CF + C\overset{*}{F})$ and $\overset{-}{W}F = \frac{1}{2}(CF - C\overset{*}{F})$. Thus f_C may be (equivalently) classified by the individual Segre types of these two (independent) maps.

Let g and g' be two smooth positive definite metrics on M which are conformally related. As remarked earlier it can then be checked that the type $(1,3)$ Weyl conformal tensor C given in (3.7) is the same whether computed with g or with g' [84, 30]. One can ask whether there is a converse to this

theorem and, after a precise formulation of such a converse, it will be shown
that for this dimension and signature a converse exists [32]. The question to be
considered is the following. Suppose (M, g) is a 4-dimensional manifold with
smooth positive definite metric g and Weyl type $(1, 3)$ conformal tensor C.
Suppose also that M admits another smooth metric g' of arbitrary signature
whose type $(1, 3)$ Weyl conformal tensor C' equals C on M and that C (and
hence C') is nowhere zero over some open dense subset of M. Are g and g'
necessarily conformally related? To consider this problem, g will be regarded
as the metric on M and will be used to raise and lower tensor indices whilst
g' will simply be another (non-degenerate) symmetric tensor on M. Consider
the type $(0, 4)$ Weyl tensor with components $C_{abcd} = g_{ae}C^e{}_{bcd}$. If C is also the
type $(1, 3)$ Weyl tensor for g', then the analogous type $(0, 4)$ tensor obtained
from C and g' with components $C'_{abcd} = g'_{ae}C^e{}_{bcd}$ must, like that from C and
g, satisfy the usual symmetries of such a tensor. Thus at any $p \in M$ where
$C(p) \neq 0$

$$g_{ae}C^e{}_{bcd} + g_{be}C^e{}_{acd} = 0, \qquad g'_{ae}C^e{}_{bcd} + g'_{be}C^e{}_{acd} = 0. \tag{3.30}$$

Then for each type $(1, 1)$ tensor $G \in V$ where $V \equiv rgf_C(p)$, so that $G^a{}_b = C^a{}_{bcd}B^{cd}$ for some $B \in \Lambda_p M$, one sees that

$$g_{ac}G^c{}_b + g_{bc}G^c{}_a = 0, \qquad g'_{ac}G^c{}_b + g'_{bc}G^c{}_a = 0, \tag{3.31}$$

that is, recalling remarks before lemma 3.6, the tensor $g'G$ is skew-symmetric
for each $G \in V$. Now if $\dim V = 1$ with V spanned by $F \in V$ then, at p, $C^a{}_{bcd}$
is a multiple of $F^a{}_b H_{cd}$ for $H \in \Lambda_p M$ and so $C_{abcd}(= g_{ae}C^e{}_{bcd}) = \lambda F_{ab}F_{cd}$
(for $0 \neq \lambda \in \mathbb{R}$ and $F_{ab} = g_{ac}F^c{}_b$) by the symmetries of C_{abcd}. But then from
(3.11), $C_{a[bcd]} = 0$ and hence $F_{a[b}F_{cd]} = 0$ which, from lemma 3.1(vii), shows
that F is simple, say $F_{ab} = p_a q_b - q_a p_b$, for 1-forms p, q which may be chosen
to be ($g-$)orthogonal. But then it is easily checked that $C^c{}_{acb} \neq 0$ which
contradicts the last equation in (3.11). [Alternatively, one could note from
a remark above that once F has been proved simple, $\overset{*}{F}$ is an independent
member of V and a contradiction follows from a remark above since then
$\dim V \geq 2$. Yet another proof follows from the fact that if $\dim V = 1$, $rgf_C(=$
$rg(\overset{+}{f}_C + \overset{-}{f}_C))$ is easily seen to lie in $\overset{+}{S}_p$ or $\overset{-}{S}_p$ and so $C_{abcd} = \lambda F_{ab}F_{cd}$ ($0 \neq$
$\lambda \in \mathbb{R}$) with $F \in \overset{+}{S}_p$ or $F \in \overset{-}{S}_p$. But then F is not simple and the relation
$C_{a[bcd]} = 0$ is contradicted.]

So $\dim V \geq 2$. Now, identifying V with the corresponding subspace of
$\Lambda_p M$, if $V \subset \widetilde{S_p}$ then either $V \subset \overset{+}{S}_p$ or $V \subset \overset{-}{S}_p$. If the former then there exists
independent $F_1, F_2 \in \overset{+}{S}_p$ which satisfy the second equation in (3.31) and so,
from lemma 3.6(iv), since g is positive definite and if g' is not proportional to
g at p, g' admits a pair of 2-dimensional eigenspaces and which are a canonical
pair of blades for each of F_1 and F_2. These eigenspace pairs must, since g' is
not proportional to g, then be the same pair and hence the canonical blade

pairs for F_1 and F_2 are the same. Since $F_1, F_2 \in \overset{+}{S}_p$, they must be proportional (lemma 3.5(v)) and a contradiction is achieved. One similarly handles $\overset{-}{S}_p$ and so V is not a subset of $\widetilde{S}_p$. In this case choose $F \in V$ with $F \notin \widetilde{S}_p$ so that $\overset{*}{F} \notin \widetilde{S}_p$ is independent of F and $\overset{*}{F} \in V$ by a result above. Then either F and $\overset{*}{F}$ are simple, or F is non-simple with $F \notin \widetilde{S}_p$, in which case one can choose simple, independent, linear combinations of F and $\overset{*}{F}$ which form a dual pair (and label them also as F and $\overset{*}{F}$). In either case and if g' is not proportional to g at p, one achieves an orthogonal pair of 2-dimensional eigenspaces of g' at p from lemma 3.6(ii). If $\dim V = 2$ then V can be taken as spanned by simple F and $\overset{*}{F}$ and one achieves

$$C_{abcd} = \alpha F_{ab}F_{cd} + \beta \overset{*}{F}_{ab}\overset{*}{F}_{cd} + \gamma(F_{ab}\overset{*}{F}_{cd} + \overset{*}{F}_{ab}F_{cd}) \qquad (3.32)$$

at p with $\alpha, \beta, \gamma \in \mathbb{R}$. Now the condition $C^c{}_{acb} = 0$ and lemma 3.1(iv) show that $\alpha = \beta = 0$ and the condition $C_{a[bcd]} = 0$ together with lemma 3.3(i) and the fact that the blades of F and $\overset{*}{F}$, being orthogonal, intersect trivially, show that $\gamma = 0$ and the contradiction that $C = 0$ is obtained. Finally suppose V is not a subset of $\widetilde{S}_p$ and $\dim V \geq 3$. One can still achieve the above, simple F and $\overset{*}{F}$ which lead to $g'(p)$, again assumed not proportional to $g(p)$ at p, having an orthogonal pair of eigenspaces together with an independent $J \in V$ which may be chosen not to be in $\widetilde{S}_p$ and J also gives rise to a 2-dimensional eigenspace or a pair of such eigenspaces of $g'(p)$. If $g'(p)$ is not a multiple of $g(p)$ this latter eigenspace, or pair of eigenspaces, must coincide with one or both of those from F and $\overset{*}{F}$ and so J is a linear combination of F and $\overset{*}{F}$ which contradicts the independence of J. Thus if $U \subset M$ is such that $p \in U \Leftrightarrow C(p) \neq 0$ with U open in M and $\mathrm{int}(M \setminus U) = \emptyset$ (so that U is open and dense in M) g' and g are proportional on U.

If $p \in M \setminus U$ and g' and g are not proportional at p there exists independent $u, v \in T_pM$ such that, at p, $g(u, u) = g(v, v) = 1$ and $g'(u, u) \neq g'(v, v)$. Then there exists a chart x containing p with domain V and smooth vector fields u' and v' on V whose components in x are, respectively, identical to the components of $u, v \in T_pM$ in x and thus $u'(p) = u$ and $v'(p) = v$. Then the vector fields $u'' = g(u', u')^{-\frac{1}{2}}u'$ and $v'' = g(v', v')^{-\frac{1}{2}}v'$ on V are such that there exists an open neighbourhood W of p, $W \subset V$, at each point of which $g(u'', u'') = g(v'', v'') = 1$ but $g'(u'', u'') \neq g'(v'', v'')$. Thus g and g' are not proportional on W. Since U is open and dense in M, $W \cap U \neq \emptyset$, contradicting the fact that g and g' are proportional on U. Thus g and g' are proportional on M. So $g' = \phi g$ on M with $\phi : M \to \mathbb{R}$. A contraction of this last equation with g^{ab} shows that ϕ is smooth and so g and g' are conformally related on M. Thus the following result has been proved [32].

Theorem 3.2 *Let M be a smooth 4-dimensional manifold admitting a smooth positive definite metric g whose Weyl conformal tensor is nowhere zero over some open dense subset of M. If g' is another smooth metric of any signature on M whose type $(1,3)$ Weyl tensor equals that of g on M, then g and g' are conformally related on M (and so g' is positive definite).*

It is remarked that, under similar restrictions, such a result fails for $\dim M \geq 5$ with metric g of any signature [32]. It similarly also fails for $\dim M = 4$ if the original metric g is of Lorentz or neutral signature and this will be established later. [Of course the Weyl tensor is identically zero for dimension 3 and is not defined for dimensions 1 and 2.]

As a final comment it is noted that M may be decomposed along similar lines with respect to the Weyl tensor to that given above in terms of curvature class for the curvature tensor (the *Weyl classes*). However, in this case the tracefree condition $C^c{}_{acb} = 0$ means that (see above proof) classes D and B are impossible at any point. A similar argument shows that class C is also impossible at any point p because if one chooses a tetrad x, y, z, w at p with x the common annihilator, the Weyl conformal tensor at p is spanned by two or three of the bivectors $y \wedge z$, $y \wedge w$ and $z \wedge w$ and then on writing out C_{abcd} in terms of symmetrised products of these bivectors, the condition $C^c{}_{acb} = 0$ is easily checked to force C to vanish at p. Thus if at $p \in M$ $C(p) \neq 0$, it is of class A. It follows that if $C(p) \neq 0$ *there are only trivial solutions for k of the equation $C^a{}_{bcd}k^d = 0$.* If $0 \neq k \in T_pM$ satisfies this latter equation it is said to *annihilate C* at p.

3.6   The Lie Algebra o(4)

In this section the well-known Lie algebra $o(4)$ will be described in terms of bivectors and which turns out to be a useful representation of it for present purposes. It is based on work in [34, 67].

Let M be a manifold of any dimension n admitting a metric g of any signature, and for $p \in M$ and $x, y \in T_pM$ let $\mathcal{G}$ be set of linear transformations on T_pM which preserve the inner product represented by the metric g in the sense that if $f \in \mathcal{G}$ then for each $x, y \in T_pM$

$$g(f(x), f(y)) = g(x, y). \tag{3.33}$$

From this one finds $g(f(x), f(x)) = g(x, x)$ and conversely, replacing x by $x + y$ in this latter formula, one can easily recover (3.33). Thus this last equation is equivalent to (3.33). Also if $y \in T_pM$ and $f(y) = 0$ then for any $x \in T_pM$, (3.33) shows that $g(x, y) = 0$ for each x and so since g is a metric, $y = 0$. Thus each $f \in \mathcal{G}$ is injective (and since they are linear maps on T_pM each member of $\mathcal{G}$ is bijective and the corresponding inverse maps are also in $\mathcal{G}$

since $g(f^{-1}(x), f^{-1}(y)) = g(f(f^{-1}(x)), f(f^{-1}(y))) = g(x, y))$. The set $\mathcal{G}$ is then easily checked to be a group under the usual composition of maps, called the *orthogonal group in n-dimensions for (the signature of) g*. For $n = 4$ and with g positive definite, one chooses a basis $\{e_a\}$ for T_pM in which the matrix representing g is $I \equiv \mathrm{diag}(1, 1, 1, 1)$. Then if the matrix representing $f \in \mathcal{G}$ is $A = (a_{ab})$ and if $x = x^a e_a$ and $y = y^a e_a$ then $f(x) = a_{ab}x^a e_b$, $f(y) = a_{ab}y^a e_b$ and (3.33) is equivalent to $a_{ba}x^b a_{cd}y^c \delta_{ad} = x^b y^c \delta_{bc}$ (repeated indices summed) for each $x, y \in T_pM$, and hence to

$$AA^T = I. \tag{3.34}$$

It follows that $\det A = \pm 1$. More generally, if a different basis for T_pM is chosen, related to the original by a non-singular matrix T and which leads to the matrix h representing $g(p)$, a repeat of the above argument shows that the matrix A representing $f \in \mathcal{G}$ satisfies $AhA^T = h$ and an isomorphic representation of this group is obtained. Of course, $\mathcal{G}$ is a subgroup of $GL(4, \mathbb{R})$, with the latter being a Lie group (chapter 2). Also, for fixed h, the map $d : GL(4, \mathbb{R}) \to M_4\mathbb{R}$ given by $A \to AhA^T - h$ is a smooth map with respect to the natural manifold topologies and $d^{-1}\{O\} = \mathcal{G}$ where $\{O\}$ represents the zero matrix in $M_4\mathbb{R}$. Thus $\mathcal{G}$ is a closed (non-discrete) subgroup of the Lie group $GL(4, \mathbb{R})$ and can thus be given a unique structure as a regular submanifold of $GL(4, \mathbb{R})$ and is hence a Lie subgroup of it (section 2.7). The Lie algebra of $GL(4, \mathbb{R})$ is $M_4\mathbb{R}$ (under the operation of matrix commutation (see, e.g.[9]). These results, appropriately modified, apply to all signatures. Returning to the positive definite case the Lie group $\mathcal{G}$ is not connected, being split by the condition $\det A = \pm 1$. Those members with $\det A = 1$ form a connected Lie group, denoted by $\mathcal{G}_o$ which is the identity component of $\mathcal{G}$. The Lie algebra of $\mathcal{G}$ (and of $\mathcal{G}_0$) is denoted by $o(4)$ and is a subalgebra of $M_4\mathbb{R}$. In fact, it can be shown (see, e.g.[9]) that the Lie algebra $o(4)$ is the 6-dimensional Lie algebra given by

$$o(4) = \{B \in M_4\mathbb{R} : Bh + (Bh)^T = 0\}. \tag{3.35}$$

Regarding B as a type $(1, 1)$ tensor this equation is just the condition that when the metric in question, $h \equiv g(p)$, is used for lowering indices, B becomes a type $(0, 2)$ skew-symmetric tensor (bivector) at p and so, with $g(p)$ understood, $o(4)$ can be regarded as the subalgebra of all such bivectors under matrix commutation.

Now one must seek all subalgebras of $o(4)$ (c.f. [34]). This will be done by taking $p \in M$ with positive definite metric $g(p)$ and taking $o(4)$ as the Lie algebra $\Lambda_p M$ under matrix commutation. Certainly $\overset{+}{S}_p$ and $\overset{-}{S}_p$ are 3-dimensional subalgebras of $o(4) = \overset{+}{S}_p + \overset{-}{S}_p$ (and will be labelled simply as $\overset{+}{S}$ and $\overset{-}{S}$ with $\widetilde{S}_p$ now labelled $\widetilde{S}$) and the projections $\pi_1 : \Lambda_p M \to \overset{+}{S}$ and $\pi_2 : \Lambda_p M \to \overset{-}{S}$ are easily checked to be Lie algebra homomorphisms with kernels $\overset{-}{S}$ and $\overset{+}{S}$, respectively. Also, as shown earlier, $\overset{+}{S}$ and $\overset{-}{S}$ are isomorphic to $o(3)$ and have no 2-dimensional subalgebras.

The trivial subalgebra of $o(4)$ is labelled S_0. The 1-dimensional subalgebras are just of the form $Sp(F)$ for some bivector F and may be subclassified according to whether F is simple (labelled S_1) or non-simple and, for the latter case, whether $F \in \overset{+}{S}_p$, $F \in \overset{-}{S}_p$ or neither and these are labelled, respectively, $\overset{+}{S}_1$, $\overset{-}{S}_1$ and $\overset{NS}{S}_1$.

Now let A be a subalgebra of $o(4)$ so that $\pi_1(A)$ and $\pi_2(A)$ are subalgebras of $\overset{+}{S}$ and $\overset{-}{S}$, respectively, of dimension 0, 1 or 3. Also, with an obvious abuse of notation, $A \cap \overset{+}{S}$ and $A \cap \overset{-}{S}$ are subalgebras of $\overset{+}{S}$ and $\overset{-}{S}$, respectively. Then if $\dim A = 2$, neither $\pi_1(A)$ nor $\pi_2(A)$ is trivial otherwise the other is 2-dimensional and a contradiction arises. So $\pi_1(A)$ and $\pi_2(A)$ are necessarily 1-dimensional subalgebras of $\overset{+}{S}$ and $\overset{-}{S}$, respectively, and writing $\pi_1(A) = Sp(F)$ and $\pi_2(A) = Sp(G)$ for $F \in \overset{+}{S}$ and $G \in \overset{-}{S}$, A is the product $Sp(F) + Sp(G) = Sp(F, G)$. Then, from lemma 3.5(iv), one may choose an orthonormal tetrad x, y, z, w with $A = Sp(x \wedge y, z \wedge w)$ and this type is denoted by S_2. If $\dim A = 3$ two possibilities are $A = \overset{+}{S}$ and $A = \overset{-}{S}$ and these are denoted by $\overset{+}{S}_3$ and $\overset{-}{S}_3$. Otherwise $\pi_1(A)$ and $\pi_2(A)$ are non-trivial subalgebras of $\overset{+}{S}$ and $\overset{-}{S}$, respectively, and, since $A \subset \pi_1(A) + \pi_2(A)$, either $\pi_1(A) = \overset{+}{S}$ or $\pi_2(A) = \overset{-}{S}$, or both. Suppose $\pi_1(A) = \overset{+}{S}$ so that π_1 is a Lie algebra isomorphism from A to $\overset{+}{S}$. If $\dim \pi_2(A) = 1$ the kernel of (the restriction to A of) π_2 would be a 2-dimensional subalgebra of A and hence of $\overset{+}{S}$ and a contradiction follows. So $\pi_2(A) = \overset{-}{S}$ and π_2 is an isomorphism between A and $\overset{-}{S}$ and $\pi_1 \circ \pi_2^{-1}$ is an isomorphism $\overset{-}{S} \to \overset{+}{S}$ which preserves the inner product P on $\Lambda_p M$ (lemma 3.5(vii)). Thus if $H \in A$ and $\pi_1(H) = F$, $\pi_2(H) = G$ for $F \in \overset{+}{S}$ and $G \in \overset{-}{S}$ then $\pi_1 \circ \pi_2^{-1}(G) = F$. So $|F| = |G|$, that is, $0 = (F + G) \cdot (F - G) = (F + G) \cdot (F + G)^*$ and then $H = F + G$ is simple (lemma 3.5(iv)). So all members of A are simple and since $\dim A = 3$ and A is a *subalgebra* (as opposed to just a subspace) it follows easily from lemma 3.2 that A is spanned by three bivectors the blades of whose duals intersect in a 1-dimensional subspace of $T_p M$ (the alternative solution from lemma 3.2 fails to be a subalgebra). One can then easily check that one may choose these duals as $w \wedge x$, $w \wedge y$ and $w \wedge z$ where $w \cdot x = w \cdot y = w \cdot z = 0$ for independent $x, y, z, w \in T_p M$ and, by an adjustment of x, y, z, one may take x, y, z, w as an orthonormal basis for $T_p M$. Then $A = Sp(x \wedge y, x \wedge z, y \wedge z)$ and this type will be denoted by S_3.

For the case when $\dim A = 4$, one has the products $A = Sp(\overset{+}{S}, G)$ and $A = Sp(F, \overset{-}{S})$, with $G \in \overset{-}{S}$ and $F \in \overset{+}{S}$ as possibilities and these are denoted by $\overset{+}{S}_4$ and $\overset{-}{S}_4$, respectively. Otherwise one must have $\pi_1(A) = \overset{+}{S}$ *and* $\pi_2(A) = \overset{-}{S}$

and (the restriction to A of) π_1 and π_2 have non-trivial kernels. Since $\dim A = 4$, a consideration of the subspaces A, $\overset{+}{S}$ and $\overset{-}{S}$ and use of the dimension formula (chapter 1) shows that $\dim A \cap \overset{+}{S} \geq 1 \leq \dim A \cap \overset{-}{S}$ and so $A \cap \overset{+}{S}$ and $A \cap \overset{-}{S}$ are not trivial. So take $0 \neq H \in A \cap \overset{+}{S}$. Then denoting the subset formed by taking the bracket of H with each member of some subset B of $\Lambda_p M$ by $[H, B]$ one gets $[H, A] = [H, \pi_1(A)]$ (since H commutes with each member of $\overset{-}{S}$) and then $[H, \pi_1(A)] = [H, \overset{+}{S}]$ and so $[H, A] = [H, \overset{+}{S}]$. But since $\overset{+}{S}$ has no 2-dimensional subalgebras the set $[H, \overset{+}{S}]$ contains two independent members a, b which, from lemma 3.4(ii), satisfy $H \cdot a = H \cdot b = 0$ and so $Sp(H, [H, \overset{+}{S}])$ is a 3-dimensional subspace of $\overset{+}{S}$. Thus $\overset{+}{S} = Sp(H, [H, \overset{+}{S}]) \subset A$. Similarly one shows that $\overset{-}{S} \subset A$ and this gives the contradiction that $\dim A = 6$. So the 4-dimensional subalgebras $\overset{+}{S}_4 \equiv Sp(\overset{+}{S}, G)$ and $\overset{-}{S}_4 \equiv Sp(F, \overset{-}{S})$, with $G \in \overset{-}{S}$ and $F \in \overset{+}{S}$ are the only possible ones. If $\dim A = 5$ the dimension formula shows that $\dim A \cap \overset{+}{S} \geq 2$ and hence $\dim A \cap \overset{+}{S} = 3$ (and similarly $\dim A \cap \overset{-}{S} = 3$) and so a contradiction follows as in the last case. This completes the list of all subalgebras of $o(4)$.

It is noted that by taking linear combinations and choosing some appropriate orthonormal basis x, y, z, w, lemma 3.5(iv) shows that the Lie algebra $\overset{+}{S}_4$ may be written as $Sp(A, B, x \wedge y, z \wedge w)$ where $A, B \in \overset{+}{S}$, and similarly for $\overset{-}{S}_4$. Subalgebra S_3 is easily seen to be isomorphic to $o(3)$ and subalgebra S_2 is the product algebra $o(2) + o(2)$. The complete list of proper subalgebras may be summarised as in Table 3.1 in which x, y, z, w is a basis of the usual form, noting that the subscript on the symbol S is the dimension of the subalgebra.

The 1-dimensional cases $\overset{+}{S}_1$, $\overset{-}{S}_1$ and $\overset{NS}{S}_1$ cannot be holonomy algebras for *metric* connections since they are spans of a non-simple bivector. This is explained below. Thus they will not be required any further in this book.

The corresponding transformations are generated by exponentiation from $\mathcal{G}_0$ as described in chapter 2. Thus, for example, S_1 gives rise to rotations in the blade of the simple bivector F.

3.7 The Holonomy Structure of (M, g)

Most of this chapter has dealt with a smooth, 4-dimensional, connected, second countable manifold M admitting a smooth, positive definite metric

TABLE 3.1: Lie subalgebras for $(+, +, +, +)$.

Type	Dimension	Basis
S_1	1	$x \wedge y$
$\overset{+}{S}_1$	1	see text
$\overset{-}{S}_1$	1	see text
$\overset{NS}{S}_1$	1	see text
S_2	2	$x \wedge y,\ z \wedge w$
S_3	3	$x \wedge y,\ x \wedge z,\ y \wedge z$
$\overset{+}{S}_3$	3	$\overset{+}{S}_p$
$\overset{-}{S}_3$	3	$\overset{-}{S}_p$
$\overset{+}{S}_4$	4	$\overset{+}{S}_p,\ G\ (G \in \overset{-}{S}_p)$
$\overset{-}{S}_4$	4	$F,\ \overset{-}{S}_p,\ (F \in \overset{+}{S}_p)$
S_6	6	$o(4)$

g and the last section explored the orthogonal algebra for T_pM with inner product $g(p)$ and listed all the proper subalgebras of $o(4)$. For such a manifold it is now possible to describe all its possible holonomy algebras. Returning to the description of holonomy groups given in chapter 2 one starts with $p \in M$ together with the set of all closed (piecewise-smooth) curves C_p starting and ending at p and the group (a subgroup of $GL(T_pM)$) of all isomorphisms τ_c for $c \in C_p$ arising as a result of parallel transport of T_pM from p back to p along c using the Levi-Civita connection ∇ compatible with g called the holonomy group of M at p. Since it is isomorphic to the holonomy group of M at any other $p' \in M$ (chapter 2), one may drop the reference to $p \in M$ and refer to the holonomy group Φ of M. This latter is a Lie group and a Lie subgroup of $GL(T_pM)$ [10]. However, with the additional information that the connection ∇ is compatible with g, each linear transformation τ_c, now must preserve the inner product arising on each copy of T_pM from $g(p)$ in the sense that for $u, v \in T_pM$ and upon parallel transport of u and v around c, the real number $g(u, v)$ is constant at each point of c. Thus Φ consists of g-orthogonal transformations and is clearly a subgroup of $\mathcal{G}$. But $\mathcal{G}$ is a Lie subgroup (and a regular submanifold) of $GL(T_pM)$ and $\Phi \subset \mathcal{G}$ and so Φ is a submanifold of $\mathcal{G}$ since $\mathcal{G}$ is regular (section 2.7). It follows that Φ is a Lie subgroup of $\mathcal{G}$ (section 2.10). Thus the holonomy algebra ϕ is a subalgebra of $o(4)$. It can be checked that $\dim \mathcal{G} = 6$. This argument also applies when the metric g has Lorentz or neutral signature (see e.g. [13]).

In chapter 2 the infinitesimal holonomy algebra ϕ'_p at $p \in M$ was introduced and is a subalgebra of ϕ for each p, and hence a subalgebra of $o(4)$. Now suppose that (M, g) has a 1-dimensional holonomy group so that, in particular, (M, g) is not flat and ϕ is spanned by a single bivector F. Then there exists $p \in M$ at which $Riem(p) \neq 0$ and so, since ϕ'_p is a subalgebra of ϕ, one must have an expression for $Riem(p)$ like $R_{abcd} = \alpha F_{ab} F_{cd}$ where $0 \neq \alpha \in \mathbb{R}$. The last equation in (3.4) then shows that $F_{a[b} F_{cd]} = 0$ and so, from lemma 3.1(vii), F is a simple bivector. It follows that, up to isomorphism, *the only* 1-*dimensional subalgebra of* $o(4)$ *in the above discussion which could be the holonomy algebra of a Levi-Civita connection on* M *is that labelled* S_1. For those other 1-dimensional subalgebras of $o(4)$, if they are holonomy algebras, they can only be such for connections which are not metric.

More information on holonomy will be given in chapter 7 where the universal covering manifold of M will be needed. Here it is sufficient to mention that if (M, g) has holonomy group Φ with associated holonomy algebra ϕ then ϕ is a subalgebra of $o(4)$ and is hence one of the subalgebras given in the last section (with the 1-dimensional exclusions mentioned above) and with ϕ'_p a subalgebra, and $rgf(p)$ a subspace, of ϕ. Each member F of the bivector representation of ϕ satisfies (3.25) for $h = g$ in lemma 3.6. These results apply, suitably modified for signature, in all the 4-dimensional cases.

3.8 Curvature and Metric

Suppose now that g and g' are smooth metrics on M with g positive definite. Suppose that their corresponding (tensor type $(1,3)$) curvature tensors $Riem$ and $Riem'$ are equal everywhere on M, that is, in any chart domain, $R^a{}_{bcd} = R'^a{}_{bcd}$. Then for $p \in M$ a consideration of the symmetries of the curvature tensor shows that, at p, $g'_{ae} R'^e{}_{bcd} + g'_{be} R'^e{}_{acd} = 0$ and so $g'_{ae} R^e{}_{bcd} + g'_{be} R^e{}_{acd} = 0$ and hence that (3.25) holds with $h = g'$ for each F in $rgf(p)$ arising from $Riem$. Thus, from lemma 3.6(v), (3.25) holds for the subalgebra $\overline{rgf(p)}$ of $o(4)$ (section 3.5). Now consider the (necessarily) open subset $A \subset M$ where the curvature class is A. If $p \in A$ and $rgf(p)$, which now has dimension ≥ 2, contains only simple members then it has dimension $\leqslant 3$ from lemma 3.2 and this lemma then shows that either $\dim rgf(p) = 2$ with curvature class C at p (and a contradiction) or $\dim rgf(p) = 3$. In this latter case lemma 3.2 leads either to $rgf(p)$ being spanned by three simple bivectors whose blades have a common non-trivial annihilator and hence one again achieves curvature class C and a contradiction, or by three simple bivectors whose blades have a common non-trivial member. In this latter case, lemma 3.6(ii) immediately shows that $T_p M$ is an eigenspace of $g'(p)$ and hence that $g' = \mu g$ at p for $0 \neq \mu \in \mathbb{R}$. So suppose $rgf(p)$ contains a non-simple member, say G. Then h admits two 2-dimensional eigenspaces from lemma 3.6(iii), (iv) and the same lemma shows that any additional independent member(s) of $rgf(p)$ will either lead to the curvature class B at p (and a contradiction) or to another eigenvector of h not in the eigenspaces already established, again forcing $g'(p)$ to be a multiple of $g(p)$. It follows that g' is necessarily a multiple of g at each point of A. Thus $g' = \lambda g$ on A for some smooth, nowhere-zero function $\lambda : A \to \mathbb{R}$, that is, g and g' are conformally related on A. Now, loosely speaking, A is the most general curvature class (and A is open in M) and if $M \setminus A$, which is a closed subset of M, has empty interior, then A is open and dense in M and, as earlier, g' and g are conformally related on M, that is, $g' = \lambda g$ on M, with $\lambda : M \to \mathbb{R}$ smooth.

Now let ∇ and ∇' denote the Levi-Civita connections associated with g and $g' = \lambda g$, respectively, and consider the Bianchi identities for *Riem* for each of ∇ and ∇' given in chapter 2 and after a contraction over the indices a and e (and using a semi-colon and a vertical stroke for a covariant derivative with respect to ∇ and ∇', respectively, and a comma for a partial derivative) one finds

$$R^a{}_{bcd;a} - R_{bd;c} + R_{bc;d} = 0, \qquad R^a{}_{bcd|a} - R_{bd|c} + R_{bc|d} = 0 \qquad (3.36)$$

where the Ricci tensor components $R_{ab} = R^c{}_{acb} = R'^c{}_{acb}$ have been introduced. The component relations between the Christoffel symbols Γ^a_{bc} for g and Γ'^a_{bc} for g' are easily calculated and are

$$P^a_{bc} \equiv \Gamma'^a_{bc} - \Gamma^a_{bc} = \frac{1}{2\lambda}(\lambda_{,c}\delta^a_b + \lambda_{,b}\delta^a_c - \lambda^a g_{bc}) \qquad (3.37)$$

where $\lambda^a = \lambda_{,b}g^{ab}$ (and P should not be confused with the bivector metric). If one subtracts the equations in (3.36) to remove the partial derivatives, one achieves

$$R^e{}_{bcd}P^a_{ea} - R^a{}_{ecd}P^e_{ba} - R^a{}_{bed}P^e_{ca} - R^a{}_{bce}P^e_{da}$$
$$+ R_{be}P^e_{dc} + R_{ed}P^e_{bc} - R_{ec}P^e_{bd} - R_{be}P^e_{cd} = 0. \qquad (3.38)$$

A substitution of (3.37) into (3.38) using the last equation in (3.4) then gives

$$-R_{cdbe}\lambda^e + R_{ec}\lambda^e g_{bd} - R_{ed}\lambda^e g_{bc} = 0. \qquad (3.39)$$

A further contraction of this last equation with g^{bd} gives $R_{ab}\lambda^b = 0$ and finally one gets $R^a{}_{bcd}\lambda^d = 0$. Previous results show that, on the subset A, (3.27) has only trivial solutions and hence $\lambda_{,a} = 0$. Thus λ is constant on each component of the subset A and so the smooth 1-form $d\lambda$ on M with components $\lambda_{,a}$ on each chart of M vanishes on A and hence on M if A is open and dense in M. In this case $\nabla' = \nabla$ and since M is connected, λ is constant on M. The following theorem has been proved.

Theorem 3.3 *Let M be a 4-dimensional manifold admitting a smooth, positive definite metric g and another arbitrary, smooth metric g' whose curvature tensors Riem agree on M. Then, if $A \neq \emptyset$, $g' = cg$ ($0 \neq c \in \mathbb{R}$) on each component of A (with c possibly being component dependent). If A is (open and) dense in M, $g' = cg$ ($0 \neq c \in \mathbb{R}$) on M and $\nabla = \nabla'$ on M.*

The first part of the proof of this theorem together with the remarks about the Weyl map following theorem 3.2 gives another proof of theorem 3.2. The techniques used here can be put in a more general setting. One is sometimes required to solve (3.25) for a symmetric (not necessarily non-degenerate) symmetric tensor h and for a certain subset of bivectors F which may be assumed to be a subalgebra of Λ_p (lemma 3.6(v)). Clearly $h = g$ is always a solution of (3.25) for all F and to seek other solutions for h one may use the results in lemma 3.6. Thus if (3.25) holds for h and the simple bivectors in $\mathrm{Sp}(x \wedge y)$ with x, y, z, w constituting an orthonormal basis at p (subalgebra type S_1) then $x \wedge y$ is an eigenspace for h and the use of the completeness relation (section 3.4) and using g to manipulate indices one finds, at p,

$$h_{ab} = ag_{ab} + bz_a z_b + cw_a w_b + d(z_a w_b + w_a z_b) \tag{3.40}$$

for $a, b, c, d \in \mathbb{R}$. If, on the other hand, (3.25) holds for h and for each F in a 2-dimensional subalgebra of the form $\mathrm{Sp}(x \wedge y, z \wedge w)$ (type S_2) then $x \wedge y$ and $z \wedge w$ are eigenspaces of h and one similarly finds using the completeness relation at p,

$$h_{ab} = ag_{ab} + b(z_a z_b + w_a w_b) \tag{3.41}$$

for $a, b \in \mathbb{R}$. If (3.25) holds for h and the 3-dimensional subalgebra of bivectors $\mathrm{Sp}(x \wedge y, x \wedge z, y \wedge z)$ (type S_3) with common annihilator w then, at p,

$$h_{ab} = ag_{ab} + bw_a w_b \tag{3.42}$$

for $a, b \in \mathbb{R}$. If h is a metric on M, the signatures of g and h may differ. If g and h above are metrics with the same tensor $Riem$ then the previous three equations would apply if the curvature class at p was class D (respectively, class B and class C).

Now suppose now that g and g' are smooth metrics on M with g positive definite and which give rise to the same Levi-Civita connection, that is, $\nabla = \nabla'$ on M. Then the holonomy algebras of ∇ and ∇' are equal and (3.25) holds for $h = g$ *and for* $h = g'$ and for each member F of this common holonomy algebra. The equality $\nabla = \nabla'$ also means that $Riem = Riem'$ and so the previous theorem holds for g and g'. But in that theorem ∇ and ∇' were not assumed equal (and need not be—see [13]).

Theorem 3.4 *Let M be a 4-dimensional manifold admitting a smooth positive definite metric g and another arbitrary, smooth metric g' whose connections ∇ and ∇' agree on M. If the holonomy algebra of ∇ is $\overset{+}{S_3}, \overset{-}{S_3}, \overset{+}{S_4}, \overset{-}{S_4}$ or $o(4)$ then $g' = cg$ where $c \in \mathbb{R}$.*

Proof The proof involves using lemma 3.6 to show that if $h = g'$ satisfies (3.25) for each F in each of the listed subalgebras then g and g' are proportional at each $p \in M$. Thus if $p \in M$ choose $F \in \overset{+}{S_3}$ to see (lemma 3.6(iv)) that g' admits a pair of 2-dimensional eigenspaces at p and which, if g and g' are not proportional at p, contain all the eigenvectors of g'. Now choose another $F' \in \overset{+}{S_3}$ independent of F and apply (3.25) at p to obtain another pair of eigenspaces of g' at p each of which is different from either of the previous two. (lemma 3.5(v)). This contradiction shows that g and g' are proportional at each $p \in M$. Thus $g' = \lambda g$ on M for a smooth function $\lambda : M \to \mathbb{R}$. Then $\nabla g = \nabla g' = 0$ shows that $d\lambda$ vanishes on M and since M is connected λ is a non-zero constant. The same proof applies to $\overset{-}{S_3}$ and then to the other listed subalgebras since each contains $\overset{+}{S_3}$ or $\overset{-}{S_3}$ as a subalgebra.   □

It is remarked here that if k is any global, smooth, type $(0,2)$, symmetric, tensor on M satisfying $\nabla k = 0$ but with k possibly degenerate and which is not proportional to g at some $p \in M$ then it is not proportional to g at *any* point of M because if $g(p) = bk(p)$, $p \in M$, $b \in \mathbb{R}$, then $\nabla(g - bk) = 0$ on M and $g - bk$ vanishes at p and hence on M. Also, since g is non-degenerate there exists, by an elementary continuity (rank-type) argument, $0 \neq a \in \mathbb{R}$ such that $g + ak$ is non-degenerate at p. But then $\nabla(g + ak) = 0$ and so $g + ak$ is non-degenerate on M. Thus the global tensor $g + ak$ is a metric on M compatible with ∇ but not conformally related to g at any point of M. In other words any global, smooth, symmetric, type $(0,2)$ tensor k on M satisfying $\nabla k = 0$ which is not proportional to g at some $p \in M$ gives rise to a *metric* on M which is not conformally related to g at any point of M and which is compatible with ∇. With regard to theorems 3.3 and 3.4, it is possible for two metrics of different signature to lead to the same tensor *Riem* or to the same Levi-Civita connection ∇. For example, suppose that M admits a global smooth 1-form field t satisfying $\nabla t = 0$. Then, in the above calculation, choose $k = t \otimes t$ so that for appropriately chosen $0 \neq a \in \mathbb{R}$, $g + ak$ is a metric on M of a different signature from g.

3.9   Sectional Curvature

In section 3.3, it was shown how the Grassmann manifold of 2-spaces at $p \in M$, now denoted for simplicity by G_p, may be identified with the manifold of projective simple bivectors at p as a 4-dimensional manifold. In this section it is geometrically convenient to think of G_p as a collection of 2-spaces but to analyse it using its (projective) bivector manifold structure.

Let $p \in M$ and consider the smooth, real-valued map σ_p on G_p, given for a non-zero, simple bivector F at p which represents a member of G_p, by

$$\sigma_p(F) = \frac{R_{abcd}F^{ab}F^{cd}}{2P_{abcd}F^{ab}F^{cd}} = \frac{R_{abcd}F^{ab}F^{cd}}{2|F|} \tag{3.43}$$

where P is the bivector metric. Since P has positive definite signature, σ_p is defined for each member of G_p, that is, for each non-zero, simple F. It is also clear that σ_p respects the equivalence relation $\sim$ on the set of non-zero bivectors $\Lambda'_p M$ given in section 3.3 and so σ_p is a smooth map on G_p. The function σ_p is called the *sectional curvature function at* p and seems to be essentially the kind of "curvature" that Riemann originally had in mind [36]. Since G_p is compact, σ_p is a bounded function on G_p (chapter 1) and since G_p is connected the range space of σ_p is a closed, bounded interval of $\mathbb{R}$. It can be interpreted in terms of the usual *Gauss curvature* of a (positive definite) 2-space as follows. One can show the existence of an open neighbourhood U of p such that the subset $N \subset U$ consisting of all points on those geodesics in U starting from p whose tangent vector at p lies in the 2-space at p represented by the blade of F is a 2-dimensional submanifold of U (and hence of M). If g' is the metric induced on N by the metric g on M, the Gauss curvature of (N, g') at p equals $\sigma_p(F)$. [It is remarked here that if g has Lorentz or neutral signature the situation is a little more complicated since it turns out that σ_p is, in general, not defined on the whole of G_p. This will be explored later.]

One can extend this definition to get the *sectional curvature function* σ on M defined for any simple bivector F at some point $p \in M$ by $\sigma(F) = \sigma_p(F)$. Thus σ is a function on the *Grassmann bundle* $\bigcup_{p \in M} G_p$. If σ_p is a constant function at p, one has either $Riem(p) = 0$ or $(R_{abcd} - 2\kappa P_{abcd})F^{ab}F^{cd} = 0$ for some constant $\kappa \neq 0$ and for each simple $F \in \Lambda_p M$. On choosing an orthonormal basis $x^1 = x$, $x^2 = y$, $x^3 = z$ and $x^4 = w$ in $T_p M$ and then a basis (of simple bivectors) $F_1 = x \wedge y$, $F_2 = x \wedge z$, $F_3 = x \wedge w$, $F_4 = y \wedge z$, $F_5 = y \wedge w$ and $F_6 = z \wedge w$ in $\Lambda_p M$ one can write the above equation as $Q_{abcd}F^{ab}F^{cd} = 0$ for all simple bivectors $F \in \Lambda_p M$ with $Q_{abcd} = R_{abcd} - 2\kappa P_{abcd}$. The basis bivectors have components $F_1^{12} = -F_1^{21} = 1, ..., F_6^{34} = -F_6^{43} = 1$ with all other components zero and hence, with the pairing of skew-symmetric tensor index pairs with block bivector indices $1, 2, 3, 4, 5, 6$ given by $12 \leftrightarrow 1$, $13 \leftrightarrow 2$, $14 \leftrightarrow 3$, $23 \leftrightarrow 4$, $24 \leftrightarrow 5$ and $34 \leftrightarrow 6$, one gets $F_1 = (1, 0, 0, 0, 0, 0), ..., F_6 = (0, 0, 0, 0, 0, 1)$. From the symmetries for R_{abcd} and P_{abcd} and using capital Latin letters for the block indices $1, ..., 6$ one may consider Q as a 6×6 symmetric matrix Q_{AB}. Then, for example, $Q_{24} = Q_{abcd}F_2^{ab}F_4^{cd}$. The above condition $Q_{abcd}F^{ab}F^{cd} = 0$ for each simple F then shows that $Q_{AA} = 0$ $(A = 1, 2, ..., 6)$ and since the bivectors with components $(1, 1, 0, 0, 0, 0), ..., (1, 0, 0, 0, 1, 0)$, etc, are also simple, $Q_{12} = Q_{13} = Q_{14} = Q_{15} = Q_{23} = Q_{24} = Q_{26} = Q_{35} = Q_{36} = Q_{45} = Q_{46} = Q_{56} = 0$. Thus all components of Q are zero except possibly Q_{16}, Q_{25} and Q_{34}. But then the condition $Q_{a[bcd]} = 0$ shows that $Q_{16} - Q_{25} + Q_{34} = 0$ and noting that $(x + y) \wedge (z + w)$ and $(x + z) \wedge (y + w)$ are also simple bivectors one finds $Q_{25} + Q_{34} = Q_{16} - Q_{34} = 0$. Thus $Q_{16} = Q_{25} = Q_{34} = 0$ and so $Q = 0$. Thus if σ_p is a constant function at p,

$$R_{abcd} = \frac{R}{6}P_{abcd} = \frac{R}{12}(g_{ac}g_{bd} - g_{ad}g_{bc}) \qquad (3.44)$$

($\kappa = \frac{R}{12}$) holds at p, and conversely, and *Riem* is said to satisfy the *constant curvature condition at p* (and the Einstein space condition $R_{ab} = \frac{R}{4}g_{ab}$ is then easily seen to hold at p). In fact, if (3.44) holds over some open, (connected) coordinate domain $U \subset M$, it is easily checked from the Bianchi identity (3.6) (contracted to $(R^b{}_a - \frac{R}{2}\delta^b_a)_{;b} = 0$) and the previous Einstein space condition that R is constant on U. This is Schur's lemma (see, for example, [37]). Then if (3.44) is true for each p in some open dense subset of M, (3.44) holds on M (by an argument similar to one given before), R is constant on M (since M is connected) and (M, g) is said to be of *constant curvature* (and (M, g) is an Einstein space).

Some years ago Kulkarni solved the following problem [38]. Suppose M is a 4-dimensional manifold admitting positive definite metrics g and g' whose sectional curvatures σ and σ' are such that (*i*) σ_p, σ'_p are identical functions on G_p for each $p \in M$ and (*ii*) the subset of M on which σ_p (and hence σ'_p) is *not* a constant function on G_p (which is necessarily open in M) is dense in M. What can one say about the metrics g and g'? The remarkable result is that $g = g'$. [Actually Kulkarni proved more than this since he proved it for $\dim M \geq 4$ and considered also the case of $\dim M = 3$.]

It is first noted that under the suppositions given above, g and g' can be shown to be conformally related [38]. To see this, briefly, note that the assumption that σ_p and σ'_p are not constant functions means from (3.43) that their equality can be written as the equality of two polynomial expressions in the coordinates of G_p and this is true for each p in some open dense subset of M. Equating coefficients then leads, after some calculation, to the fact that g' and g are conformally related on this open dense subset of M and hence, by a proof given earlier, on M. Thus $g' = \phi g$ on M for some real-valued, nowhere-zero, smooth, function ϕ on M. The tensors arising from g' will be distinguished from those of g by the use of a prime and C and C' will then denote their respective type $(1, 3)$ Weyl conformal tensors.

Lemma 3.7 *Under the assumptions above one has on M*

   (*i*) $g' = \phi g$,   (*ii*)$R'_{abcd} = \phi^2 R_{abcd}$,   (*iii*) $R'^a{}_{bcd} = \phi R^a{}_{bcd}$,
   (*iv*) $R'_{ab} = \phi R_{ab}$,   (*v*) $R' = R$,   (*vi*) $C' = \phi C$.

Proof Part (*i*) has already been given. Next let $X \subset M$ be precisely the (closed) subset of M on which σ_p (and hence σ'_p) are constant functions, so that $M \setminus X$ is open and dense in M (since $\text{int} X = \emptyset$). Then if $p \in X$, it follows from (3.7) and (3.44) after a simple calculation using the fact that (M, g) and (M, g') each satisfy the Einstein space condition at p, that $C(p) = C'(p) = 0$. Also if for $p \in M$ $Riem(p) = 0$, then σ_p (and hence σ'_p) is a constant function (the zero function) on G_p and so $p \in X$ (and $Riem'(p) = 0$ by a similar argument to that given above for the tensor Q). So those points where $Riem'$ and $Riem$ vanish are thus included in X. Now from (3.43) and part (*i*) it follows that for each $p \in M$ and with $W_{abcd} \equiv (R'_{abcd} - \phi^2 R_{abcd})$,

$$W_{abcd}F^{ab}F^{cd} \equiv (R'_{abcd} - \phi^2 R_{abcd})F^{ab}F^{cd} = 0 \qquad (3.45)$$

for every simple bivector F. Another argument identical to one given above then shows that $W \equiv 0$ and so result (ii) is proved. The results (iii), (iv) and (v) then follow since $g'^{ab} = \phi^{-1}g^{ab}$. When these results are substituted into (3.7) the final result (vi) follows. [It is important to note here that since g' and g are conformally related, $C' = C$ on M but C (and hence C') may be zero. Result (vi) holds in *all* cases.]□

A slightly different, more direct approach from [38] will now be followed. Let U denote the open subset of M on which C (and hence C', since g' and g are conformally related on M) are non-zero and let V denote the open subset of M on which the 1-form $d\phi$ does not vanish. Then, from lemma 3.7(vi), $\phi = 1$ on U and so $U \cap V = \emptyset$. Also ϕ is a non-zero constant on each component of $\text{int}(M \setminus V)$ and, since g and g' are related by a *constant* conformal factor on (each component of) $\text{int}(M \setminus V)$, $Riem' = Riem$ on $\text{int}(M \setminus V)$. Thus since $Riem$ and $Riem'$ do not vanish on the open subset $M' = M \setminus X$, $\phi = 1$ on $M' \cap \text{int}(M \setminus V)$ from (iii) above. Let Y be the closed subset of all points of M at which $\phi = 1$, so that $U \subset Y$.

This allows a disjoint decomposition of M as $M = V \cup \text{int}Y \cup K$ (since if $\text{int}Y \neq \emptyset \neq V$, $(\text{int}Y) \cap V = \emptyset$) and where K is a closed subset of M defined by the disjointness of the decomposition. Now the subset K satisfies $\text{int}K = \emptyset$. This follows because any non-empty open subset $U' \subset K$ would satisfy $U' \cap V = \emptyset$ (because of the disjointness of the decomposition) and so $U' \subset (M \setminus V)$ and hence, since U' is open, $U' \subset \text{int}(M \setminus V)$. Thus ϕ is constant on each component of U'. Further, since $\text{int}X = \emptyset$, U' is not contained in X and $Riem$ and $Riem'$ are equal and non-zero and $\phi = 1$ (lemma 3.7) at each point of the non-empty open subset $U' \cap (M \setminus X)$ of U', this latter set then being contained in Y and hence in $\text{int}Y$. From this it follows that $U' \cap (\text{int}Y) \neq \emptyset$ contradicting the disjointness of the above decomposition and so $U' = \emptyset$. In summary, $M = V \cup \text{int}Y \cup K$ is a disjoint decomposition with $\phi = 1$ and $g' = g$ on $\text{int}Y$ and with V an open subset on which $d\phi$ is nowhere zero. Since $U \cap V = \emptyset$ the open subset V is conformally flat for g' and g. The subset K is closed with empty interior.

The open subset V can now be explored further, assuming it is not empty. (if $V = \emptyset$, $\phi = 1$ on $M \setminus K$ and hence on M and so $g' = g$ on M.) On V, $C = C' = 0$ and so, from the first of these and equations (3.8) and (3.9),

$$R_{abcd} = \frac{1}{2}(\widetilde{R}_{ac}g_{bd} - \widetilde{R}_{ad}g_{bc} + \widetilde{R}_{bd}g_{ac} - \widetilde{R}_{bc}g_{ad}) + \frac{R}{6}P_{abcd}. \tag{3.46}$$

Next consider the Bianchi identities in the conformally flat case, one for each of the connections ∇ and ∇', for g and g', respectively, with the symbols ; and | denoting covariant derivatives with respect to ∇ and ∇', respectively. These identities can be obtained from (3.6) and (3.46) after a somewhat tedious but straightforward calculation, recalling the identities $2R^a{}_{b;a} = R_{,b}$ (and hence $4\widetilde{R}^a{}_{b;a} = R_{,b}$) and a contraction with g^{ac}. One finds using a prime to denote quantities formed from g'

$$R_{ca;b} - R_{cb;a} = \frac{1}{6}(g_{ac}R_{,b} - g_{cb}R_{,a}), \qquad R'_{ca|b} - R'_{cb|a} = \frac{1}{6}(g'_{ac}R'_{,b} - g'_{cb}R'_{,a}).$$

$$(3.47)$$

One can now substitute results (iv) and (v) of lemma 3.7 into the second of (3.47) and subtract from it ϕ times the first of (3.47) to remove the partial derivatives, Then the differences between the Christoffel symbols arising from ∇' and ∇ give rise to the tensor P in (3.37) (not to be confused with the bivector metric) and on substituting all this into (3.47) one finds

$$\phi_{,b}R_{ac} - \phi_{,a}R_{bc} = \phi R_{ae}P^e_{bc} - \phi R_{be}P^e_{ac} = R_{be}\phi^e g_{ac} - R_{ae}\phi^e g_{bc} \qquad (3.48)$$

where $\phi^a = g^{ac}\phi_{,c}$. If one contracts (3.48) with g^{ac} one sees that $4R_{ab}\phi^b = R\phi_a$ and hence that the tracefree Ricci tensor satisfies $\widetilde{R}_{ab}\phi^b = 0$. Substituting this back into (3.48) gives $\widetilde{R}_{ac}\phi_b - \widetilde{R}_{bc}\phi_a = 0$ from which one finds $\widetilde{R}_{ab} = \psi\phi_a\phi_b$ for some smooth real-valued function ψ on V. It follows that $\psi(\phi_a\phi^a) = 0$ on V. Now suppose that $p \in V$ and $(\phi_a\phi^a)(p) \neq 0$. Then $\phi_a\phi^a$ is non-zero over some non-empty open subset W of M contained in V and ψ vanishes on W. But then $\widetilde{Ricc}$ vanishes on W and so, from (3.9), E vanishes on W and so from (3.8) it follows that (3.44) holds on W showing that $W \subset X$. Since X has interior W must be empty and then $\phi_a\phi^a = 0$ on V. Since, by definition of the set V, $d\phi$ is nowhere zero on V this gives a contradiction to the signature of g and so, in the positive definite case, $V = \emptyset$ and $M = \text{int}Y \cup K$ is a disjoint decomposition with $\phi = 1$ and $g' = g$ on $\text{int}Y$. Since K is closed with empty interior $\phi = 1$ on M and so $g' = g$ on M. [It is remarked at this point that the initial assumption that g' was also of positive definite signature can actually be removed. This is because since g is positive definite, σ_p is defined on the whole of G_p for each $p \in M$. Thus it is implicitly assumed that σ'_p is also so defined. As will be seen in the next two chapters, this (and the conditions of the theorem) force g' to be positive definite.]

The following result is thus proved.

Theorem 3.5 *(Kulkarni, 1970 [38])*

Suppose M is a 4-dimensional manifold admitting a smooth, positive definite metric g and another smooth metric g' of arbitrary signature and whose sectional curvatures σ and σ' are such that (i) σ_p and σ'_p are identical functions on G_p for each $p \in M$ and (ii) σ_p (and hence σ'_p) are not constant functions on G_p for each p in an open dense subset of M. Then $g' = g$ on M.

The necessity of assuming that the interior of that subset of M of points at which σ_p and σ'_p are constant functions, that is, those points of "constant curvature", is empty was stressed and exemplified by Kulkarni [38]. It is also remarked that if F and $\overset{*}{F}$ are dual simple bivectors representing a pair of orthogonal 2-spaces at p one has from (3.8), (3.17) and (3.19) and noting that $|F| = |\overset{*}{F}|$ (lemma 3.4(iii))

$$\sigma_p(\overset{*}{F}) - \sigma_p(F) = \frac{R_{abcd}\overset{*}{F}{}^{ab}\overset{*}{F}{}^{cd} - R_{abcd}F^{ab}F^{cd}}{2P_{abcd}F^{ab}F^{cd}} \tag{3.49}$$

$$= \frac{({}^*R^*_{abcd} - R_{abcd})F^{ab}F^{cd}}{2P_{abcd}F^{ab}F^{cd}} = -\frac{E_{abcd}F^{ab}F^{cd}}{P_{abcd}F^{ab}F^{cd}} \tag{3.50}$$

which shows that if (M, g) satisfies the Einstein space condition at p, so that $E(p) = 0$, $\sigma(F) = \sigma(\overset{*}{F})$ for each such F. A similar argument to that given above for the tensor W gives the converse of this result (cf, [64]).

Theorems 3.2 to 3.5 reveal strong relations between the metric, the Levi-Civita connection, the curvature tensor, the sectional curvature function, the holonomy group and the Weyl conformal tensor for 4-dimensional manifolds admitting a positive definite metric. It will be seen later that similar, but slightly less strong, results apply in the cases of Lorentz and neutral signature.

3.10 The Ricci Flat Case

Now consider the situation when the 4-dimensional manifold with smooth, positive definite metric, (M, g), is *Ricci flat*, that is, when the Ricci tensor is identically zero on M. In this case the Ricci scalar and the tensor E are identically zero on M, $R \equiv 0$ and $E \equiv 0$ on M, and then (3.8) shows that the Riemann tensor and Weyl tensor are equal, $C = Riem$, on M. It now follows from section 3.5 that, at each $p \in M$, $\overset{+}{S}_p$ and $\overset{-}{S}_p$ are invariant subspaces of the curvature map f (and are orthogonal with respect to the bivector metric P). Thus, choosing a basis F_i for $\overset{+}{S}_p$ and a basis G_i for $\overset{-}{S}_p$ ($i = 1, 2, 3$), one may write out $Riem(p)$ in this bivector basis using the abbreviated form used earlier as

$$Riem(p) = \sum_{i,j=1}^{3} (\alpha_{ij} F_i F_j + \beta_{ij} G_i G_j) \tag{3.51}$$

where α_{ij} and β_{ij} are symmetric arrays of real numbers. Thus the range space $rgf(p)$ of the curvature map at p admits a basis consisting of members of $\widetilde{S}_p$. [Of course, this result applies also to the Weyl tensor in *all* cases, not just the Ricci flat case—see section 3.5.] It also follows from section 3.5 that if (M, g) is Ricci flat and $p \in M$ the curvature class at p is either O or A. A study of the curvature map f reveals the following results.

Theorem 3.6 *Let* (M, g) *be a 4-dimensional, positive definite, Ricci-flat manifold with Levi-Civita connection* ∇ *and let* $p \in M$.

(i) *Suppose (M, g) is not flat. Then if $Riem(p) \neq 0$, $rankf_p \geq 2$, the holonomy algebra ϕ cannot be of type S_1, S_2 or S_3 and ∇ determines the metric up to a constant conformal factor.*

(ii) *If (M, g) is non-flat, Riem determines g up to a constant conformal factor and determines ∇ uniquely.*

Proof For (i) if $Riem(p) \neq 0$ the curvature class at p is, from a remark above, class A and hence $\dim rgf(p) \geq 2$. But $rgf(p)$ is a non-trivial subspace of the infinitesimal holonomy algebra at p, and hence of ϕ, and so this curvature class restriction shows that ϕ cannot be of type S_1, S_2 or S_3. Thus the holonomy algebra at p is of type $\overset{+}{S_3}$, $\overset{-}{S_3}$, $\overset{+}{S_4}$, $\overset{-}{S_4}$ or $S_6(= o(4))$ and theorem 3.4 completes the proof.

For part (ii) one notes that if g and g' are smooth metrics on M with the same tensor $Riem$ and with g positive definite then, since g is Ricci flat and non-flat, g' is also Ricci-flat and non-flat and so from (3.8) g and g' have the same Weyl conformal tensor which does not vanish over any non-empty, open subset of M (by the non-flat condition). Theorem 3.2 then shows that g and g' are conformally related, $g' = \lambda g$, for some smooth, real function λ on M. Then the argument leading to theorem 3.3 shows that $R_{abcd}\lambda^d = 0$ ($\lambda^a = g^{ab}\lambda_{,b}$) holds on M and it follows that $\lambda_{,a}$ vanishes over the open dense subset of M where $Riem = C$ is not zero. Since M is connected, λ is thus constant on M and the proof is complete. □

It is remarked here that if the pair (M, g) is non-flat and Ricci-flat, $C = Riem$ on M and if the open, dense subset of points of M at which $Riem$ (and hence C) is non-zero is labelled U the sectional curvature function σ_p is nowhere a constant function on U since, from (3.44), such a condition would force the contradiction $Riem = 0$ at each point of U (since $R \equiv 0$ on M). Thus one has from the work of section 3.9 the much tidier result in the Ricci-flat case.

Theorem 3.7 *If M is a 4-dimensional manifold with smooth, positive definite metric g which is non-flat and Ricci-flat and if M admits another smooth metric g' of arbitrary signature and with the same sectional curvature function as g at each $p \in M$, then $g' = g$.*

There is another straightforward result which is, in a sense, a trivial variant of a theorem due to Brinkmann [39].

Theorem 3.8 *If M is a 4-dimensional manifold which admits conformally related, smooth, positive definite metrics g and g' each of which is Ricci flat and non-flat. Then $g' = \lambda g$ for some constant λ.*

Proof Write $g' = \lambda g$ for some smooth function $\lambda : M \to \mathbb{R}$. Then the respective type $(1, 3)$ Weyl conformal tensors C and C' for g and g' are equal,

$C' = C$, on M (which, by the non-flat and Ricci-flat conditions, are each non-zero on some open, dense subset $U \subset M$). The Ricci-flat condition then shows that the respective curvature tensors are equal, $Riem' = Riem$. The work leading to theorem 3.3 above then shows that in any chart domain $R_{abcd}k^d = 0$ where $k_a = \lambda_{,a}$, and hence, again from the Ricci flat condition, $C_{abcd}k^d = 0$. The result now follows. $\qquad\square$

Chapter 4

Four-Dimensional Lorentz Manifolds

4.1 Lorentz Tangent Space Geometry

This chapter will be devoted to the study of the geometry of a 4-dimensional Lorentz manifold, (M, g), with metric g of signature $(-, +, +, +)$ and with M being smooth, connected, Hausdorff and second countable and g smooth. Let $p \in M$ with tangent space $T_p M$ at p. As before, for $u, v \in T_p M$, the inner product $g(u, v)$ is denoted by $u \cdot v$ and $|u| = g(u, u)$. For this signature one has spacelike, timelike and null vectors in $T_p M$. Timelike and spacelike vectors are sometimes collectively referred to as *non-null* and a non-null vector u satisfying $|u| = \pm 1$ is called a unit vector. For a non-zero $u \in T_p M$ the 1-dimensional subspace of $T_p M$ (direction) spanned by u is called spacelike (respectively, timelike or null) if u is spacelike (respectively, timelike or null). Of course one may choose a basis in $T_p M$ such that the metric at p takes the Sylvester canonical form $\mathrm{diag}(-1, 1, 1, 1)$. This is referred to as a *Minkowski basis* and enables a quick proof of the facts that two timelike vectors, a timelike and a null vector, or two independent null vectors can never be orthogonal. For the first of these let $u, v \in T_p M$ be timelike and choose a Minkowski basis for which u (assumed unit) satisfies $u = (1, 0, 0, 0)$ and $v = (a, b, c, d)$ with $a, b, c, d \in \mathbb{R}$ with $a^2 > b^2 + c^2 + d^2$, Then $u \cdot v = 0$ gives $a = 0$ and so $b = c = d = 0$ and hence the contradiction $v = 0$. For the second and third, choose v with components as above and $k \in T_p M$ null with $k = (1, 1, 0, 0)$. Then $v \cdot k = 0 \Rightarrow a = b \Rightarrow |v| \geq 0$ and so either v is spacelike or a multiple of k. Two useful bases for $T_p M$ are a (pseudo-)orthonormal tetrad of unit vectors $t, x, y, z \in T_p M$ with $|x| = |y| = |z| = -|t| = 1$ and all other inner products between these basis members zero, and a *(real) null tetrad* $l, n, x, y \in T_p M$ with l and n null vectors satisfying $l \cdot n = 1$, $|x| = |y| = 1$ and all other inner products between basis members zero. Two such bases are said to *correspond* if $\sqrt{2} z = l + n$ and $\sqrt{2} t = l - n$. The collection of all null vectors in $T_p M$ is called the *null cone* at p. From the above, one has useful *completeness relations* relating a basis and the metric at any point and given by $g_{ab} = -t_a t_b + x_a x_b + y_a y_b + z_a z_b = l_a n_b + n_a l_b + x_a x_b + y_a y_b$. Conversely these last two relations imply that the collections (t, x, y, z) and (l, n, x, y) are, respectively, an orthonormal and a null basis (see section 3.4).

DOI: 10.1201/ 9781003023166-4

If a 2-dimensional subspace (2-space) U of T_pM is such that each of its non-zero members is spacelike it is called *spacelike*, if U contains exactly two distinct null directions it is called *timelike* and if U contains a unique null direction it is called *null*. [If U contains three distinct null directions spanned by null vectors $u, v, w \in T_pM$, one of u, v, w must be a non-trivial linear combination of the other two. The null condition then forces the contradiction that these other two are orthogonal.] It follows that these are the only possibilities for U. Then it is easily checked that if U is spacelike it may be spanned by an orthogonal pair of spacelike vectors. If U is timelike it contains non-orthogonal (independent) null vectors l and n which may be scaled so that $l \cdot n = 1$ and U also contains the (orthogonal) spacelike and timelike members $l \pm n$. The only *orthogonal pairs* of independent vectors of a timelike U consist of a spacelike and a timelike vector (all other possibilities are easily checked to forbid the presence of timelike vectors in U) and the null directions spanned by l and n are unique up to interchange and referred to as the *principal null directions* of U. If U is null it contains a null vector l unique up to a scaling (and called the *principal null direction* of U) and all other non-zero members of U are either proportional to l or spacelike and orthogonal to l. To see this let U be spanned by $l, k \in T_pM$. Then k is not null and if $l \cdot k \neq 0$ one can find $0 \neq \lambda \in \mathbb{R}$ such that $k + \lambda l$ is null and a contradiction follows. This completes the proof. Thus if u, v are an orthogonal pair of independent members of a null 2-space U exactly one of them is a multiple of l otherwise each member of the 2-space would be spacelike. The orthogonal complement $U^\perp$ of a timelike 2-space U is spacelike, and vice versa and the members of U and $U^\perp$ collectively span T_pM. For $p \in M$ and a subspace $U \subset T_pM$ of dimension ≤ 3 the metric $g(p)$ naturally induces a mapping $U + U \to \mathbb{R}$ given for $u, v \in U$ by $(u, v) \to g(p)(u, v)$. This need not be an inner product on U (it may not be non-degenerate) but if it is, it is called the *induced metric* on U. The induced metric on a spacelike 2-space is Euclidean whilst the metric induced on a timelike 2-space is Lorentzian. There is no induced metric on a null 2-space. The orthogonal complement $U^\perp$ of a null 2-space U is also null and the principal null directions of U and $U^\perp$ coincide. To see this let l be null and x spacelike with $l \cdot x = 0$ and $U = l \wedge x$. Then clearly $U^\perp = l \wedge y$ with $l \cdot y = x \cdot y = 0$. So y is spacelike and $U^\perp$ null. The span of U and $U^\perp$ is 3-dimensional. This completes the classification of 2-spaces of T_pM. In fact, for each type one may choose a real null tetrad l, n, x, y such that U can be represented as $l \wedge x$ (null), $x \wedge y$ (spacelike) and $l \wedge n$ (timelike) and conversely any 2-space of one of these forms is of the indicated type.

Let U is a 3-dimensional subspace (a 3-space) of T_pM. Then U is called *spacelike* if all its non-zero members are spacelike (and then its 1-dimensional orthogonal complement, or *normal*, is, from the Lorentz signature, a timelike direction). U is called *timelike* if it contains infinitely many null directions (and then it contains also timelike and spacelike members and its orthogonal complement (normal) is a spacelike direction). U is called *null* if it contains a unique null direction (and then all its other non-zero members are spacelike and orthogonal to it and its orthogonal complement (normal) is the same null direction). This null direction is sometimes referred to as its *principal null direction*. The metric induced on a spacelike (respectively timelike) 3−space is positive definite (respectively, Lorentz). There is no metric induced on a null 3−space. This completes the classification of 3−spaces of T_pM. In the tetrads used so far typical examples of 3−spaces are $Sp(x, y, z)$ (spacelike), $Sp(x, y, t)$ (timelike) and $Sp(l, x, y)$ (null).

It is stressed here that if U is a spacelike or timelike 2-space at p then U and $U^\perp$ are *complementary* in the sense that the span of $U \cup U^\perp$ equals T_pM. However, if U is a null 2-space then the span of $U \cup U^\perp$ gives a null 3−space of T_pM whose principal null direction equals the (common) principal null direction of U and $U^\perp$.

4.2 Classification of Second Order Tensors

It is important to know the algebraic structure of the main tensors which can occur in Lorentzian geometry. In this section, the classification of second order symmetric and skew-symmetric tensors will be considered. In addition to the algebraic and geometrical aspects of such classifications they have also been found very useful in the physics of Einstein's general relativity theory. First it is convenient to gather together some general elementary results on the algebraic theory of symmetric tensors (recalling the convention adopted for "complex" eigen structures in section 3.3).

Lemma 4.1 *Let (M, g) be a 4-dimensional manifold admitting a Lorentz metric g and let S be a real, non-zero, symmetric, type $(0, 2)$ tensor at $p \in M$ with associated linear map f.*

(i) The map f must admit an invariant 2-space whose orthogonal complement is then also invariant. If $U \subset T_pM$ is an invariant 2-space for f then if U is spacelike it contains two independent (real) spacelike eigenvectors of f (or S). If U is null the principal null direction of U is an eigendirection of f. There may or may not be another eigendirection of f in U. If U is timelike with principal null directions spanned by null vectors l and n, then either l and n span eigendirections of f with equal eigenvalues, or U contains two independent real, orthogonal, non-null eigendirections with distinct eigenvalues, or U gives rise to a conjugate pair of complex eigendirections (which may, after an appropriate choice of l and n, be taken as spanned by $l \pm in$), or exactly one of l and n spans the only eigendirection of f. If f admits a complex eigenvector the invariant 2-space spanned by its real and imaginary parts is timelike. Thus f admits (up to complex scalings) at most one conjugate pair of (complex) eigendirections and must admit a real eigenvector.

(ii) Any null eigenvector of f is necessarily real and any eigenvector of f corresponding to a non-simple elementary divisor is null (and hence real). Thus non-null eigenvectors of f correspond to simple elementary divisors. Conversely any (real) null eigenvector corresponds either to a non-simple elementary divisor or arises from an eigenvalue degeneracy (that is, the associated eigenspace has dimension ≥ 2).

Proof The proof of the first part of (i) was given in chapter 3. The next part of (i) follows from the principal axes theorem since f restricts to a linear map $U \to U$ and the metric g, restricted to U, is positive definite. If U is null then U and $U^\perp$ are invariant and have a common principal null direction which then becomes an eigendirection of f. If U is timelike and spanned by null vectors $l, n \in T_pM$ with $l \cdot n = 1$ then, since $S_{ab}l^a n^b = S_{ab}n^a l^b$ one has $f(l) = al + bn$ and $f(n) = cl + an$ $(a, b, c \in \mathbb{R})$. Suppose n is an eigenvector of f so that $c = 0$ and $f(n) = an$. Then either n is the only real independent eigenvector in U or any other real independent eigenvector in U is of the form $l + \lambda n$ $(\lambda \in \mathbb{R})$ and satisfies $f(l + \lambda n) = \mu(l + \lambda n)$ $(\mu \in \mathbb{R})$. Thus $a = \mu$ and $b + \lambda a = \mu \lambda$ which leads to $b = 0$ and to l also being an eigenvector with the same eigenvalue a as n. If neither l nor n is an eigenvector, $\lambda \neq 0$, $b \neq 0 \neq c$ and the above calculation leads to $\lambda^2 = \frac{b}{c}$ and hence to a pair of independent eigenvectors which, depending in the sign of $\frac{b}{c}$, are either real with distinct eigenvalues and hence orthogonal (section 3.3) and non-null or complex conjugates. If they are complex conjugates they can be taken as $l \pm i\nu n$ $(\nu \in \mathbb{R}, \nu > 0$ and $\nu^2 = -\lambda^2)$ and then as $\nu^{-\frac{1}{2}}(l \pm i\nu n)$. Thus they may be taken as $l' \pm in'$ with $l' = \nu^{-\frac{1}{2}}l$ and $n' = \nu^{\frac{1}{2}}n$ and so $l' \cdot n' = 1$. The next part follows since if $r \pm is$ are complex eigenvectors $(0 \neq r, s \in T_pM)$ with distinct eigenvalues $a \pm ib, b \neq 0$, then $(r + is) \cdot (r - is) = 0$ implies that $|r| + |s| = 0$ and so the span of r and s is timelike. Since f admits either a null invariant 2-space or an orthogonal timelike/spacelike pair of invariant 2-spaces the rest of part (i) follows easily by a simple counting of real eigenvalues of f.

Some parts of this proof were mentioned in [49] but with few details given. Here a full proof will be given. If $r + is$ is a complex null eigenvector of f, $(r, s \in T_pM)$, with eigenvalue $a + ib$ $(a, b, \in \mathbb{R}, b \neq 0)$ then so is $r - is$ with eigenvalue $a - ib$ (which is different from $a + ib$ since $b \neq 0$). Hence $|r \pm is| = 0$ and $(r + is) \cdot (r - is) = 0$. So $|r| = |s| = 0$ and $r \cdot s = 0$ which is a contradiction for Lorentz signature. So $b = 0$ and any null eigenvector must be real. Now suppose that k is a real or complex eigenvector of f at p corresponding to a non-simple elementary divisor. Then selecting a Jordan basis (chapter 1) one has vectors k and k' satisfying $f(k) = \alpha k$ and $f(k') = \alpha k' + k$ $(\alpha \in \mathbb{C})$. The symmetry of S then shows that $k \cdot f(k') = k' \cdot f(k)$ and so k is null (and hence real). Now suppose that k is a real null eigenvector of f at p with $f(k) = \alpha k$ $(\alpha \in \mathbb{R})$ such that α is associated with a simple elementary divisor and is not degenerate. Then clearly k is, by non-degeneracy, orthogonal to each eigenvector of f (including itself) and hence the eigenvectors of f cannot form a basis for (the complexification of) T_pM. It follows that f admits a non-simple elementary divisor whose associated eigenvalue $\gamma \neq \alpha$ and whose associated eigenvector u is null, and hence real, from the previous argument, and satisfies $k \cdot u = 0$. Thus k and u are real, orthogonal, independent null vectors contradicting Lorentz signature. $\qquad \square$

One can now complete the classification of symmetric tensors of second-order in the 4-dimensional case of Lorentz signature. For this it is convenient to prove the following lemma for the 3−dimensional Lorentz case where the notation and definitions are carried forward in a consistent manner from the 4-dimensional situation and where the signature is $(-, +, +)$. The definitions of spacelike, timelike and null 2-spaces are as in the 4-dimensional case. The orthogonal complement of such a 2-space is a direction. In the expression for a Segre symbol a positive integer entry always refers to a real eigenvalue whilst a complex conjugate pair of eigenvalues is denoted by the pair entry $z\bar{z}$. (This rule will be slightly modified in the case of neutral signature in chapter 5.)

Lemma 4.2 *Let M be a 3−dimensional manifold on which there is a Lorentz metric g of signature $(-, +, +)$. Let S be a non-zero second-order symmetric tensor at $p \in M$ with associated linear map f. The only possibilities for the Jordan/Segre type of S are $\{111\}$, $\{z\bar{z}1\}$, $\{21\}$ and $\{3\}$ together with their (possible) degeneracies and each can occur.*

Proof Quite generally f either admits a real eigenvector together with a complex conjugate pair of complex eigenvectors whose real and imaginary parts span a timelike 2-space (and hence its spacelike orthogonal complement is invariant and gives rise to a (real) eigendirection), or three independent real eigenvectors associated with simple elementary divisors, or a real null eigenvector associated with a non-simple elementary divisor. That this latter case must result in a real null eigenvector and that the invariant 2-space in the first possibility is timelike follow as in the 4-dimensional case above. The first of these possibilities leads to a possible Segre type $\{z\bar{z}1\}$. The second leads to the possibility $\{111\}$ or some degeneracy of this type whilst the third leads to the possibilities $\{21\}$ or $\{3\}$. Now suppose $p \in M$ and (using the obvious notation from the 4-dimensional case) that $t, x, z \in T_pM$ is a pseudo-orthonormal basis at p with $|x| = |z| = -|t| = 1$ and l, n, x is a null basis at p with $\sqrt{2}l = z + t$ and $\sqrt{2}n = z - t$ null, $l \cdot n = 1$ and all other inner products between them being zero. Using these one can construct general canonical forms for the four types claimed in the lemma as follows, where $\rho_1, \rho_2, \rho_3, \lambda, \mu \in \mathbb{R}$ with $\lambda \neq 0 \neq \mu$.

$$\rho_1(l_a n_b + n_a l_b) + \rho_2(l_a l_b \pm n_a n_b) + \rho_3 x_a x_b, \qquad (4.1)$$

$$\rho_1(l_a n_b + n_a l_b) + \lambda l_a l_b + \rho_2 x_a x_b, \qquad (4.2)$$

$$\rho_1(l_a n_b + n_a l_b) + \mu(l_a x_b + x_a l_b) + \rho_1 x_a x_b. \qquad (4.3)$$

In (4.1) with the $+$ option the eigenvectors are $l \pm n$ with eigenvalues $\rho_1 \pm \rho_2$ and x with eigenvalue ρ_3 and the Segre type is $\{111\}$ or some degeneracy of this type, and with the $-$ option they are $l \pm in$ with eigenvalues $\rho_1 \pm i\rho_2$ and x with eigenvalue ρ_3 and the Segre type is $\{z\bar{z}1\}$. In (4.2) the eigenvectors are l with eigenvalue ρ_1 and x with eigenvalue ρ_2 and Segre type $\{21\}$ or its degeneracy and in (4.3) the eigenvector is l with eigenvalue ρ_1. In (4.2) one may scale l (with a compensating scaling of n) so that $\lambda = \pm 1$ and in (4.3) one may similarly scale l so that $\mu = 1$. In the Segre type $\{111\}$ case one may write it conveniently in terms of the orthonormal triad t, x, y as

$$(\rho_2 - \rho_1)t_a t_b + (\rho_2 + \rho_1)z_a z_b + \rho_3 x_a x_b. \tag{4.4}$$

This completes the proof.$\square$

Theorem 4.1 *Let M be a 4-dimensional manifold on which there is a Lorentz metric g of signature $(-, +, +, +)$. Let S be a non-zero second order symmetric tensor at $p \in M$ with associated linear map f. The only possibilities for the Jordan/Segre type of S are $\{1111\}$, $\{z\bar{z}11\}$, $\{211\}$ and $\{31\}$ together with each of their degeneracies, and all of these types can occur. It follows that a real eigenvector is always admitted.*

Proof From lemma 4.1 if f admits a complex eigenvector it is non-null with real and imaginary parts spanning a timelike, invariant 2-space U for f and then $U^{\perp}$ is invariant and spacelike. Then the action of f on $U^{\perp}$ is diagonalisable over $\mathbb{R}$ and the Segre type of f is $\{z\bar{z}11\}$. or its degeneracy. Otherwise all eigenvectors of f are real (and at least one exists from lemma 4.1(i)). If f admits a timelike eigenvector k, the $3-$dimensional orthogonal complement V of k is spacelike and invariant for f and the action of f on V is diagonalisable over $\mathbb{R}$. The resulting Segre type of f is $\{1111\}$ or some degeneracy of this type (and conversely). If f admits a spacelike eigenvector k, the $3-$dimensional orthogonal complement V of k is invariant for f and the metric induced on V is Lorentz. It follows from lemma 4.2 that the Segre type of f restricted to V is as given there. Thus the Segre type of f is $\{1111\}$, $\{z\bar{z}11\}$, $\{211\}$ or $\{31\}$ or some degeneracy of one of these types. Finally suppose all eigenvectors of f are (real and) null. If there are at least two independent such eigenvectors, l and n then, because of the Lorentz signature, one has $l \cdot n \neq 0$ and the 2-space U they span is a timelike eigenspace giving rise to a contradiction since non-null eigenvectors arise in $U^{\perp}$. So suppose there exists exactly one real null eigenvector k so that the Segre type of f is $\{4\}$. In a Jordan basis k, r, s, q at p one then has (chapter 1) $f(k) = ak$, $f(r) = ar + k$, $f(s) = as + r$ and $f(q) = aq + s$ and using the symmetry relations $k \cdot f(q) = q \cdot f(k)$, $k \cdot f(s) = s \cdot f(k)$ and $r \cdot f(s) = s \cdot f(r)$ gives k, r null and $k \cdot r = 0$ and a contradiction. The general canonical forms for each type can be obtained directly from (4.1), (4.2) and (4.3) in terms of a null basis l, n, x, y and are

$$\rho_1(l_a n_b + n_a l_b) + \rho_2(l_a l_b \pm n_a n_b) + \rho_3 x_a x_b + \rho_4 y_a y_b, \tag{4.5}$$

$$\rho_1(l_a n_b + n_a l_b) + \lambda l_a l_b + \rho_2 x_a x_b + \rho_3 y_a y_b, \tag{4.6}$$

$$\rho_1(l_a n_b + n_a l_b) + \mu(l_a x_b + x_a l_b) + \rho_1 x_a x_b + \rho_2 y_a y_b. \tag{4.7}$$

In (4.5) with the $+$ option the Segre type is $\{1111\}$ or one of its degeneracies with eigenvectors $l \pm n$ (eigenvalues $\rho_1 \pm \rho_2$), x (ρ_3) and y (ρ_4) and with the $-$ option it is $\{z\bar{z}11\}$ or its degeneracy with eigenvectors $l \pm in$ (eigenvalues $\rho_1 \pm i\rho_2$), x (ρ_3) and y (ρ_4). In (4.6) the Segre type is $\{211\}$ with eigenvectors l (ρ_1), x (ρ_2) and y (ρ_3) and in (4.7) it is $\{31\}$ with eigenvectors l (ρ_1) and y (ρ_2) or, in each case, one of its degeneracies. Again one can scale l so that $\lambda = \pm 1$ in (4.6) and $\mu = 1$ in (4.7). In the corresponding orthonormal basis t, x, y, z one then gets for the general Segre type $\{1111\}$ above

$$(\rho_2 - \rho_1)t_a t_b + (\rho_2 + \rho_1)z_a z_b + \rho_3 x_a x_b + \rho_4 y_a y_b. \tag{4.8}$$

$\square$

There are a number of other ways of achieving such a classification but it seems that the above is perhaps the simplest and quickest. One can achieve the same results using techniques of algebraic geometry [41], spinors [42], a direct computation in a null basis at p [17] or by use of the tensor denoted by E introduced in chapter 3 [13, 44]. Another more general approach is discussed in [49] (see also [45]) and a summary of such methods may be found in [13]. The idea of using invariant 2-spaces for this end was first raised in [43] and developed in a different direction in [13].

Turning attention now to skew-symmetric second order tensors, let $0 \neq F \in \Lambda_p M$ be a bivector at p. Then F is either simple or non-simple (chapter 3). If F is simple then F is called *spacelike, timelike or null* if its blade is, respectively, spacelike, timelike or null and any null direction in the blade of F is referred to as a *principal null direction* of F. If F is spacelike one may then choose an orthonormal basis t, x, y, z at p such that $F = x^a y^b - y^a x^b$ (sometimes written $x \wedge y$) and if F is timelike a similar choice reveals $F^{ab} = t^a z^b - z^a t^b$, $(t \wedge z)$ or, in a corresponding null basis, $F^{ab} = l^a n^b - n^a l^b$, $(l \wedge n)$. If F is null a null basis may be chosen at p such that $F^{ab} = l^a x^b - x^a l^b$, $(l \wedge x)$. Thus for the spacelike F above, F admits a complex conjugate pair of complex eigenvectors $x \pm iy$ with eigenvalues $\pm i$ and a 0−eigenspace spanned by z, t and thus has Segre type $\{z\bar{z}(11)\}$. For the timelike F one has null eigenvectors l, n with respective eigenvalues 1 and -1 and a 0−eigenspace spanned by x and y. The Segre type is thus $\{11(11)\}$. If F is null and, in a null basis x, y, l, n, represented by $F^{ab} = l^a x^b - x^a l^b$, it is easily checked that $F^a{}_b l^b = 0$, $F^a{}_b x^b = l^a$, $F^a{}_b n^b = -x^b$ and $F^a{}_b y^b = 0$. The obvious Jordan basis here shows the Segre type to be $\{(31)\}$ with zero eigenvalue. This exhausts the possibilities if F is simple.

Now suppose that F is non-simple. Then all its eigenvalues (real or complex) are non-zero and all eigenvectors (real or complex) are null (section 3.3). Suppose all eigenvalues are real. Since $F^c{}_c = 0$ their sum is zero. So there exists eigenvalues $a, b \in \mathbb{R}$ with $a \neq b$ and $F^a{}_b k^b = ak^a$ $F^a{}_b q^b = bq^a$ with $k, q \in T_p M$ independent and both null (and since $k \cdot q \neq 0$, $b = -a$). Then $k \wedge q$ is a timelike invariant 2-space for F and its orthogonal complement is also invariant and spacelike. It then follows by an identical argument to that given in section 3.4 that F necessarily admits a conjugate pair of complex eigenvalues. Thus the case of all the eigenvalues being real is forbidden. So F admits a complex eigenvalue $a \pm ib$ with $x \pm iy$ the corresponding (null) eigenvector with $x, y \in T_p M$ independent. Then $|(x + iy)| = 0$ and so $|x| = |y|$ and $x \cdot y = 0$. This means that the 2-space $x \wedge y$, which is an invariant 2-space for F, is spacelike. Hence its timelike orthogonal complement U is also invariant and if spanned by null vectors l and n (with, say, $l \cdot n = 1$) one gets $F^a{}_b l^b = cl^a + dn^a$ ($c, d \in \mathbb{R}$). It follows, since F is skew-symmetric, that $F_{ab} l^a l^b = 0$, hence $d = 0$ and so $F^a{}_b l^b = cl^a$ (and similarly $F^a{}_b n^b = -cl^a$ since F is skew-symmetric) with $c \neq 0$. Thus l and n are real eigenvectors of F with eigenvalues differing only in sign. The condition $F^c{}_c = 0$ then shows that $a = 0$. With the scaling $|x| = |y| = 1$, l, n, x, y form a null tetrad and one achieves a canonical form for a non-simple bivector

$$F^{ab} = c(l^a n^b - n^a l^b) + b(x^a y^b - y^a x^b) \tag{4.9}$$

with Segre type $\{z\bar{z}11\}$. In this case the null directions spanned by l and n are referred to as *principal null directions* of F and the uniquely determined pair of 2-spaces $l \wedge n$ and $x \wedge y$ are called the *canonical blades* of F. This completes the classification for bivectors in Lorentz signature. Dealing with the non-simple case is usually achieved after introducing complex bivectors [40] but the above proof removes the necessity for this and is, in any case, more direct. These results are summarised in the following theorem.

Theorem 4.2 *Let M be a 4-dimensional manifold on which there is a Lorentz metric g of signature $(-, +, +, +)$. Let F be a non-zero, second order, skew-symmetric tensor at $p \in M$. The only possibilities for the Jordan/Segre type of F are (for F simple) $\{11(11)\}$, $\{z\bar{z}(11)\}$ and $\{(31)\}$, with no further degeneracies permitted, and $\{z\bar{z}11\}$ (for F non-simple) and again no further degeneracies are permitted.*

4.3 Bivectors in Lorentz Signature

It is convenient in this section to develop further the theory of bivectors in Lorentz signature. For this signature the dual of a bivector F satisfies $\overset{**}{F} = -F$ and one has the property that F and $\overset{*}{F}$ are always independent (that is, the subspaces $\overset{+}{S}_p$ and $\overset{-}{S}_p$ of $\Lambda_p M$ defined in the last chapter are trivial). The process of "transferring" the duality operation then shows that for bivectors F, G, one has $F \cdot \overset{*}{G} = \overset{*}{F} \cdot G$ and also $\overset{*}{F} \cdot \overset{*}{G} = F \cdot \overset{**}{G} = -F \cdot G$ and then $|F| = -|\overset{*}{F}|$. Further, F is simple $\Leftrightarrow \overset{*}{F}$ is simple (chapter 3) and the blades of

F and $\overset{*}{F}$ are orthogonal. Thus if F is spacelike (respectively, timelike) then its dual $\overset{*}{F}$ is timelike (respectively, spacelike) whilst if F is null $\overset{*}{F}$ is null. More precisely, one may start from the bases l, n, x, y, and t, x, y, z, and choose a basis for $\Lambda_p M$ of the form $l \wedge n$, $x \wedge y$, $l \wedge x$, $l \wedge y$, $n \wedge x$ and $n \wedge y$, or of the form $x \wedge y$, $x \wedge z$, $y \wedge z$, $x \wedge t$, $y \wedge t$ and $z \wedge t$. The members here are sometimes referred to as *basis bivectors* (for the corresponding basis) of $\Lambda_p M$. Then one may assume the orientation chosen so that $(l \wedge x)^* = -l \wedge y$, $(l \wedge y)^* = l \wedge x$, $(l \wedge n)^* = x \wedge y$, $(x \wedge y)^* = -l \wedge n$, $(n \wedge x)^* = n \wedge y$ and $(n \wedge y)^* = -n \wedge x$ from which it follows that $(z \wedge t)^*(= -(l \wedge n)^*) = -x \wedge y$. A simple bivector F is spacelike (respectively, timelike, null) if $|F| > 0$ (respectively, $|F| < 0$, $|F| = 0$). Noting that lemma 3.1 holds for all signatures one may add to the results of that lemma the following results for Lorentz signature.

Lemma 4.3 *For a non-trivial bivector* $F \in \Lambda_p M$, $|F| = 0$ *if and only if either* F *is null or* F *is non-simple and satisfies (4.9) with* $b = \pm c$. *In addition, the following conditions are equivalent.*

(*i*) F *is null,*

(*ii*) *There exists* $0 \neq k \in T_p M$ *such that* $F_{ab}k^b = \overset{*}{F}_{ab}k^b = 0$. *The direction spanned by* k *is necessarily unique and null, being the principal null direction of* F *and* $\overset{*}{F}$,

(*iii*) $|F| = \overset{*}{F} \cdot F = 0$ *(that is,* $F_{ab}F^{ab} = \overset{*}{F}_{ab}F^{ab} = 0$*).*

Proof The first part is straightforward. For the remainder, if (*i*) holds F is null (and hence simple) and from the above $|\overset{*}{F}| = 0$ since $|F| = 0$ and $\overset{*}{F}$ is simple since F is. Thus from the first part $\overset{*}{F}$ is null and the orthogonality of their blades shows that $\overset{*}{F}$ has the same principal null direction as F which is the intersection of their blades. So (*ii*) clearly holds with k spanning the common principal null direction of F and $\overset{*}{F}$. It is also clear that (*ii*) implies that F and $\overset{*}{F}$ are simple with blades intersecting in the direction spanned by k. Thus k is null and so (*i*) holds. Then if (*i*) (or (*ii*)) is true, F is null and simple and hence (*iii*) holds, and conversely, (*iii*) implies F is simple and null and so (*i*) holds. $\square$

In the last chapter the subsets $\overset{+}{S}_p$ and $\overset{-}{S}_p$ for $p \in M$ proved useful in positive definite signature. However, since for a (real) bivector F the bivectors F and $\overset{*}{F}$ are independent in Lorentz signature, these corresponding subsets now are trivial. This is because the linear duality map on $\Lambda_p M$ satisfies $\overset{**}{F} = -F$ and so its only eigenvalues are $\pm i$. One can explore this further by first extending $\Lambda_p M$ to its complexification, thought of as the $6-$dimensional complex vector space $\widetilde{\Lambda_p M}$ of all complex bivectors at p. Then extend the duality operator $*$ to $\widetilde{\Lambda_p M}$ by defining for $W = F + iG \in \widetilde{\Lambda_p M}$ $(F, G \in \Lambda_p M)$,

$\overset{*}{W} \equiv \overset{*}{F} + i\overset{*}{G}$. Then define subsets $\overset{+}{S}_p$ and $\overset{-}{S}_p$ of $\widetilde{\Lambda_p M}$ by

$$\overset{+}{S}_p = \{W : \overset{*}{W} = -iW\}, \qquad \overset{-}{S}_p \equiv \{W : \overset{*}{W} = iW\}, \qquad (4.10)$$

referred to, respectively, as the subsets of *self dual* and *anti-self dual* complex bivectors at p. It is clear that $\overset{+}{S}_p \cap \overset{-}{S}_p = \{0\}$. Further, any complex bivector W may be written as $W = \frac{1}{2}[(W + i\overset{*}{W}) + (W - i\overset{*}{W})]$ with $W + i\overset{*}{W} \in \overset{+}{S}_p$ and $W - i\overset{*}{W} \in \overset{-}{S}_p$ and this decomposition of W into a sum of members of $\overset{+}{S}_p$ and $\overset{-}{S}_p$ is unique. Thus, as vector spaces, $\widetilde{\Lambda_p M}$ is isomorphic to $\overset{+}{S}_p + \overset{-}{S}_p$. So any member of $\overset{+}{S}_p$ may be written as $F + i\overset{*}{F}$ and any member of $\overset{-}{S}_p$ may be written as $G - i\overset{*}{G}$ for unique, (real) $F, G \in \Lambda_p M$. Then starting from a (real) null basis l, n, x, y for $T_p M$ as above define a conjugate pair of complex vectors $m, \bar{m}$ by $\sqrt{2}m = x + iy$ and $\sqrt{2}\bar{m} = x - iy$. So m and $\bar{m}$ are complex null vectors and $l, n, m, \bar{m}$ form a basis for the complexification of $T_p M$ called a *complex null tetrad* at p. In fact, the only non-vanishing inner products between the members of this complex null tetrad are $l \cdot n = m \cdot \bar{m} = 1$. It is noted here that since complex null vectors are involved, independent (complex) null vectors may be orthogonal, for example, $l \cdot m = l \cdot \bar{m} = n \cdot m = n \cdot \bar{m} = 0$. Now consider the three complex bivectors given by $F + i\overset{*}{F}$ for the successive choices $F = l \wedge x$, $F = n \wedge x$ and $F = l \wedge n$. These lead to a basis for $\overset{+}{S}_p$ with members

$$V_{ab} \equiv 2l_{[a}\bar{m}_{b]}, \qquad U_{ab} \equiv 2n_{[a}m_{b]}, \qquad M_{ab} \equiv 2l_{[a}n_{b]} + 2\bar{m}_{[a}m_{b]}. \qquad (4.11)$$

Their conjugates $\bar{V}, \bar{U}$ and $\bar{M}$ then yield a basis for $\overset{-}{S}_p$. Thus $\overset{+}{S}_p$ and $\overset{-}{S}_p$ are 3–dimensional subspaces of $\widetilde{\Lambda_p M}$ and, in fact, subalgebras of the Lie algebra $\widetilde{\Lambda_p M}$ (when the latter has the obvious "complexified" Lie product from $\Lambda_p M$) since it is easily checked that $[V, U] = -M$, $[V, M] = -2V$ and $[U, M] = 2U$ and, by conjugation, $[\bar{V}, \bar{U}] = -\bar{M}$, $[\bar{V}, \bar{M}] = -2\bar{V}$ and $[\bar{U}, \bar{M}] = 2\bar{U}$. The basis members also satisfy the following conditions using an obvious extension of the bivector metric P to complex bivectors,

$$|U| = |V| = V \cdot M = U \cdot M = 0, \qquad U \cdot V = 2, \qquad M \cdot M = -4, \qquad (4.12)$$

and similarly, by conjugation, for $\bar{V}, \bar{U}$ and $\bar{M}$. It is also easily checked that if $A \in \overset{+}{S}_p$ and $B \in \overset{-}{S}_p$ then $A \cdot B = 0$ and $[A, B] = 0$. Thus $\widetilde{\Lambda_p M}$ is Lie algebra isomorphic to $\overset{+}{S}_p + \overset{-}{S}_p$. It is also remarked here for later use that if $A, B \in \overset{+}{S}_p$ are independent then $[A, B] \neq 0$. To see this let $A = aU + bV + cM$ and $B = a'U + b'V + c'M$ for $a, b, c, a', b', c' \in \mathbb{C}$. Then $[A, B] = 0 \Leftrightarrow ab' - ba' = ac' - ca' = cb' - bc' = 0$ These equations can be solved by noting their symmetry in the unknowns a, b, c, a', b', c'. Thus if $a = 0 \neq a'$, then $b = c = 0$

and a contradiction arises. Then if $a = 0 = a'$, either b, c, b', c' are all non-zero and so $\frac{b'}{b} = \frac{c'}{c}$ which makes A and B proportional and gives a contradiction, or (say) $b = 0$ which forces $b' = 0$ and another contradiction. Thus each of $a, ..., c'$ is non-zero and $\frac{a'}{a} = \frac{b'}{b} = \frac{c'}{c}$ and again A and B are proportional.

Similar comments apply to $\bar{S}_p$.

The next result collects together some further results about real bivectors in Lorentz signature.

Lemma 4.4 *For real bivectors F, G and H,*

(i) $H^{ac}\overset{*}{F}_{bc} - \overset{*}{F}^{ac}H_{bc} = \frac{1}{2}(F \cdot H)\delta^a_b$,

(ii) $[\overset{*}{F}, \overset{*}{G}] = -[F, G]$ *and* $[F, \overset{*}{G}] = [\overset{*}{F}, G] = [F, G]^*$ *($\Rightarrow [F, \overset{*}{F}] = 0$),*

(iii) *For a fixed bivector $F \neq 0$, if a bivector X satisfies $[X, F] = 0$, then* $X = aF + b\overset{*}{F}$ *for $a, b \in \mathbb{R}$,*

(iv) *If $\{0\} \neq W \subset \Lambda_p M$ is a subspace with the "dual invariant" property that $A \in W \Leftrightarrow \overset{*}{A} \in W$ (equivalently, W admits a basis the duals of whose members constitute a basis for W) then W is even-dimensional.*

Proof For part (i), on writing out the left hand side using the definition of the duality operator and using standard formulae for handling the alternating symbol products for Lorentz signature [33] one easily achieves the desired result.

For part (ii) let $X \equiv F + i\overset{*}{F} \in \overset{+}{S}_p$ and $Y \equiv G - i\overset{*}{G} \in \bar{S}_p$ so that, with a bar denoting conjugation, $2F = X + \bar{X}$ and $2i\overset{*}{F} = X - \bar{X}$ and similarly $2G = Y + \bar{Y}$ and $2i\overset{*}{G} = Y - \bar{Y}$. Then a remark above gives $[X, Y] = 0$ and expanding this gives $[\overset{*}{F}, \overset{*}{G}] = -[F, G]$ and $[F, \overset{*}{G}] = [\overset{*}{F}, G]$. Then one computes $[F, G]$, $[F, G]^*$ and $[F, \overset{*}{G}]$ using the results of this paragraph, noting that $[X, \bar{Y}] \in \overset{+}{S}_p$ and $[\bar{X}, Y] \in \bar{S}_p$, to get $[\overset{*}{F}, G] = [F, G]^*$. The final part of (ii) follows by putting $F = G$ in an appropriate previous result.

For part (iii) Suppose $[X, F] = 0$. Then from part (ii) $[\overset{*}{F}, \overset{*}{X}] = [F, \overset{*}{X}] = [\overset{*}{F}, X] = 0$ and so with $G \equiv F + i\overset{*}{F} \in \overset{+}{S}_p$ and $H \equiv X + i\overset{*}{X} \in \overset{+}{S}_p$ one computes that $[G, H] = 0$. A remark above shows that G and H are (complex) proportional and so X is a linear combination of F and $\overset{*}{F}$.

For part (iv) let $A \in W$. Then since A and $\overset{*}{A}$ are independent $\dim W \geq 2$. Now suppose $A, \overset{*}{A}$ and B are independent members of W and consider the equation $aA + b\overset{*}{A} + cB + d\overset{*}{B} = 0$ for $a, b, c, d \in \mathbb{R}$ not all zero. Taking the dual of this gives $a\overset{*}{A} - bA + c\overset{*}{B} - dB = 0$ and so, by independence of $A, \overset{*}{A}$ and B, $c \neq 0 \neq d$. Eliminating $\overset{*}{B}$ from these equations gives the contradiction that $c^2 + d^2 = 0$. Thus $A, \overset{*}{A}, B$ and $\overset{*}{B}$ are independent and $\dim W \geq 4$. Then if $A, \overset{*}{A}, B, \overset{*}{B}$ and C are independent members of W a consideration of the equation $aA + b\overset{*}{A} + cB + d\overset{*}{B} + eC + f\overset{*}{C} = 0$ for $a, b, c, d, e, f \in \mathbb{R}$ not all zero and a similar argument to the above shows that $A, \overset{*}{A}, B, \overset{*}{B}, C$ and $\overset{*}{C}$ are independent and the proof is complete. [It is remarked that if W is dual invariant and $\dim W = 2$ or $\dim W = 6$, W is a subalgebra of $\Lambda_p M$ (lemma 4.4) but this is not necessarily true if $\dim W = 4$. To see this consider, for a l, n, x, y, the dual invariant subspace $W \equiv \mathrm{Sp}(l \wedge x, l \wedge y, n \wedge x, n \wedge y)$ which is not a subalgebra of $\Lambda_p M$ since $[l \wedge x, n \wedge x.] = -l \wedge n.$] $\qquad\square$

A little more can be said about the 6−dimensional complex vector space of all complex bivectors at $p \in M$, $\widetilde{\Lambda_p M}$, and, in particular, about its subspaces $\overset{+}{S}_p$ and $\overset{-}{S}_p$. For members of $\widetilde{\Lambda_p M}$ the terms simple and non-simple will be used as they were for real bivectors, noting that the blades are now, in general, spanned by *complex* vectors. For simple such bivectors one can say more and for this another definition is needed. A simple member of $\widetilde{\Lambda_p M}$ is called *totally null* if each member of its blade is null (and hence, by taking obvious linear combinations, any two of its blade members are orthogonal). Thus the bivectors V and U in (4.11) are totally null. The following result can then be proved [46].

Lemma 4.5 *If Q is a non-zero, complex (not proportional over $\mathbb{C}$ to a real) bivector then Q is a simple member of $\overset{+}{S}_p$ or $\overset{-}{S}_p$ if and only if Q is totally null. If Q is totally null its blade admits a unique (up to a complex scaling), real, necessarily null member.*

Proof Suppose Q is a simple member of $\overset{+}{S}_p$ or $\overset{-}{S}_p$. Then $Q = (x+iy) \wedge (r + is)$ for $x, y, r, s \in T_p M$. Since Q is simple so is its dual, $\overset{*}{Q}$, which is $\pm iQ$, and hence this has the same blade as Q. But the blades of Q and $\overset{*}{Q}$ are orthogonal and it follows that each member of the blade of Q is null and so Q is totally null. Conversely, if Q is totally null, it is simple and the blades of Q and $\overset{*}{Q}$ are the same and so $\overset{*}{Q} = \lambda Q$ ($\lambda \in \mathbb{C}$). But then the result $\overset{**}{Q} = -Q$ shows that $\lambda = \pm i$ and so Q is a simple member of $\overset{+}{S}_p$ or $\overset{-}{S}_p$.

If Q is totally null and if its blade contains at least two real (up to complex scalings) members they are null and orthogonal and a contradiction to the Lorentz signature arises. Thus there is at most one such member in its blade. So assume that in the expression for Q above x, y, r and s are each non-zero with $x + iy$ and $r + is$ null and orthogonal and with x and y independent to avoid $x + iy$ being complex proportional to a real vector (and similarly r and s independent). This gives $|x| = |y|$, $|r| = |s|$, $x \cdot y = r \cdot s = 0$ and $x \cdot r - y \cdot s = 0 = x \cdot s + y \cdot r$. Because of the Lorentz signature the conditions $|x| = |y|$ and $x \cdot y = 0$ force x, y to be spacelike and similarly r, s are spacelike. Now consider the condition $x \cdot r - y \cdot s = 0 = x \cdot s + y \cdot r$. The numbers $x \cdot r$ and $x \cdot s$

cannot both be zero since then $y \cdot r = y \cdot s = 0$ and then x, y, r, s are mutually orthogonal and hence independent and yield the contradiction of a basis for $T_p M$ consisting of four orthogonal spacelike vectors. If neither $x \cdot r$ nor $x \cdot s$ is zero one can easily replace r, s by independent linear combinations r', s' of r, s still satisfying $|r'| = |s'|$ and $r' \cdot s' = 0$ and with $x \cdot r' = 0$. Thus, dropping primes, one has $|x| = |y|$, $|r| = |s|$, $x \cdot y = r \cdot s = 0$, $x \cdot r = y \cdot s = 0$ and $x \cdot s = -y \cdot r = \alpha \neq 0$. If the collection of vectors given, for $a = 1, ..., 4$, by $e_a = x, y, r, s$ is independent (and with $|x| = |y| = \mu > 0$ and $|r| = |s| = \nu > 0$) and hence a basis for $T_p M$ at p the corresponding metric coefficients $g_{ab} = g(e_a, e_b)$ at p are easily checked to have determinant $(\alpha^2 - \mu\nu)^2$ which is positive, contradicting Lorentz signature. Thus the collection x, y, r, s is a dependent set of vectors in $T_p M$. It follows that there exists $a, b, c, d \in \mathbb{R}$ not all zero such that $bx + ay + dr + cs = 0$ and so the member $(a + ib)(x + iy) + (c + id)(r + is)$ is a real member of the blade of Q and since Q is totally null this real vector is null. The uniqueness follows. $\qquad\square$

A totally null bivector Q whose blade contains the real null vector l can be written as $Q = l \wedge z$ for a complex null vector z orthogonal to l. Extending l to a complex null tetrad $l, n, m, \bar{m}$, it is easily checked that z is a linear combination of l, m and $\bar{m}$ and so, since z is null and $m \cdot \bar{m} \neq 0$, Q is either a multiple of $V \in \overset{+}{S}_p$ in (4.11) or its conjugate $\bar{V} \in \overset{-}{S}_p$. Such totally null bivectors are usually called *complex null bivectors*. Thus for any complex null bivector V there exists a unique (up to a scaling) real null vector l satisfying $V_{ab} l^b = 0$. The null direction spanned by l is called the *principal null direction* of V. The only complex null bivectors whose principal null direction is (spanned by–these words will sometimes be understood) l are, up to (complex) multiples, the bivectors V and $\bar{V}$ in (4.11). If, however, Q and Q' are independent complex null bivectors with respective (distinct) principal null directions l and n one may form a complex null tetrad $l, n, m, \bar{m}$ to see that Q is a (complex) multiple of $l \wedge m \in \overset{-}{S}_p$ or $l \wedge \bar{m} \in \overset{+}{S}_p$ and that Q' is a (complex) multiple of $n \wedge m \in \overset{+}{S}_p$ or $n \wedge \bar{m} \in \overset{-}{S}_p$. It follows from this that if Q and Q' are either both in $\overset{+}{S}_p$ or both in $\overset{-}{S}_p$ their blades intersect only in the trivial subspace (since $l, n, m, \bar{m}$ are an independent set) whereas if $Q \in \overset{+}{S}_p$ and $Q' \in \overset{-}{S}_p$, or vice versa, their blades intersect in a null vector which may be real or complex and is real if and only if the two complex null bivectors are conjugates, sharing the same principal null direction.

Lemma 4.5 leads to the following decomposition of a complex null bivector $Q = A + iB$ at p, $(A, B \in \Lambda_p M)$. If $Q \in \overset{+}{S}_p$ the condition $\overset{*}{Q} = -iQ$ immediately gives $B = \overset{*}{A}$ and so $Q = A + i\overset{*}{A}$. If l is the real null vector in the blade of Q, $Q = l \wedge z$ for a complex null vector $z = r + is$, $r, s \in T_p M$ with

$|r| = |s|, r \cdot s = l \cdot r = l \cdot s = 0$. Again r and s are spacelike and it is easily checked that, with an appropriate orientation in the $r \wedge s$ plane, $Q = (l \wedge r) + i(l \wedge s)$ with $(l \overset{*}{\wedge} r) = l \wedge s$. Clearly the real bivectors $l \wedge r$ and $l \wedge s$ are null with common principal null direction spanned by l which is the principal null direction of Q.

Similar comments apply if $Q \in \bar{S}_p$. Conversely any complex bivector $A + iB$ where A and B are real null bivectors with common principal null direction spanned by l and satisfying $A = l \wedge r$, $B = l \wedge s$, $|r| = |s|$ and $l \cdot r = l \cdot s = r \cdot s = 0$, is a complex null bivector in $\overset{+}{S}_p$ or $\bar{S}_p$. It is noted that if $X \in \overset{+}{S}_p$ ($\bar{S}_p$ is similar) with $X = F + i\overset{*}{F}$ ($F \in \Lambda_p M$) then $|X| = 2|F| + 2i F \cdot \overset{*}{F}$ and so F is simple if and only if $|X| \in \mathbb{R}$ and (from lemma 4.3) F is null if and only if $|X| = 0$. Thus $X \in \overset{+}{S}_p$ is a complex null bivector if and only if $|X| = 0$.

There is a result which is analogous to the triple vector product in 3–dimensional vector analysis. It was given without proof in [47] and will here be proved. One notes that the definition of bivector inner product in [47] differs by a factor 2 from that given here and this explains the factor 2 (rather than 4) in the equation below.

Lemma 4.6 *Let F, G and H be (real) bivectors. Then*

$$2[F, [G, H]] = (F \cdot H)G - (F \cdot G)H - (F \cdot \overset{*}{H})\overset{*}{G} + (F \cdot \overset{*}{G})\overset{*}{H}. \qquad (4.13)$$

Proof Starting from a real null basis l, n, x, y construct the basis bivectors $F_1 = l \wedge x \equiv l^a x^b - x^a l^b$, $F_2 = n \wedge x$, $F_3 = l \wedge n$, $\overset{*}{F}_1 = -(l \wedge y)$, $\overset{*}{F}_2 = n \wedge y$ and $\overset{*}{F}_3 = x \wedge y$. One notes that $F_1 \cdot F_2 = -\overset{*}{F}_1 \cdot \overset{*}{F}_2 = 2$, $|F_3| = -|\overset{*}{F}_3| = -2$ whilst $[F_1, F_2] = -F_3$, $[F_1, F_3] = -F_1$ and $[F_2, F_3] = F_2$. (thus the collection F_1, F_2, F_3 constitute a subalgebra of $\Lambda_p M$, but the collection $\overset{*}{F}_1, \overset{*}{F}_2, \overset{*}{F}_3$ do not—see lemma 4.4(ii)). One can now show that (4.13) holds when F, G and H are any combinations of F_1, F_2 and F_3. Use of lemma 4.4(ii) and the result $\overset{*}{P} \cdot Q = -P \cdot \overset{*}{Q}$ can then be used to check the result when F, G, H are any combination of $F_1, F_2, ..., \overset{*}{F}_3$. For example, $[\overset{*}{F}, [\overset{*}{G}, \overset{*}{H}]] = -[\overset{*}{F}, [G, H]] = -[F, [G, H]]^*$ and $[\overset{*}{F}, [G, \overset{*}{H}]] = [\overset{*}{F}, [G, H]^*] = -[F, [G, H]]$.

Then writing out each bivector in (4.13) in terms of the basis bivectors F_1, $F_2, ..., \overset{*}{F}_3$ and using linearity gives the desired result. $\qquad \square$

For future use it is convenient at this point to consider the Lorentz equivalent of lemma 3.6.

Lemma 4.7 *Suppose that h is a non-zero, symmetric tensor and F a bivector at $p \in M$ which satisfy*

$$h_{ac} F^c{}_b + h_{bc} F^c{}_a = 0. \qquad (4.14)$$

(i) *If F is simple then the blade of F is an eigenspace of h (with respect to g),* (*ii*) *If* F *is non-simple its canonical blades are eigenspaces of* h *but with the resulting eigenvalues possibly distinct,*

(*iii*) *For given* h *if (4.14) holds for bivectors* F *and* G *it holds for* $[F, G]$.

Proof For part (*i*) one could proceed as in the analogous lemma in the last chapter. In each case the blade of $\overset{*}{F}$ is the $0-$eigenspace of F and hence invariant for h, as is its orthogonal complement, that is, the blade of F. If F is spacelike choose a null tetrad l, n, x, y so that $F = x \wedge y$. It follows that $x \wedge y$ is an invariant 2-space of h and has an induced positive definite metric. Hence x and y may be chosen as orthogonal eigenvectors of h with real eigenvalues. A back substitution then shows that the eigenvalues of x and y are equal. If F is null, say $F = l \wedge x$ one similarly sees that $l \wedge x$ and $l \wedge y$ are invariant for h and hence that l is an eigenvector of h. Again a back substitution completes the proof. For F timelike, say $F = l \wedge n$, and for F non-simple (part (*ii*)), say $F = a(l \wedge n) + b(x \wedge y)$, similar ideas give the desired result. For (*iii*) the proof is as in chapter 3 and so for a given h the solutions of (4.14) for F form a subalgebra of $\Lambda_p M$. $\qquad \square$

4.4 The Lorentz Algebra o(1,3) and Lorentz Group

Let M be a 4-dimensional manifold admitting a Lorentz metric g, let $p \in M$ and let $\mathcal{L}$ denote the collection of all linear maps $f : T_p M \to T_p M$ which preserve the metric $g(p)$, that is, $g(p)(f(u), f(v)) = g(p)(u, v)$ for each $u, v \in T_p M$. The work in chapter 3 on the Lie algebra $o(4)$, is easily modified to show that $\mathcal{L}$ is a group under the usual composition of maps and, in fact, a 6-dimensional Lie group called the *Lorentz group*. Its Lie algebra, $o(1, 3)$, the Lorentz algebra, is denoted by L and is isomorphic to the 6-dimensional Lie algebra, under the bracket operation, of all bivectors at p. Each member of $\mathcal{L}$ is a bijective map and the above definition of $\mathcal{L}$ may be replaced by the equivalent one $g(p)(f(u), f(u)) = g(p)(u, u)$ for each $u \in T_p M$. Of course, one may always choose coordinates in some neighbourhood of p so that the components of $g(p)$ take the Sylvester form η with components $\eta_{ab} = \mathrm{diag}(-1, 1, 1, 1)$. In such coordinates at p, called *Minkowski coordinates at* p, the resulting pair $T_p M (= \mathbb{R}^4)$ together with η with components η_{ab} is referred to as *Minkowski space* and η_{ab} as the *Minkowski metric* on $T_p M$. From section 3.6 one sees that the matrix A representing a Lorentz transformation satisfies $A \eta A^T = \eta$ and hence that $\det A = \pm 1$.

For $p \in M$ consider the 4-dimensional subset $S \subset T_p M \setminus \{0\}$ of all spacelike vectors at p. This is an open and hence regular, 4-dimensional submanifold of $T_p M$ and it is connected. To see this let u, v be spacelike vectors at p and choose Minkowski coordinates so that $u = (0, 1, 0, 0)$ and $v = (d, a, b, c)$ with

$-d^2 + a^2 + b^2 + c^2 > 0$. Then construct obvious smooth paths from $(0, 1, 0, 0)$ to $(0, a, b, c)$ and from $(0, a, b, c)$ to (d, a, b, c) which pass only through spacelike vectors. Thus S is path connected and hence connected.

Next consider the subset $T \subset T_pM \setminus \{0\}$ of all timelike vectors at p represented in Minkowski coordinates there. This is an open (hence regular) 4-dimensional submanifold of T_pM and so gets subspace topology from T_pM but it is not connected. To see this let T' (respectively, T'') denote the subset of T whose members are such that their first ("time") component in this coordinate system, which is necessarily non-zero, is positive (respectively, negative). Then one can easily construct a smooth path between any two members of T' (respectively, T'') which passes only through members of T' (respectively, T''). Thus T' and T'' are connected. However, T is not connected since any path connecting a member of T' to a member of T'' must pass through one with zero time component and is then not in T. It follows that T has two components T' and T'' and which are then open submanifolds of $T_pM \setminus \{0\}$. Hence the product manifold $T \times T$ has four components $T' \times T'$, $T'' \times T''$, $T' \times T''$ and $T'' \times T'$. Now consider the smooth map $\sigma : T \times T \to \mathbb{R} \setminus \{0\}$ given by $\sigma(u, v) = \eta(u, v)$ whose range does not include 0 (since two timelike vectors can never be orthogonal) but which clearly takes any positive or negative value. Further, $u \in T' \Rightarrow -u \in T''$ (and vice versa). So σ maps $T' \times T'$ and $T'' \times T''$ to the negative real numbers and $T' \times T''$ and $T'' \times T'$ to the positive reals. Then the relation $\sim$ on T given by $u \sim v \Leftrightarrow \eta(u, v) < 0$ can be shown to be an equivalence relation with precisely two equivalence classes T' and T''.

Again using Minkowski coordinates consider the smooth real-valued map on $T_pM \setminus \{0\}$ given by $f : (d, a, b, c) \to -d^2 + a^2 + b^2 + c^2$. Arguments similar to ones given previously (see also chapter 2) then show that the set of all null vectors (the *null cone*) N at p is the 3−dimensional regular submanifold $f^{-1}\{0\}$ of the manifold $T_pM \setminus \{0\}$. Each member of this submanifold has $d \neq 0$. A similar argument to that above shows that N with its subspace topology from $T_pM \setminus \{0\}$ is not connected but has two components N' and N'' and a continuity argument shows that this labelling may be chosen such that $k \in N' \Leftrightarrow \eta(u, k) < 0$ for each $u \in T'$ and $k \in N'' \Leftrightarrow \eta(u, k) < 0$ for each $u \in T''$. Thus the set $C \equiv T \cup N$ may be partitioned as $C = C' \cup C''$ where $C' = T' \cup N'$ and $C'' = T'' \cup N''$. The members of one of these partitions are called *future pointing* and the other *past pointing*. If $f \in \mathcal{L}$ then f is a continuous map on T_pM and from the definition of $\mathcal{L}$, either f maps each of C' and C'' into itself or it maps C' into C'' and vice versa. In the first case f is called *future preserving* and in the second, *future reversing*. Thus $\mathcal{L}$ may be decomposed into four disjoint subsets $\mathcal{L} = \mathcal{L}_+^\uparrow \cup \mathcal{L}_+^\downarrow \cup \mathcal{L}_-^\uparrow \cup \mathcal{L}_-^\downarrow$ where $\uparrow$ ($\downarrow$) refer to the future preserving (reversing) properties and $\pm$ to the sign of $\det A$. Of these the most important subset is $\mathcal{L}_+^\uparrow$, which is actually a Lie subgroup of $\mathcal{L}$, and is labelled $\mathcal{L}_0$ and referred to as the *proper Lorentz group*.

As in the positive definite case, it can be shown (for details see [13]) that $\mathcal{L}$ is a closed (not open and non-discrete) subgroup of $GL(4, \mathbb{R})$ and hence admits the structure of a regular submanifold, and hence of a Lie subgroup, of $GL(4, \mathbb{R})$ of dimension 6 with Lie algebra L. It can also be checked that $\mathcal{L}_0$ is a connected, open Lie subgroup of $\mathcal{L}$ which contains the identity and is thus (chapter 2) the *identity component* of $\mathcal{L}$ and has Lie algebra L. Its cosets in $\mathcal{L}$ are $\mathcal{L}_0$, $\mathcal{L}_+^{\downarrow}$, $\mathcal{L}_-^{\uparrow}$ and $\mathcal{L}_-^{\downarrow}$.

In this section the subalgebras of $o(1,3)$ will be computed. There are many ways of doing this. One interesting method is to work almost entirely with complex algebras as is done in [47]. However, this approach will not be followed here but rather a direct method involving only real Lie algebras will be considered. (However, it is acknowledged that what is to follow is influenced and guided by [47]). The notation used in the classifying of these subalgebras is taken from [48] and the full list will be tabulated. The work on bivectors in the last section and also in chapter 3 will be useful in what is to follow.

Let $V \subset L$ be a subalgebra of L. If $\dim L = 0$ (the trivial case) the resulting subalgebra is denoted by R_1. If $\dim L = 1$ then L is of the form $Sp(F)$ where F is a bivector. Thus one has four possibilities; when F is timelike, null, spacelike or non-simple. These subalgebra types are labelled, respectively, R_2, R_3, R_4 and R_5. The case R_5 cannot occur for the holonomy algebra of a *metric* connection since it is the span of a non-simple bivector (as explained in section 3.7).

Now suppose that $\dim V = 2$ and that $V = Sp(A, B)$ for bivectors A, B. If V is Abelian, $[A, B] = 0$ and so, from lemma 4.4(iii) above, B is a linear combination of A and $\overset{*}{A}$ and so $V = Sp(A, \overset{*}{A})$. Thus there are only two possibilities here (up to isomorphism—this will always be understood); when A is spacelike (and $\overset{*}{A}$ timelike) and when A (and hence $\overset{*}{A}$) is null. (The case when A is non-simple is the same as for A spacelike, from (4.9)). Thus for some real null tetrad l, n, x, y, the possibilities are $V = Sp(l \wedge n, x \wedge y)$ (labelled R_7) and $V = Sp(l \wedge x, l \wedge y)$, labelled R_8. Now suppose that V is not Abelian so that $[A, B] = C \neq 0$. Then since V is a subalgebra, $C = aA + bB$ for $a, b \in \mathbb{R}$. If $a \neq 0 \neq b$, V is spanned by A and C and $[A, C] = bC$. It follows that one may always choose A, B such that $V = Sp(A, B)$ with $[A, B] = \mu B$ ($0 \neq \mu \in \mathbb{R}$). Then from lemma 3.4$(ii)$, $|B| = A \cdot B = 0$. Also, from lemma 4.4(ii), $[\overset{*}{A}, \overset{*}{B}] = -[A, B] = -\mu B$ and so $B \cdot \overset{*}{B} = 0$, that is, B is null from lemma 4.3. Thus one must have $V = Sp(A, B)$, $[A, B] = \mu B$ and B null. So choose a real null tetrad l, n, x, y so that $B = l \wedge x$. Then in matrix language $AB - BA = \mu B$ and a contraction first with l and then with y gives, using the same notation, $B(Al) = B(Ay) = 0$. It follows that Al and Ay annihilate B and so lie in the blade of $\overset{*}{B} = -l \wedge y$. Now write out A in terms of the basis bivectors from this real null tetrad to see that $A = al \wedge x + bl \wedge y + cl \wedge n$ $(a, b, c \in \mathbb{R})$. Thus $V = Sp(l \wedge x, bl \wedge y + cl \wedge n)$ with the non-Abelian condition giving $c \neq 0$. Now change the real null tetrad l, n, x, y to the (easily checked to be a) real null tetrad l', n', x', y' where $l' = l$,

$x' = x$, $y' = y - \frac{b}{c}l$, $n' = n + \frac{b}{c}y - \frac{b^2}{2c^2}l$. Then $V = Sp(l' \wedge x', l' \wedge n')$. This subalgebra is labelled R_6.

Now suppose $\dim V = 3$, let $\overset{*}{V}$ be the span of the duals of the members of V (so that $\overset{*}{V}$ is a 3–dimensional subspace of L) and consider $V \cap \overset{*}{V}$. If $0 \neq A \in V \cap \overset{*}{V}$ then $A \in V$ ($\Rightarrow \overset{*}{A} \in \overset{*}{V}$) and $A \in \overset{*}{V}$ ($\Rightarrow A = \overset{*}{B}, B \in V, \Rightarrow \overset{*}{A} = -B \in V$). So A and $\overset{*}{A}$ are members of $V \cap \overset{*}{V}$ and hence, since A and $\overset{*}{A}$ are independent, it follows from lemma 4.4(iv) that $\dim(V \cap \overset{*}{V})$ is 2 or 0 (and that, although $\overset{*}{V}$ may not be a subalgebra, $V \cap \overset{*}{V}$ is an (Abelian) subalgebra from lemma 4.4(ii)). If $\dim(V \cap \overset{*}{V}) = 2$ one can, from the above argument, choose $V \cap \overset{*}{V} = Sp(C, \overset{*}{C})$ for some non-zero bivector C and hence $V = Sp(C, \overset{*}{C}, D)$ for some bivector D independent of C and $\overset{*}{C}$. It follows that V admits a 2-dimensional Abelian subalgebra, $Sp(C, \overset{*}{C})$. From the 2-dimensional cases above this subalgebra must be of type R_7 or R_8. Next one has

$$[D, C] = aD + bC + c\overset{*}{C}, \qquad [D, \overset{*}{C}] = dD + eC + f\overset{*}{C}, \qquad (4.15)$$

for $a, b, c, d, e, f \in \mathbb{R}$. Lemma 4.4($ii$) then gives $[D, \overset{*}{C}] = [D, C]^*$ and so, from (4.15), $a = d = 0$, $b = f$ and $e = -c$ (because D and $\overset{*}{D}$ are independent of C and $\overset{*}{C}$—see the proof of lemma 4.4(iv)). If $Sp(C, \overset{*}{C})$ is of type R_7 a null tetrad may be chosen so that $C = l \wedge n$ and $\overset{*}{C} = x \wedge y$. But then (4.15) with $a = d = 0$, $b = f$ and $e = -c$ shows that $[D, C]$ and $[D, \overset{*}{C}]$ are linear combinations of only $l \wedge n$ and $x \wedge y$ and it is easily checked by computing some straightforward Lie brackets that D must also be a linear combination of $l \wedge n$ and $x \wedge y$ and the independence of D, C and $\overset{*}{C}$ is contradicted. If $Sp(C, \overset{*}{C})$ is of type R_8 a null tetrad may be chosen so that is $C = l \wedge x$ and $\overset{*}{C} = l \wedge y$ and a similar argument shows that D is a linear combination of $l \wedge x$, $l \wedge y$, $l \wedge n$ and $x \wedge y$. Thus by taking linear combinations of the basis members for V one may take $D = a(l \wedge n) + b(x \wedge y)$ ($a, b \in \mathbb{R}, a^2 + b^2 \neq 0$). There are three choices given by $a = 0 \neq b$, $a \neq 0 = b$ and $a \neq 0 \neq b$. The first two of these give rise to the (non-isomorphic) types labelled R_{11} ($V = Sp(l \wedge x, l \wedge y, x \wedge y)$) and R_9 ($V = Sp(l \wedge x, l \wedge y, l \wedge n)$) whilst the third gives rise to an infinite collection of non-isomorphic types labelled collectively as R_{12} and which are distinguished by the non-zero ratio $\omega \equiv \frac{b}{a}$ ($V = Sp(l \wedge x, l \wedge y, l \wedge n + \omega(x \wedge y))$).

Now suppose that $V \cap \overset{*}{V} = \{0\}$ and let $V = Sp(P, Q, R)$ for bivectors P, Q and R, so that $\overset{*}{V} = Sp(\overset{*}{P}, \overset{*}{Q}, \overset{*}{R})$. Then $\dim(V + \overset{*}{V}) = \dim V + \dim \overset{*}{V} = 6$ and so $P, Q, R, \overset{*}{P}, \overset{*}{Q}$ and $\overset{*}{R}$ are independent. If $X, Y \in V$ lemma 4.6 shows that

$$2[X, [X, Y]] = (X \cdot Y)X - |X|Y - (X \cdot \overset{*}{Y})\overset{*}{X} + (X \cdot \overset{*}{X})\overset{*}{Y}. \qquad (4.16)$$

Since $[X, [X, Y]] \in V$ it follows that, since $V \cap \overset{*}{V} = \{0\}$, $X \cdot \overset{*}{X} = X \cdot \overset{*}{Y} = 0$ for each $X, Y \in V$ and so every member of V is simple (lemma 3.1 and any member of V is orthogonal to any member of $\overset{*}{V}$, that is V and $\overset{*}{V}$ are orthogonal complements in $\Lambda_p M$. Since $\dim V = 3$ and each member of V is simple, lemma 3.2 shows that either the blades of P, Q and R, or those of $\overset{*}{P}, \overset{*}{Q}$ and $\overset{*}{R}$, have a single common direction, say, k. In the latter case (the former case is similar), k is orthogonal to the blades of P, Q and R and hence these blades collectively span a $3-$space $W \subset T_p M$ with normal k. If k is null it lies in W and, further, W contains two independent orthogonal spacelike unit vectors, say, x and y, orthogonal to k and hence V contains the dual pair $k \wedge x$ and $k \wedge y$ which contradicts $V \cap \overset{*}{V} = \{0\}$. So k is either spacelike or timelike. For these cases one may choose an orthonormal tetrad t, x, y, z such that $k = x$ or $k = t$ and (up to isomorphism) $\overset{*}{V}$ is either $Sp(x \wedge y, x \wedge z, x \wedge t)$ or $Sp(x \wedge t, y \wedge t, z \wedge t)$. These lead to $V = Sp(y \wedge z, y \wedge t, z \wedge t)$ and $V = Sp(x \wedge y, x \wedge z, y \wedge z)$. These are the types labelled R_{10} and R_{13}, respectively, with the latter isomorphic to $o(3)$ (and it is noted that in each of these cases $\overset{*}{V}$ is *not* a subalgebra). This completes the case when $\dim V = 3$.

Now suppose $\dim V = 4$ and again introduce $\overset{*}{V}$. From equation (1.1) applied to V and $\overset{*}{V}$ as subspaces of $\Lambda_p M$ one finds $2 \leqslant \dim(V \cap \overset{*}{V})$ and so $2 \leqslant \dim(V \cap \overset{*}{V}) \leqslant 4$. Also $\dim(V \cap \overset{*}{V})$ must be even from Lemma 4.4(iv). So suppose $\dim(V \cap \overset{*}{V}) = 2$ which implies that $\dim Sp(V, \overset{*}{V}) = 6$. Now $V \cap \overset{*}{V}$ is an (Abelian) subalgebra of V of the form $Sp(C, \overset{*}{C})$ for some bivector C and so is of the type R_7 or R_8. Thus, in the first case one may choose a real null tetrad l, n, x, y and write $V = Sp(A, B, l \wedge n, x \wedge y)$ for bivectors A, B and, since $\dim Sp(V, \overset{*}{V}) = 6$, $\overset{*}{A}$ and $\overset{*}{B}$ are not in V. Now

$$[A, l \wedge n] = aA + bB + c(l \wedge n) + d(x \wedge y) \tag{4.17}$$

and $[A, l \wedge n] = -[A, (x \wedge y)^*] = -[A, x \wedge y]^*$ by lemma 4.4(ii). It follows that $[A, l \wedge n]^* \in V$ and then, since $\overset{*}{A}$ and $\overset{*}{B}$ are not in V, that $a = b = 0$ and hence that $[A, l \wedge n]$ is a linear combination of $l \wedge n$ and $x \wedge y$. But A is a linear combination of the bivector basis members generated by the above null tetrad and a short computation then shows that this linear combination cannot depend on $l \wedge x, l \wedge y, n \wedge x$ or $n \wedge x$. Similar comments apply to B. Thus A and B are linear combinations of $l \wedge n$ and $x \wedge y$ and a contradiction follows. Now suppose $Sp(C, \overset{*}{C})$ is of the type R_8. A similar calculation shows that A and B are each linear combinations of $l \wedge x, l \wedge y, l \wedge n$ and $x \wedge y$ and so, since $\dim V = 4$, $V = Sp(l \wedge x, l \wedge y, l \wedge n, x \wedge y)$. This implies $\dim(V \cap \overset{*}{V}) = 4$ and a contradiction follows. Now suppose that $\dim(V \cap \overset{*}{V}) = 4$, that is, $V = \overset{*}{V}$. Then $V = Sp(A, B, \overset{*}{A}, \overset{*}{B})$ for independent bivectors A and B. Suppose $[A, B] = 0$.

Use of lemma 4.4 then shows that V is Abelian and $\overset{*}{V}$, now a subalgebra, is also Abelian. It follows that $A + i\overset{*}{A}$ and $B + i\overset{*}{B}$ are independent members of $\overset{+}{S}_p$ and have zero bracket and this is a contradiction (see section 4.3 following (4.15)). So $[A, B] \neq 0$. Use of lemma 4.4 again shows that the set of all commutators of all members of V is a 2-dimensional subalgebra Z of V of the form $Sp(C, \overset{*}{C})$ with $C = [A, B]$ and that $C \cdot A = C \cdot \overset{*}{A} = C \cdot B = C \cdot \overset{*}{B} = 0$ and similarly with C replaced by $\overset{*}{C}$. Thus $Z \in V^{\perp}$ and hence $Z \subset V \cap V^{\perp}$ and it follows that $C \cdot C = C \cdot \overset{*}{C} = 0$ and hence, from lemma 4.3, that C (and $\overset{*}{C}$) are null bivectors. Choosing a null tetrad l, n, x, y with $C = l \wedge x$ and $\overset{*}{C} = -l \wedge y$ one finds, since $C \cdot A = C \cdot \overset{*}{A} = C \cdot B = C \cdot \overset{*}{B} = 0$, that $V = Sp(l \wedge x, l \wedge y, l \wedge n, x \wedge y)$, which is labelled R_{14}.

Finally suppose that $\dim V = 5$. Let O be a 3−dimensional Lie subalgebra of the Lorentz algebra L of type R_{13} and isomorphic to $o(3)$, discussed earlier. If O is not a subalgebra of V, $\dim(V + O) = 6$ and (1.1) applied to V and O shows that $\dim(V \cap O) = 2$. Since there are no 2-dimensional subalgebras of O, it follows that each such subalgebra O is contained in V. Since any simple, spacelike bivector is a member of a subalgebra like O and since one may form a basis for $\Lambda_p M$ consisting of simple, spacelike bivectors, [for example, $x \wedge y \pm \frac{1}{2} x \wedge t$, $x \wedge z \pm \frac{1}{2} z \wedge t$, $y \wedge z \pm \frac{1}{2} y \wedge t$], at least one of which is not in V, this gives a contradiction. Thus there are no 5−dimensional subalgebras of L. This completes the classification of the subalgebras of L. One has 13 proper subalgebras, labelled R_2–R_{14}, the trivial subalgebra R_1 and the full Lorentz algebra L, sometimes labelled R_{15}. These are summarised in Table 4.1 in which l, n, x, y and x, y, z, t are the usual bases and $0 \neq \omega \in \mathbb{R}$. (This table is taken from [48, 13].) As explained earlier the 1-dimensional type R_5 cannot arise for the holonomy algebra of a *metric* connection since it is the span of a *non-simple* bivector (see chapter 3).

All the connected Lie subgroups of $\mathcal{L}_0$ can now be found by exponentiation, as described in chapter 2. [In fact $\mathcal{L}_0$ is an *exponential* Lie group [47] in that each $f \in \mathcal{L}_0$ is the exponential of some bivector in L. However, not all of its subgroups are exponential.] From the above classification of the subalgebras of L one can write down a typical member of a connected subgroup $\mathcal{H} \subset \mathcal{L}_0$ by noting (chapter 2) that $\mathcal{H}$ corresponds to a unique subalgebra $H \subset L$ and that each member of $\mathcal{H}$ is then a finite product of exponentials of members of H. If $F \in L$ is spacelike, say $F = x \wedge y$ in some null tetrad l, n, x, y, exponentiation yields a transformation $f = exp(tF) \in \mathcal{L}_0$ given by

$$l' = l, \qquad n' = n, \qquad x' = \cos tx - \sin ty, \qquad y' = \cos ty + \sin tx. \quad (4.18)$$

If $F = l \wedge n$ is timelike one gets

$$l' = e^t l, \qquad n' = e^{-t} n, \qquad x' = x, \qquad y' = y. \quad (4.19)$$

TABLE 4.1: Lie subalgebras for $(-, +, +, +)$.

Type	Dimension	Basis
R_1	0	0
R_2	1	$l \wedge n$
R_3	1	$l \wedge x$
R_4	1	$x \wedge y$
R_5	1	$l \wedge n + \omega(x \wedge y)$
R_6	2	$l \wedge n, l \wedge x$
R_7	2	$l \wedge n, x \wedge y$
R_8	2	$l \wedge x, l \wedge y$
R_9	3	$l \wedge n, l \wedge x, l \wedge y$
R_{10}	3	$l \wedge n, l \wedge x, n \wedge x$
R_{11}	3	$l \wedge x, l \wedge y, x \wedge y$
R_{12}	3	$l \wedge x, l \wedge y, l \wedge n + \omega(x \wedge y)$
R_{13}	3	$x \wedge y, y \wedge z, x \wedge z$
R_{14}	4	$l \wedge x, l \wedge y, l \wedge n, x \wedge y$
R_{15}	6	L

If $F = l \wedge y$ is null one has

$$l' = l, \qquad n' = n - ty - \frac{1}{2}t^2 l, \qquad x' = x, \qquad y' = y + tl. \tag{4.20}$$

where in each case $t \in \mathbb{R}$ and satisfies $0 \leq t < 2\pi$ in (4.18). If $F = l \wedge n + \omega(x \wedge y)$ is non-simple one gets a combination of (4.18) and (4.19). Transformations like (4.18) are just *rotations* in the $x \wedge y$ plane whilst those in (4.19) are usually referred to as *boosts* in the $l \wedge n$ plane. The transformation in (4.20) is a *null rotation* (about l). The transformations generated by a non-simple bivector are called *screw motions*. The Lie algebra R_{13} is just $o(3)$ (fixing the timelike vector t), R_{10} can be checked to be the Lie algebra $o(1, 2)$, R_{11} is the Lie algebra which leads to those members of $\mathcal{L}_0$ which fix the null vector l and R_{14} is the important *null rotation subgroup*–the members of $\mathcal{L}_0$ which fix the *direction* spanned by l. The transformations arising from R_{14} can be represented elegantly in terms of a complex null tetrad $l, n, m, \bar{m}$ by [40]

$$l' = e^\lambda l, \qquad m' = e^{i\theta}(m - e^\lambda \bar{B} l), \tag{4.21}$$
$$n' = e^{-\lambda} n + Bm + \bar{B}\bar{m} - e^\lambda |B|^2 l,$$

where $\lambda, \theta \in \mathbb{R}, 0 \leqslant \theta < 2\pi$ and $B \in \mathbb{C}$. The condition $e^\lambda > 0$ reflects the fact that these transformations are future preserving. Judicious choices of λ, θ and B in (4.21) can be used to recover transformations arising from the subalgebras $R_2 - R_9$, R_{11} and R_{12}. The transformations (4.21) all fix the direction spanned by l and (4.21) with $\lambda = 0$ are those transformations which fix l. If n_1 and n_2 span distinct null directions, neither of which is that spanned by l, they may be mapped onto each other using some member of (4.21). To see this construct a null basis $l, n_1, m, \bar{m}$ and let $n_2 = al + bn_1 + Bm + \bar{B}\bar{m}$ for $a, b \in \mathbb{R}$

and $B \in \mathbb{C}$. Since n_2 is null one finds $ab = -|B|^2$ with $ab \neq 0$ since n_2 spans a direction distinct from those spanned by l and n_1. Then n_2 is as the last equation in (4.21) with $b = e^{-\lambda}$.

Since $\mathcal{L}_0$ is exponential and since any non-zero bivector admits at least one and at most two independent null eigenvectors, any $f \in \mathcal{L}_0$ which is not the identity map fixes at least one, and at most two, null directions. Each of these may be false if f is taken from $\mathcal{L} \setminus \mathcal{L}_0$ [13].

4.5   The Curvature and Weyl Conformal Tensors

For the curvature tensor $Riem$ one may, as in the last chapter, construct the linear curvature map f on bivectors (and taking advantage of the existence of the metric to "abuse" indices according to $f : F^{ab} \to R^a{}_{bcd}F^{cd}$). In this case, however, the bivector metric P is not positive definite and diagonalisation of f does not follow. Once again one may classify $Riem(p)$ for $p \in M$ into the five classes based on $rgf(p)$ and given in the last chapter. They are defined as in the positive definite case but with the following caveats. Class D may be subdivided into the classes when the necessarily simple bivector spanning $rgf(p)$ is spacelike, timelike or null. (That it must be simple was explained in the last chapter and is due to the curvature symmetry $R_{a[bcd]} = 0$). Class C may be subdivided into the cases when the unique (up to scaling) annihilating vector is spacelike, timelike or null. In class B the spanning bivectors for $rgf(p)$ form a 2-dimensional Abelian subalgebra with no common annihilator and this fixes a subalgebra of type R_7 from the last section. If the curvature class is C and $rgf(p)$ is a subalgebra, it is necessarily of the type R_6 or R_{10} (spacelike annihilator), R_{13} (timelike annihilator) or R_8 or R_{11} (null annihilator). That the classification thus achieved is disjoint and exhaustive is proved in exactly the same way as in the positive definite case (this was the reason for the general, metric-independent way of phrasing the curvature class definitions). Again one allows the symbols A, B, C, D and O to denote also those subsets of M where the curvature is of that class to get the disjoint decomposition of M as $M = A \cup B \cup C \cup D \cup O$. As in the positive definite case one can show using an almost identical proof that A and $A \cup B$ are open in M. The same argument as before then shows that $A \cup B \cup C$ and $A \cup B \cup C \cup D$ are also open in M and that the decomposition of theorem 3.1 holds with $\text{int} Z = \emptyset$. The equation $R^a{}_{bcd}k^d = 0$ for $k \in T_pM$ has non-trivial solutions for k at p if and only if $p \in C \cup D \cup O$.

Now consider the Weyl conformal tensor C on M. For Lorentz signature one has $\epsilon = -1$ in the appropriate equations of chapter 3 and so for C (and also for the tensor E) one has

$$^*C = C^*, \qquad ^*C^* = -C, \qquad ^*E = -E^*, \qquad ^*E^* = E. \tag{4.22}$$

Again one can write out the type $(0,4)$ tensor C_{abcd} as a sum of symmetrised products of members of a basis for $\Lambda_p M$ (as one can for the tensor R_{abcd}) and then one can introduce the *Weyl map* f_C as before and given by $f_C : F^{ab} \to C^a{}_{bcd}F^{cd}$ (or by the usual abuse of notation and retaining the same symbol for the map $f_C : F^{ab} \to C^{ab}{}_{cd}F^{cd}$). These maps are self-adjoint (since $C_{abcd} = C_{cdab}$) and have the same rank, the latter being called the *Weyl rank* at p. Thus one is considering, just as for the curvature tensor in chapter 3, the Weyl tensor in the 6×6 matrix form C_{AB} as a real, symmetric 6×6 matrix. It follows that the rank of f_C (or of the matrix C_{AB}) is, when non-zero, an even integer. To see this note that there exists non-zero bivectors F and A such that

$$f_C(F) = A = C_{abcd}F^{cd} \text{ and hence } f_C(\overset{*}{F}) = \overset{*}{A} \text{ (since } C_{abcd}\overset{*}{F}{}^{cd} = C^*_{abcd}F^{cd}$$

$=^* C_{abcd}F^{cd} = (C_{abcd}F^{cd})^* = \overset{*}{A}_{ab}$ with A and $\overset{*}{A}$ independent. Thus rank$f_C \geq$ 2. If rank$f_C \geq 3$ there exists a non-zero bivector H with $f_C(H) = B$ (and hence $f_C(\overset{*}{H}) = \overset{*}{B}$) for some bivector B which is independent of A and $\overset{*}{A}$. It then follows, as in the proof of lemma 4.4(iv) that $A, \overset{*}{A}, B, \overset{*}{B}$ are independent bivectors. So rank$f_C \neq 3$. A similar argument shows that rank$f_C \neq 5$ and so *the rank of f_C is even* (and rgf_C has the "dual invariant property").

In order to construct the sets $\overset{+}{S}_p$ and $\overset{-}{S}_p$ in the Lorentz case one had to go to the vector space of all complex bivectors $\widetilde{\Lambda_p M}$ at p from which they emerged as 3−dimensional complex subspaces of $\widetilde{\Lambda_p M}$. This leads to a different method for decomposing the Weyl tensor. First one has the following relations (in addition to those already found) and which can be computed from [14]. They are $C^*_{abcd} = C^*_{cdab}$, $C^*_{a[bcd]} = 0$ and $C^{*c}{}_{acb} = 0$ and instead of the real decomposition of C as in the positive definite case one may perform a complex decomposition of C which emerged out of the original research of Petrov [49, 50]. This work, inspired by its potential importance in Einstein's general relativity theory, led to a classification of C in the case of Lorentz signature, known as the *Petrov classification* and developed in [49, 50, 86, 40, 51, 52, 64, 87] amongst many others (and summaries may be found in [16, 13]). This classification is more complicated than in the positive definite case mainly because of the fact that the bivector metric P is no longer positive definite and also because of the existence of null vectors and bivectors. However, the latter objects add a richness to the classification and the complexification of C simplifies the situation.

To achieve this classification one first constructs the *complex Weyl tensor* $\overset{+}{C}$ at p with components $\overset{+}{C}_{abcd}$ and defined by $\overset{+}{C}_{abcd} = C_{abcd} + iC^*_{abcd}$. This has the easily checked properties

$$\overset{+}{C}_{abcd} = -\overset{+}{C}_{bacd} = -\overset{+}{C}_{abdc}, \quad \overset{+}{C}_{abcd} = \overset{+}{C}_{cdab}, \quad \overset{+}{C}_{a[bcd]} = 0, \quad \overset{+}{C}^c{}_{acb} = 0, \quad (4.23)$$

and it also has the "self dual" property $^*\overset{+}{C} = -i\overset{+}{C}$. It is clear that $C(p) = 0 \Leftrightarrow \overset{+}{C}(p) = 0$. Essentially the algebraic classification of C is just the algebraic eigenvector/eigenvalue problem for the linear map f_C at p. A convenient method of effecting this classification starts with the idea of an eigenbivector (and in this respect the work in section 4.3 above is important). It is first recalled that the Weyl tensor requires the metric g for its existence and so one has the bivector metric P and which may be used to raise and lower skew-symmetric pairs on real bivectors, a procedure which may (and will) be extended to complex bivectors. Then a bivector $F \in \widetilde{\Lambda_p M}$ is called an *eigenbivector* of $C(p)$ (respectively, of $\overset{+}{C}(p)$) if the first (respectively, the second) equation below holds for $\lambda, \mu \in \mathbb{C}$

$$C_{abcd}F^{cd} = \lambda P_{abcd}F^{cd} = \lambda F_{ab}, \qquad \overset{+}{C}_{abcd}F^{cd} = \mu P_{abcd}F^{cd} = \mu F_{ab}, \quad (4.24)$$

and then λ (respectively, μ) is the associated *eigenvalue*. Thus if $F \in \widetilde{\Lambda_p M}$ then, using the shorthand notation given in chapter 3, one has $^*(CF)$ $=^* CF = C^*F = C\overset{*}{F}$ and so if $F \in \overset{+}{S}_p$ (respectively $\overset{-}{S}_p$), $CF \in \overset{+}{S}_p$ (respectively $\overset{-}{S}_p$). Thus decomposing C as

$$C = \frac{1}{2}(C + iC^*) + \frac{1}{2}(C - iC^*) \equiv C_1 + C_2, \qquad (4.25)$$

one sees that the obvious self-adjoint maps f_{C_1} and f_{C_2} associated with C_1 and C_2 (just as f_C was with C) are such that f_{C_1} maps $\overset{+}{S}_p$ into itself and $\overset{-}{S}_p$ to the zero bivector whilst f_{C_2} maps $\overset{-}{S}_p$ into itself and $\overset{+}{S}_p$ to the zero bivector and that $f_C = f_{C_1} + f_{C_2}$. The complex tensor C_1 is just half the original $\overset{+}{C}$ and $2C_2$ is sometimes denoted by $\overset{-}{C}$ (and equals the conjugate of $\overset{+}{C}$). Also if F, G and H form a basis for $\overset{+}{S}_p$, the conjugates $\bar{F}, \bar{G}$ and $\bar{H}$ form a basis for $\overset{-}{S}_p$ and $C_1 F = CF$, $C_1 \bar{F} = 0$, $C_2 F = 0$ and $C_2 \bar{F} = C\bar{F} = \overline{CF}$, etc. It follows that the Jordan forms of C_1 (as a linear map on $\overset{+}{S}_p$) and C_2 (as a linear map on $\overset{-}{S}_p$) are the same (with eigenvalues differing only by conjugation) and hence all the algebraic information about C is contained in the action of C_1 (that is, the restriction of $\overset{+}{C}$) on $\overset{+}{S}_p$.

Since $\overset{+}{S}_p$ is a 3–dimensional vector space over $\mathbb{C}$ there are three possible Jordan forms for (the restrictions to $\overset{+}{S}_p$) of C (or $\overset{+}{C}$) represented in Segre notation by $\{111\}$, $\{21\}$ and $\{3\}$. These are the respective *Petrov types* **I**, **II** and **III**. The degeneracy $\{(11)1\}$ is Petrov type **D** and the degeneracy $\{(21)\}$ is Petrov type **N**. (Historically, the symbol **D** refers to the degeneracy in the Segre type and the **N** stands for "null" since this type was associated with possible null radiation fields.) It is noted that the tracefree condition in (4.23) shows that the sum of the eigenvalues in each case is zero. Thus in types **III** and **N** all eigenvalues are zero. Petrov type **O** refers to the vanishing of C at

p. A consideration of the tracefree condition on $\overset{+}{C}$ shows that the map $f_{+\atop C}$ on $\overset{+}{S_p}$ has rank 2 or 3 for type **I**, rank 3 for type **II** and **D**, rank 2 for type **III**, rank 1 for type **N** and rank 0 for type **O**. Lemma 4.1(ii) also applies to the complex situation here and so the (complex) eigenbivector F corresponding to the non-simple elementary divisor in types **II**, **N** or **III** satisfies $|F| = 0$ and is hence a complex null bivector (see after lemma 4.5). The eigenvalues arising from this classification are sometime referred to as *Petrov* or *Weyl invariants*.

Since $\overset{+}{C}$ is, in essence, the restriction to $\overset{+}{S_p}$ of C one may decompose $\overset{+}{C}$ as a sum of symmetrised products of the basis members U, V and M given in (4.11). Then one applies one of the equivalent conditions $\overset{+}{C}_{a[bcd]} = 0$ and $\overset{+}{C}{}^c{}_{acb} = 0$ to get, after a straightforward calculation at p, the elegant expression [40]

$$\overset{+}{C}_{abcd} = C^1 V_{ab}V_{cd} + C^2(V_{ab}M_{cd} + M_{ab}V_{cd}) + C^3(V_{ab}U_{cd} + U_{ab}V_{cd} + M_{ab}M_{cd})$$
$$+ C^4(U_{ab}M_{cd} + M_{ab}U_{cd}) + C^5 U_{ab}U_{cd} \tag{4.26}$$

for $C^1, ..., C^5 \in \mathbb{C}$. Then, for example, if the Petrov type at p is **N** the eigenbivector corresponding to the non-simple elementary divisor must be complex null (and its eigenvalue zero). Choosing it as V, (4.26) gives $C^3 = C^4 = C^5 = 0$. Similar remarks apply if the Petrov type at p is **III**. These two cases can be distinguished by noting that for type **N**, another complex eigenbivector $Q \in \overset{+}{S_p}$ must exist with zero eigenvalue. Writing $Q = aU + bV + cM$ $(a, b, c \in \mathbb{C})$ the equation $\overset{+}{C}Q = 0$ then gives $aC^1 - 2cC^2 = 0 = aC^2$. One cannot have $a = c = 0$ and so $a = 0 \Rightarrow C^2 = 0$, whilst $a \neq 0 \Rightarrow C^2 = 0$ ($\Rightarrow C^1 = 0$ and the contradiction $\overset{+}{C} = 0$.) Thus $C^2 = 0$ for type **N** (and $C^2 \neq 0$ for type **III**). If the Petrov type at p is **II** again the eigenbivector corresponding to the non-simple elementary divisor is complex null (and chosen to be V) but the eigenvalue is non-zero. One gets $C^4 = C^5 = 0$ and the above eigenvalue is $2C^3$.whilst the other eigenvalue is $-4C^3$. It will be shown below that one may choose the basis $l, n, m, \bar{m}$ at p so that $C^2 = 0$. If the Petrov type at p is **D** one has two complex eigenbivectors R and S with equal (non-zero) eigenvalues α and one complex eigenbivector Q with (non-zero) eigenvalue -2α. If $|R| \neq 0 \neq |S|$ one may choose independent linear combinations R' and S' of them within the α−eigenspace such that $|R'| = |S'| = 0$. If $|R| = 0 \neq |S|$ then if $R \cdot S = 0$ one has $R \cdot S = R \cdot Q = S \cdot Q = 0$ (the latter two since $\alpha \neq -2\alpha$). But then R, S, Q form a basis for $\overset{+}{S_p}$ with $R \cdot S = R \cdot Q = R \cdot R = 0$ which is a contradiction. So $R \cdot S \neq 0$ and again one may find a linear combination S' of R and S, independent of R and satisfying $|S'| = 0$. It follows that the α−eigenspace may be spanned by two complex null bivectors $R, S \in \overset{+}{S_p}$ satisfying $R \cdot S \neq 0$ and hence they may be taken as V and U above (section 4.3). Then Q is uniquely determined up to a complex scaling since it is (bivector-) orthogonal to V and U and hence may be taken as M above. Thus, from (4.26), $C^1 = C^2 = C^4 = C^5 = 0$. So for the Petrov types $\mathbf{N}, \mathbf{III}, \mathbf{II}$ and $\mathbf{D}$ one may choose a complex null tetrad $l, n, m, \bar{m}$ at p so that $\overset{+}{C}$ takes the respective canonical forms

$$\overset{+}{C}_{abcd} = C^1 V_{ab} V_{cd}, \qquad (\mathbf{N}) \tag{4.27}$$

$$\overset{+}{C}_{abcd} = C^1 V_{ab} V_{cd} + C^2 (V_{ab} M_{cd} + M_{ab} V_{cd}), \qquad (\mathbf{III}) \tag{4.28}$$

$$\overset{+}{C}_{abcd} = C^1 V_{ab} V_{cd} + C^3 (V_{ab} U_{cd} + U_{ab} V_{cd} + M_{ab} M_{cd}), \qquad (\mathbf{II}) \tag{4.29}$$

$$\overset{+}{C}_{abcd} = C^3 (V_{ab} U_{cd} + U_{ab} V_{cd} + M_{ab} M_{cd}), \qquad (\mathbf{D}) \tag{4.30}$$

where, in each case, $C^1, C^2, C^3 \in \mathbb{C}$. Conversely, each of the expressions above are of the required algebraic Petrov type. Next, the general equation for a change of complex null tetrad from $l, n, m, \bar{m}$ to $l', n', m', \bar{m}'$ given in (4.21) can be used to see how such a change affects a basis change for $\overset{+}{S}_p$ from V, U, M to V', U', M' where V', U' and M' are given in terms of the new basis l', n', m' and $\bar{m}'$ as V, U and M were in terms of the original basis. This gives, after a calculation,

$$V' = e^\lambda e^{-i\theta} V, \qquad M' = 2e^\lambda \bar{B} V + M, \tag{4.31}$$
$$U' = e^\lambda \bar{B}^2 e^{i\theta} V + \bar{B} e^{i\theta} M + e^{-\lambda} e^{i\theta} U.$$

Finally one may substitute this last equation into (4.26) (written with primes on the bivector basis members and coefficients) to see how the coefficients in (4.26) are affected by this change. A lengthy but straightforward calculation gives

$$C^1 = e^{2\lambda} e^{-2i\theta} C'^1 + 4e^{2\lambda} \bar{B} e^{-i\theta} C'^2 + 6e^{2\lambda} \bar{B}^2 C'^3$$
$$+ 4e^{2\lambda} \bar{B}^3 e^{i\theta} C'^4 + e^{2\lambda} \bar{B}^4 e^{2i\theta} C'^5, \tag{4.32}$$
$$C^2 = e^\lambda e^{-i\theta} C'^2 + 3e^\lambda \bar{B} C'^3 + 3e^\lambda \bar{B}^2 e^{i\theta} C'^4 + e^\lambda \bar{B}^3 e^{2i\theta} C'^5, \tag{4.33}$$
$$C^3 = C'^3 + 2\bar{B} e^{i\theta} C'^4 + \bar{B}^2 e^{2i\theta} C'^5, \tag{4.34}$$
$$C^4 = e^{-\lambda} e^{i\theta} C'^4 + e^{-\lambda} \bar{B} e^{2i\theta} C'^5, \tag{4.35}$$
$$C^5 = e^{-2\lambda} e^{2i\theta} C'^5. \tag{4.36}$$

From this equation, and considering Petrov type $\mathbf{N}$ as in (4.27), so that $C^2 = C^3 = C^4 = C^5 = 0$, one may make a basis change (4.21) with $\lambda = 1$ and with an appropriate choice of θ to make C^1 real in (4.27). For Petrov type $\mathbf{III}$ one may first change basis and choose θ to make C^2 real and then (keeping C^2 real) choose B to set $C^1 = 0$. For Petrov type $\mathbf{II}$ one may arrange that C^1 is real.

The Petrov classification can be described in another way by following the penetrating observations initiated by Bel [52] and others [53, 54, 40, 51]. They discovered that the Petrov type at p can be characterised by a certain number of real null directions at p which lie in a special way with respect to the Weyl tensor. The geometrical relations controlling this phenomenon are referred to as the *Bel criteria*. To see how this works suppose that $C(p) \neq 0$, consider the following two equations (which are easily shown to be equivalent).

$$k_{[e}\overset{+}{C}_{a]bc[d}k_{f]}k^b k^c = 0, \qquad \overset{+}{C}_{abcd}k^b k^c = k_a q_d + q_a k_d, \qquad (4.37)$$

for a (non-zero, real) $k \in T_p M$ and a complex 1–form q at p. If $q \neq 0$ the condition $g^{ad}\overset{+}{C}_{abcd} = 0$ shows that $k \cdot q = 0$ and then a contraction of (the second of) (4.37) with k^a shows that k is necessarily null. Also it was shown above that $\overset{+}{C}_{abcd}\bar{V}^{cd} = 0$ (since $\bar{V} = l \wedge m \in \bar{S}_p$) and so, choosing a complex null tetrad $l, n, m, \bar{m}$ with $k = l$ and using this last relation in (4.37), one sees by contracting the second of (4.37) with m^d that $q \cdot m = 0$ and so q is a linear combination of l and m and hence null. On the other hand, if $q = 0$, one has $C_{abcd}k^b k^d =^* C_{abcd}k^b k^d = 0$. The second of these is $(\epsilon^{cdrs}C_{abrs})k^b k_c = 0$ which implies $\epsilon^{cdrs}T_{arsc} = 0$ where $T_{arsc} = C_{abrs}k^b k_c$. It follows that $\epsilon^{rscd}T_{arsc} = 0$ and hence that $T_{a[rsc]} = 0$, that is, $k^b C_{ab[rs}k_{c]} = 0$. A contraction of this equation with k^c and use of the first equation above shows that either $C_{abcd}k^a = 0$ or k is null. If $C_{abcd}k^d = 0$ then $^*C_{abcd}k^d = 0$, that is, $C^*_{abcd}k^d = 0$ and so $\epsilon^{rscd}C_{abrs}k_d = 0$ and hence $C_{ab[cd}k_{e]} = 0$ and a contraction of this equation with k^e again shows that k is null. Thus the equivalent conditions (4.37) force k to be null and then k is said to span a *principal null direction* (*pnd*) for $C(p)$ or $\overset{+}{C}(p)$.

Now, again with $C(p) \neq 0$, consider the following two equations at p (again easily checked to be equivalent).

$$k_{[e}\overset{+}{C}_{a]bcd}k^b k^c = 0, \qquad \overset{+}{C}_{abcd}k^b k^c = K k_a k_d, \qquad (4.38)$$

for a (non-zero, real) $k \in T_p M$ and $K \in \mathbb{C}$. It is noted that if $K \neq 0$ a contraction of (the second of) (4.38) with k^a reveals that k is necessarily null. This result also follows if $K = 0$ by the argument just given for the case $q = 0$ in (4.38). Thus the equivalent conditions (4.38) force k to be null and then k is said to span a *repeated principal null direction* (*repeated pnd*) for $C(p)$ or $\overset{+}{C}(p)$. The term "repeated" will be explained later.

Thus a repeated pnd is a pnd, but not necessarily conversely. Let n be null and choose a complex tetrad $l, n, m, \bar{m}$ at p. On using (4.26) and performing some simple contractions one sees that (4.38) holds for $k = n$, that is, n is a pnd of C at p, if and only if $C^1 = 0$. This allows one to count the number of pnds which can exist for $C(p) \neq 0$ by fixing some null direction spanned by l and using (4.21) to seek solutions of (4.38) for n by seeking solutions for $\bar{B}$ of the equation $C^1 = 0$ in (4.26) using (4.37), each one of which will give a solution (up to a scaling) for n. It follows that if $C^5 \neq 0$ one has a quartic equation in $\bar{B}$ from (4.37) and hence at least one and at most four pnds exist. If $C^5 = 0 \neq C^4$ one gets a cubic equation for $\bar{B}$, that is, for n but then l is another distinct pnd. Thus there is at least one and at most four pnds in all cases. [It is noted that the solutions are *counted properly* and some of them may coincide. This is the reason for the term "repeated" pnd] If l (respectively, n) is a repeated pnd then, from (4.26), $C^4 = C^5 = 0$ (respectively, $C^1 = C^2 = 0$).

For the individual Petrov types listed above, one can check that if $C(p)$ is of type **N**, l is the only pnd and is (quadruply) repeated pnd. Similarly, if $C(p)$ is of type **III** (with C^1 set to zero as was explained earlier) l is a (triply) repeated pnd and n is a (non-repeated) pnd. If $C(p)$ is of type **II**, l is a (doubly) repeated pnd and it is easily checked that there are two other distinct (non-repeated) pnds. If $C(p)$ is of type **D**, l and n are each (doubly) repeated pnds and there are no non-repeated pnds. Now from the above description of the algebraic types for $\overset{+}{C}(p) \neq 0$ it can be seen that if $\overset{+}{C}$ admits a repeated pnd, say l, it may be reduced to one of the types **N**, **III**, **II** or **D** and so, defining these types to be *algebraically special*, one has the result that $C(p)$ or $\overset{+}{C}(p)$ is algebraically special if and only if it admits a repeated pnd and hence it is of Petrov type **I** if and only if it has four (non-repeated) pnds and is then referred to as *algebraically general*.

The above Bel criteria are stated in terms of $\overset{+}{C}$. However, they can be restated, with very little change, in terms of the real tensor C. This makes them more accessible and, in fact, one of the main reasons for expressing the criteria in this latter form is its usefulness in calculation especially in general relativity theory. For example, if $p \in M$ and $C(p) \neq 0$, the Petrov type at p is **N** if and only if there exists $0 \neq k \in T_pM$ such that either $\overset{+}{C}_{abcd}k^d = 0$, or $C_{abcd}k^d = 0$, or $*C_{abcd}k^d = 0$, or $\overset{+}{C}_{ab[cd}k_{e]} = 0$, or $C_{ab[cd}k_{e]} = 0$, or $*C_{ab[cd}k_{e]} = 0$. The vector k is, in each case, null, unique up to scaling and spans the quadruply repeated pnd of $C(p)$. The other Petrov types are similar and are discussed in [13].

Another type of study of the Weyl conformal tensor can be found in [80].

As before, using a Petrov symbol to denote precisely those points of M where the Weyl tensor has that Petrov type, one has $M = \mathbf{I} \cup \mathbf{II} \cup \mathbf{D} \cup \mathbf{III} \cup \mathbf{N} \cup \mathbf{O}$. To describe this decomposition in more detail one must consider the characteristic polynomial Q arising from $\overset{+}{C}$, the latter regarded as the linear map $f_{\overset{+}{C}}$ on the 3−dimensional complex vector space $\overset{+}{S}_p$ given by $F^{ab} \rightarrow \overset{+}{C}_{abcd}F^{cd}$. Let P_n denote the set of all polynomials with coefficients in $\mathbb{C}$ and of degree $\leq n$. P_n can be regarded as a manifold of dimension $2n+2$ according to the chart scheme

$$Q = c_n z^n + ... + c_1 z + c_0 \longleftrightarrow (a_0, b_0, ..., a_n, b_n) \tag{4.39}$$

for $Q \in P_n$ and $c_k = a_k + ib_k$, $(a_k, b_k \in \mathbb{R})$, $(0 \leq k \leq n)$. If $\lambda \in \mathbb{C}$ is a *simple* root of Q, so that $Q(\lambda) = 0$ then there exists a smooth map h from some open neighbourhood U of Q in $\mathbb{R}^{2n+2}$ to $\mathbb{C}$ such that $h(Q) = \lambda$ and that $h(Q')$ is a root of Q' for each $Q' \in U$ (see, for example, [29]). Thus a simple root of Q depends smoothly on the polynomial coefficients. Since g, and hence $\overset{+}{C}$, are smooth on M, the characteristic polynomial of $\overset{+}{C}$ gives rise to a smooth map $S : V : M \to \mathbb{R}^8$ where V is some coordinate neighbourhood in M. Then there exists a smooth map h from some open neighbourhood W of the characteristic polynomial Q' for $\overset{+}{C}$ at p to $\mathbb{C}$ such that $h(Q') = \lambda$ and $h(Q)$ is a root of Q for each $Q \in W$. Thus $U_1 \equiv S^{-1}W$ is an open coordinate neighbourhood of p and $S(p)$ is the characteristic polynomial of $\overset{+}{C}$ at p. Now consider the smooth map $h \circ S$ on U_1 so that $q \in U_1 \Rightarrow S(q)$ is the characteristic polynomial at q and $h(S(q)) = (h \circ S)(q)$ is a root of $S(q)$ with $h \circ S : U_1 \to \mathbb{C}$. It follows that the smooth complex function $h \circ S$ on U_1 gives rise to a smooth eigenvalue of $\overset{+}{C}$. Thus if $p \in \mathbf{I}$, $S(p) = Q'$ has three simple (distinct) roots in $\mathbb{C}$ and it follows that Q has three simple, distinct roots in some open neighbourhood of p. Thus $\mathbf{I}$ is an open subset of M.

One may then refine the decomposition of M above in terms of open subsets of M. For this it is recalled that, to allow use of the rank theorem, the rank of $f_{\overset{+}{C}}$ is 2 or 3 when restricted to $\mathbf{I}$, 3 on $\mathbf{II}$ or $\mathbf{D}$, 2 on $\mathbf{III}$, 1 on $\mathbf{N}$ and zero on $\mathbf{O}$. (When considered as a real tensor C the rank of f_C is twice the above rank for each of the Petrov types; cf the earlier remark that rank f_C is even.)

Theorem 4.3 *Let M be a smooth, connected 4-dimensional manifold admitting a Lorentz metric. One has the following disjoint decomposition of M in terms of the Petrov types of C (or $\overset{+}{C}$).*

$$M = \mathbf{I} \cup int\mathbf{II} \cup int\mathbf{D} \cup int\mathbf{III} \cup int\mathbf{N} \cup int\mathbf{O} \cup \mathbf{X} \qquad (4.40)$$

where int denotes the interior operator in the manifold topology of M and $\mathbf{X}$, which is determined by the disjointness of the decomposition, is a closed subset of M satisfying $int\mathbf{X} = \emptyset$. The subset $\mathbf{X}$ is empty if and only if the Petrov type is the same at each point of M.

Proof It was shown above that $\mathbf{I}$ is open in M. The above remarks on rank show that the subsets $\mathbf{I} \cup \mathbf{II} \cup \mathbf{D} \cup \mathbf{III}$ and $\mathbf{I} \cup \mathbf{II} \cup \mathbf{D} \cup \mathbf{III} \cup \mathbf{N}$ are open in M. In addition, the subset $\mathbf{I} \cup \mathbf{II} \cup \mathbf{D}$ is also open since $\mathbf{I}$ is and since the rank of $f_{\overset{+}{C}}$ is equal to 3 on $\mathbf{II} \cup \mathbf{D}$. It remains only to show that $int\mathbf{X} = \emptyset$. So let $W \subset \mathbf{X}$ be open so that, by disjointness, $W \cap \mathbf{I} = \emptyset$. Now $\mathbf{I} \cup \mathbf{II} \cup \mathbf{D}$ is open and so $W \cap (\mathbf{I} \cup \mathbf{II} \cup \mathbf{D}) = W \cap (\mathbf{II} \cup \mathbf{D})$ is open. If this latter subset is non-empty it cannot be completely contained in $\mathbf{D}$ since, being open, it would lead to $W \cap int\mathbf{D} \neq \emptyset$, contradicting the disjointness of the decomposition. So

$W \cap \mathbf{II} \neq \emptyset$ and there exists $p \in W \cap \mathbf{II}$. Since $W \cap (\mathbf{II} \cup \mathbf{D})$ is open one has, from the above, the existence of a simple root of the characteristic equation at p and hence a corresponding smooth root γ of this equation in some neighbourhood V of p with $V \subset W \cap (\mathbf{II} \cup \mathbf{D})$. Since the Petrov type of $\overset{+}{C}$ at each point of V is $\mathbf{II}$ or $\mathbf{D}$, the tracefree condition on $\overset{+}{C}$ shows that the roots of the characteristic polynomial on V are γ and $-\frac{1}{2}\gamma$ and they are smooth. Now consider the smooth matrix function $Z = (\overset{+}{C} - \gamma I)(\overset{+}{C} + \frac{1}{2}\gamma I)$ on V where I is the unit 3×3 matrix. A consideration of the *minimal* polynomial associated with $\overset{+}{C}$ on V shows that Z vanishes at points where the Petrov type is $\mathbf{D}$ since all elementary divisors are simple for this Petrov type, but not where it is $\mathbf{II}$ since for this type $-\frac{1}{2}\gamma$ corresponds to a non-simple elementary divisor (chapter 1). Thus $Z(p) \neq 0$ and so there exists an open neighbourhood V' of p where Z does not vanish. It follows that $W \cap \mathbf{II}$ is non-empty and open, contradicting $W \cap \text{int}\mathbf{II} = \emptyset$ by disjointness of the decomposition. It follows that $W \cap (\mathbf{II} \cup \mathbf{D}) = \emptyset$ and W is disjoint from $\mathbf{I}, \mathbf{II}$ and $\mathbf{D}$. Now suppose that $W \cap \mathbf{III} \neq \emptyset$. Now $W \cap \mathbf{III} = W \cap (\mathbf{I} \cup \mathbf{II} \cup \mathbf{D} \cup \mathbf{III})$ is open and immediately one gets the contradiction $W \cap \text{int}\mathbf{III} \neq \emptyset$ so that $W \cap \mathbf{III} = \emptyset$. Similarly one gets $W \cap \mathbf{N} = W \cap \mathbf{O} = \emptyset$ and thus $W = \emptyset$. Hence $\text{int}\mathbf{X} = \emptyset$. The final sentence of the theorem follows from the fact that M is connected. □

This decomposition shows that each point of the open dense subset $M \setminus \mathbf{X}$ of M lies in an open subset of M on which the *Petrov type is constant*. From this it can be shown [29] that the eigenvalues are *locally smooth* and the eigenbivectors of $\overset{+}{C}$ may be chosen to be locally smooth. In practice this is convenient and necessary for local calculations involving calculus in both classical geometry and general relativity theory.

4.6   Curvature Structure

Again suppose that $\dim M = 4$ and that g is a smooth Lorentz metric on M with curvature tensor *Riem*. If $0 \neq \alpha \in \mathbb{R}$, the Lorentz metric αg on M is also smooth and has the same curvature tensor *Riem*. Now suppose that g' is another smooth metric on M of arbitrary signature and which has the same curvature tensor *Riem* as g. What can one say about g'? As discussed in the last chapter, one necessarily has the conditions of lemma 4.7 satisfied for each bivector $F \in rgf(p)$ at each $p \in M$, where f is the curvature map. The structure of $rgf(p)$ is given by the curvature class of *Riem*(p) as detailed earlier this chapter. According to this classification M may be decomposed into open subsets of fixed curvature class as $M = A \cup \text{int}B \cup \text{int}C \cup \text{int}D \cup \text{int}O \cup Z$ where A, B, C, D and O are the subsets of M where the curvature class is, respectively, A, B, C, D and O (with A open in M) and where Z is a closed subset of M with empty interior. If A is open and dense in M (the "general situation") similar techniques to those in section 3.8 show that for $p \in A$, $rgf(p)$ must have dimension ≥ 2 and if this dimension is 2 it must contain a non-simple member, say, F (to avoid annihilators—see lemma 3.2), and a member G independent of F which may, by taking linear combinations, also be chosen non-simple and which, to avoid the contradiction that $p \in B$, may be chosen so that none of the canonical blades of F coincides with either of the canonical blades of G. Then lemma 4.7 shows that T_pM is an eigenspace for h. Similar comments apply if $\dim rgf(p) > 3$. If $\dim rgf(p) = 3$, $rgf(p)$ may contain a non-simple member (and the above argument applies again) or it may contain only simple members (see lemma 3.2) and again T_pM is an eigenspace for h. Recalling the result that the equation $R^a{}_{bcd}k^d = 0$ has only trivial solutions for $k \in T_pM$ at points of the subset A, one thus has the following result [13, 55] (see also [56, 18, 19])

Theorem 4.4 *Let M be a 4-dimensional manifold and let g be a smooth Lorentz metric on M with curvature tensor Riem. Suppose that A is an open dense subset of M. Suppose also that g' is a smooth metric on M of arbitrary signature which has the same curvature tensor Riem as g does on M. Then $g' = h$ satisfies lemma 4.7 for each $F \in rgf(p)$ and for each $p \in A$ and so, since A is dense in M, $g' = \alpha g$ for $0 \neq \alpha \in \mathbb{R}$. The Levi-Civita connections for g and g' are the same and g' has Lorentz signature.*

If A is not dense in M, so that at least one of int$B$, int$C$, int$D$ and int$O$ is not empty one can, omitting the case int$O \neq \emptyset$, derive expressions relating g and g' on each of these subsets in the same manner as that in section 3.8. In these cases, g and g' may have different signatures. [It is noted that lemma 4.7 still holds even if h is not non-degenerate.]

As shown earlier (chapter 3) the holonomy group Φ of (M, g) is now a Lie subgroup of $\mathcal{L}$ and so the holonomy algebra ϕ is a Lie subalgebra of $o(1, 3)$. Since Φ is a metric holonomy group the subalgebra ϕ is a one of the subalgebras $R_1 - R_{15}$ from Table 4.1 with R_5 omitted (see section 3.7).

Theorem 4.5 *Let M be a manifold admitting smooth Lorentz metrics g and g' with respective Levi-Civita connections ∇ and ∇'. Suppose $\nabla = \nabla'$. Then the holonomy algebras of ∇ and ∇' are the same $(=\phi)$ and if ϕ is of type R_9, R_{12}, R_{14}, or R_{15}, $g' = \lambda g$ $(\lambda \in \mathbb{R}.)$*

Proof The proof follows immediately from the definitions of the appropriate subalgebras for each type given in section 4.4, use of lemma 4.7 and the calculations in section 3.8. $\qquad\square$

Now consider the Weyl conformal tensor C and associated Weyl map f_C for (M, g). As mentioned earlier, conformally related metrics on M have the same (tensor-) type $(1, 3)$ conformal Weyl tensor. Now with the original metric g of Lorentz signature given suppose, conversely, that g' is another smooth metric of arbitrary signature on M whose Weyl conformal (type $(1,3)$) tensor (C') equals that of g on M. What can one say about the relationship between g and g'? One can examine the range space of f_C at $p \in M$ through the Petrov canonical forms of C (with respect to g) and then use lemma 4.7. It is then recalled from the previous chapter that one may, at each $p \in M$, find a Weyl class for the Weyl conformal tensor at $p \in M$ in exactly the same way as one originally found the curvature class at p. It is then clear, following a simple consideration of the rank of f_C and the tracefree condition $C^c{}_{acb} = 0$ on C, that classes B and D are again impossible, that class C can only arise when the rank of f_C is 2 with a null annihilator and then the Petrov type is $\mathbf{N}$ (section 4.5) and that, otherwise, one has class A or O. Further, and in an appropriately chosen real null tetrad l, n, x, y at p, the Petrov types listed above for g show that $rg f_C$ is spanned by $l \wedge x$ and $l \wedge y$ for $p \in \mathbf{N}$ and contains $l \wedge x$, $l \wedge y$ $l \wedge n$ and $x \wedge y$ for all other Petrov types at p except type $\mathbf{0}$. Use of lemma 4.7 with $h = g'$ shows, using the decomposition for (M, g) in terms of its Petrov types, as in theorem 4.4, shows that at each $p \in M \setminus (\mathbf{N} \cup \mathbf{O})$, $l \wedge x$, $l \wedge y$, $l \wedge n$ and $x \wedge y$ are eigenspaces of g' with respect to g and hence $g' = \phi g$ for some function ϕ on $M \setminus (\mathbf{N} \cup \mathbf{O})$ and that for $p \in \mathbf{N}$, $l \wedge x$ and $l \wedge y$ are eigenspaces for g and hence $g'_{ab} = \alpha g_{ab} + \beta l_a l_b$ for functions α and β on $\mathbf{N}$. The functions ϕ, α and β are easily seen to be smooth. Further, since the Petrov type is a statement between the Weyl conformal tensor and the bivector metric constructed from the metric giving rise to it, it can be checked from the above statements that g' has Lorentz signature and the Petrov types of g and g' are identical on $M \setminus (\mathbf{N} \cup \mathbf{O})$ and also on $\mathbf{N}$. Thus if $\text{int}(\mathbf{N} \cup \mathbf{O}) = \emptyset (\Rightarrow \text{int} \mathbf{N} = \text{int} \mathbf{O} = \emptyset)$ g and g' are conformally related on the open dense subset $M \setminus X$ of theorem 4.4 and hence on M, $g' = \phi g$, with ϕ a smooth map $M \to \mathbb{R}$. One has the following theorem.

Theorem 4.6 *Let M be a 4-dimensional manifold and let g be a smooth Lorentz metric on M with type $(1,3)$ Weyl conformal tensor C. Suppose g' is another smooth metric of arbitrary signature on M which has the same type $(1,3)$ Weyl conformal tensor as g on M. If, when M is decomposed with respect to the Petrov type of g, $int(\mathbf{N} \cup \mathbf{O}) = \emptyset$ then g and g' are conformally related on M.*

The clause $\text{int} \mathbf{O} = \emptyset$ is obviously necessary whilst the clause $\text{int} \mathbf{N} = \emptyset$ is (less obviously) also necessary. To see this, suppose that (M, g) is such that it is vacuum (that is, $Ricc = 0$ on M) with nowhere zero curvature tensor and admits a global function u such that the covector field $l_a \equiv u_{,a}$ is nowhere zero, null and parallel, $l_{a;b} = 0$. [Such pairs (M, g) with all these imposed conditions exist–see, for example, [13].] Then consider the metric g' such that, in any local coordinate domain, $g'_{ab} = g_{ab} + \lambda(u) l_a l_b$ for some smooth function λ. The metric g' is easily seen to be of Lorentz signature (since in any real, null tetrad l, n, x, y (with respect to g) based on l at any $p \in M$, $g'(l, l) = 0$, $g'(x, x) = g'(y, y) = 1$ and $g'(l, x) = g'(l, y) = 0$). The Ricci identity then gives for the curvature from g, $R_{abcd} l^d = 0$ and, since $Ricc = 0$, this leads to $C_{abcd}l^d = 0$. Thus, from the Bel criteria, (M, g) is of Petrov type **N** everywhere. One can now show [13] that the type $(1, 3)$ curvature tensors of g and g' are equal on M and hence g' is also a vacuum metric on M. It then easily follows that g' is also of Petrov type **N** on M and, if λ is chosen to be nowhere zero on M, g and g' are not conformally related on M. Further the type $(1, 3)$ Weyl tensors of g and g' are equal on M (but, if $\dot{\lambda}$ is not zero at some point of M, ∇ and ∇' differ). Thus the need for clauses in the above theorems.

4.7 Sectional Curvature

The sectional curvature function for positive definite metrics was discussed in the last chapter. Here a study of this function will be undertaken for a 4-dimensional manifold M admitting a Lorentz metric g. One considers the Grassmann manifold of all 2-spaces at $p \in M$, now denoted by G_p and identified as the (diffeomorphic) manifold of projective simple bivectors at p. The formal definition of the sectional curvature function $\sigma_p : G_p \to \mathbb{R}$ at p is

$$\sigma_p(F) = \frac{R_{abcd}F^{ab}F^{cd}}{2P_{abcd}F^{ab}F^{cd}} = \frac{R_{abcd}F^{ab}F^{cd}}{2|F|} \qquad (4.41)$$

where F_{ab} is any non-zero, simple bivector whose blade is the 2-space F in G_p and P is the bivector metric at p. The definition is clearly independent of the representative bivector chosen for F. The problem now is that the denominator in the above definition of σ_p may vanish and will do so if and only if the simple bivector in the denominator satisfies $|F| = 0$, that is, if and only if it is null (lemma 4.3). More formally, if one denotes the subsets of the Grassmann manifold G_p consisting of all spacelike (respectively, timelike, null) 2-spaces at p by S_p^2 (respectively, T_p^2, N_p^2) and which can be shown to be submanifolds of G_p of dimension 4, 4 and 3, respectively [13], one has the disjoint union $G_p = S_p^2 \cup T_p^2 \cup N_p^2$ with $F \in S_p^2 \Leftrightarrow |F| > 0$, $F \in T_p^2 \Leftrightarrow |F| < 0$ and $F \in N_p^2 \Leftrightarrow |F| = 0$. The submanifold N_p^2 is the topological boundary of S_p^2 and also of T_p^2 and is closed and not open. Thus σ_p is only defined on the 4-dimensional open submanifold $\bar{G}_p \equiv G_p \setminus N_p^2 = S_p^2 \cup T_p^2$ of G_p consisting of all non-null 2-spaces in G_p and is smooth (in fact, analytic) there. Further, $\bar{G}_p$ is (open and) dense in G_p. It is clear that $\bar{G}_p$ is not connected and also is not compact (otherwise it would be a closed subspace of the Hausdorff space G_p and this would contradict the fact that N_p^2 is not open). [This should be compared to the situation in the last chapter when g was positive definite and where σ_p was defined on the whole of the compact, connected space G_p.]

Consider the situation when the map σ_p is a constant (continuous) function on $\bar{G}_p$, say mapping each member of $\bar{G}_p$ to $K \in \mathbb{R}$. Then one may trivially extend it to a constant (continuous) function on G_p, mapping each member of G_p to K and one achieves the constant curvature condition at p (see section 3.9). But suppose that σ_p is not a constant function on $\bar{G}_p$. In this case one may, by considering sequences in $\bar{G}_p$ which converge to a limit in N_p^2 [57] (or see, for example, [58, 60, 61]), show the rather useful result that if σ_p can be continuously extended to *any* member of N_p^2 it can be continuously extended to the whole of N_p^2 and is then a constant function on G_p. Thus if σ_p is not a constant function on $\bar{G}_p$ it cannot be continuously extended to *any* member of N_p^2. [This follows a weaker result in [62].] This allows the following deduction to be made. Suppose that for (M, g) the function σ_p for g is given and is not a constant function at some $p \in M$. Then the (complement of the) domain space of σ_p determines the subset N_p^2 at p on which σ_p is not defined and about which the following geometrical remarks may be made. Suppose F and G represent distinct members of N_p^2. (All bivectors will be assumed to be in their tensor type $(2, 0)$ form to avoid any confusion when a second metric is introduced, and will be identified with the null 2-spaces which they represent.) There are three cases arising here. The first is when F and G have the same principal null direction, say, spanned by $l \in T_pM$. Then clearly, the bivectors $F + \lambda G$ for each $\lambda \in \mathbb{R}$ are members of N_p^2 (with the same principal null direction). The second is when F and G have distinct principal null directions but whose blades intersect (in a necessarily spacelike direction). In this case the bivector $F + \lambda G$ for $0 \neq \lambda \in \mathbb{R}$ is simple but not null since it can be spanned by a pair of orthogonal non-null vectors (section 4.1). The third is when F and G have distinct principal null directions and whose blades do not intersect. Then $F + \lambda G$ for $0 \neq \lambda \in \mathbb{R}$ is clearly non-simple and hence not in N_p^2. It follows that for the bivectors $F + \lambda G$, $0 \neq \lambda \in \mathbb{R}$ to be in N_p^2 the first case must hold and the blades of F and G have a common principal null direction (spanned by) l. Let $N_p(l)$ denote this collection of (all) null 2-spaces at p with principal null direction l and which represents the collection $F + \lambda G$ in N_p^2. Now suppose this metric g on M is changed to a smooth, Lorentz metric g' on M such that g and g' have the same *non-constant* sectional curvature function at p, that is, $\sigma_p = \sigma_p'$. The equality of these non-constant functions leads to the equality of the (complements of) their domain spaces and hence to the *same* collection of 2-spaces, N_p^2 at p and which are now null for $g'(p)$. Thus the collection $N_p(l)$ determines, by their common intersections, common null directions for $g(p)$ and $g'(p)$. Thus the collection of null vectors for $g(p)$ and $g'(p)$ coincide, that is, $g(p)$ and $g'(p)$ are proportional. If this is true at each p in some open dense subset of M, g and g' are conformally related on M. [This clarifies the proof in [58] which was spoiled by typos.]

The following result now arises in a similar way to the one given in chapter 3. For σ_p a constant function on $\bar{G}_p$ one has $Q_{abcd}F^{ab}F^{cd} = 0$ for $Q_{abcd} = R_{abcd} - \frac{R}{6}P_{abcd}$ and for all simple, non-null bivectors F, that is, for each $F \in \bar{G}_p$. Also if two Lorentz metrics g and g' with sectional curvature functions σ_p and σ_p', respectively, have the same sectional curvatures, one has, from (4.41), $Q_{abcd}F^{ab}F^{cd} = 0$ for $Q_{abcd} = \phi^2 R_{abcd} - R'_{abcd}$ and all $F \in \bar{G}_p$. But then $\bar{G}_p$ is open and dense in G_p and so these equations for Q hold for each $F \in G_p$, that is, for any simple bivector. Then the argument given in chapter 3 shows that if σ_p is a constant function *Riem* takes the constant curvature form at p, $R_{abcd} = \frac{R}{6} P_{abcd}$, and if, for $p \in M$ $\sigma_p = \sigma'_p$ and with these functions not constant, $\phi^2 R_{abcd} = R'_{abcd}$. Thus the following lemma holds.

Lemma 4.8 *Let M be a 4-dimensional manifold admitting smooth Lorentz metrics g and g' whose sectional curvature functions are equal at each point of M but are such that this common function is not a constant function on G_p for each p in some open dense subset of M. Then at any $p \in M$*
(i) $g' = \phi g$,   (ii) $R'_{abcd} = \phi^2 R_{abcd}$,   (iii) $R'^a{}_{bcd} = \phi R^a{}_{bcd}$,
(iv) $R'_{ab} = \phi R_{ab}$,   (v) $R' = R$,   (vi) $C' = \phi C$
for some smooth function ϕ on M.

It is stressed here that part (vi) above for the type $(1,3)$ Weyl conformal tensor is deduced from the other parts, as before. Of course, since g and g' are conformally related, $C' = C$, but one may have $C' = C = 0$. Now using the same notation as in section 3.9, let $X \subset M$ denote the (necessarily closed) subset of M on which the sectional curvature is a constant function (so that $M \setminus X$ is the open dense subset W of M on which the sectional curvature function is not constant) and, using primes to denote tensors associated with g', let U denote the open subset of M on which C (and hence C') is not zero. Also let V denote the open subset of M on which the $1-$form $d\phi$ does not vanish and let Y denote the closed subset of M where $\phi = 1$. Thus all points where *Riem* and *Riem'* vanish are contained in X. One then gets the disjoint decomposition $M = V \cup \text{int} Y \cup K$ where K is the closed subset of M defined by the disjointness of the decomposition, so that, as before, $\text{int} K = \emptyset$. In this decomposition, $g' = g$ on $\text{int} Y$ and, whereas in the positive definite case $V = \emptyset$, now V is an open, possibly non-empty, subset of M where $d\phi$ does not vanish and on which g and g' are conformally related, conformally flat Lorentz metrics. If $V = \emptyset$ one sees that $g' = g$ on the open dense subset $\text{int} Y$ and hence on M. So suppose that the set V is non-empty. This subset V will now be investigated following [58, 59]. It is pointed out here that similar results were arrived at, using different techniques, independently in [63].

On V $d\phi$, with components ϕ_a, is nowhere zero and each of g and g' is conformally flat, $C = C' = 0$ on V. Use of the expressions for the tensors *Riem*, $C(= 0)$ and E for g from chapter 3 then give, for g

$$R_{abcd}(= g_{ae} R^e{}_{bcd}) = \frac{1}{2} [\tilde{R}_{ac} g_{bd} - \tilde{R}_{ad} g_{bc} + \tilde{R}_{bd} g_{ac} - \tilde{R}_{bc} g_{ad}] + \frac{R}{6} P_{abcd} \quad (4.42)$$

and similarly for g'.

One has two conformally flat structures (V, g) and (V, g') with $g' = \phi g$ for $\phi : V \to \mathbb{R}$ nowhere zero and smooth and with ∇ and ∇' denoting the respective Levi-Civita connections for g and g'. Denoting (as usual) their respective covariant derivatives in components by a semi-colon and a stroke, and a partial derivative by a comma, one may write down the (conformally flat) Bianchi identities for (V, g) and (V, g')

$$R_{ca;b} - R_{cb;a} = \frac{1}{6}[g_{ac}R_{,b} - g_{cb}R_{,a}], \qquad R'_{ca|b} - R'_{cb|a} = \frac{1}{6}[g'_{ac}R'_{,b} - g'_{cb}R'_{,a}].$$
(4.43)

Now evaluate the second in (4.43) using the previous lemma and subtract from it ϕ times the first in (4.43). The partial derivatives disappear and terms like $P^a_{bc} = \Gamma'^a_{bc} - \Gamma^a_{bc}$, which are given in section 3.8, can then be used to get

$$\phi_b R_{ac} - \phi_a R_{bc} = \phi R_{ae}P^e_{bc} - \phi R_{be}P^e_{ac} = R_{be}\phi^e g_{ac} - R_{ae}\phi^e g_{bc} \qquad (4.44)$$

where $\phi_a \equiv \phi_{,a}$ and $\phi^a \equiv g^{ab}\phi_b$. A contraction with g^{ac} gives $\tilde{R}_{ab}\phi^b = 0$ and a back substitution into the last equation gives $\tilde{R}_{ac}\phi_b - \tilde{R}_{bc}\phi_a = 0$ and so $\tilde{R}_{ab} = \psi\phi_a\phi_b$ for some smooth function $\psi : V \to \mathbb{R}$. It follows that $\psi(\phi^a\phi_a) = 0$ on V. If at some $p \in V$ $\psi(p) = 0$, then $\tilde{Ricc}(p) = 0$ and so $E(p) = 0$ and, since $C(p) = 0$, *Riem* takes the constant curvature form at p and σ_p is a constant function at p. It follows, by assumption, that ψ cannot vanish over any non-empty open subset of V and thus $\phi^a\phi_a = 0$, that is, since $d\phi$ never vanishes on V, ϕ_a is null on V with respect to both g and g'. The earlier expression for P^a_{bc} then shows that $\phi_{a;b} = \phi_{a|b} + \phi^{-1}\phi_a\phi_b$. Taking a further ∇-covariant derivative gives

$$\phi_{a;bc} = (\phi_{a|b})_{;c} - \phi^{-2}\phi_a\phi_b\phi_c + \phi^{-1}\phi_a\phi_{b;c} + \phi^{-1}\phi_{a;c}\phi_b \qquad (4.45)$$

and so, since $\phi^a\phi_{a|b} = 0$,

$$(\phi_{a|b})_{;c} - (\phi_{a|b})_{|c} = \phi_{a|e}P^e_{bc} + \phi_{e|b}P^e_{ac} = (2\phi)^{-1}[\phi_{a|c}\phi_b + \phi_{c|b}\phi_a + 2\phi_{a|b}\phi_c].$$
(4.46)

The last two equations then combine to give

$$\phi_{a;bc} - \phi_{a;cb} = \phi_{a|bc} - \phi_{a|cb} + (2\phi^{-1})[\phi_{a;c}\phi_b - \phi_{a;b}\phi_c]. \qquad (4.47)$$

Now one uses the Ricci identities for ∇ and ∇' in the last equation and contracts with ϕ^b to get

$$\phi_d R^d{}_{abc}\phi^b = \phi_d R'^d{}_{abc}\phi^b = \phi\phi_d R^d{}_{abc}\phi^b \qquad (4.48)$$

where the fact that $\phi^a\phi_a = 0$ and $\phi_{a;b} = \phi_{b;a}$ on V means that $\phi^a\phi_{a;b} = \phi^a\phi_{b;a} = 0$. Equation (4.42) and the condition $\phi^a\phi_a = 0$ on V then give $R_{abcd}\phi^b\phi^d = -\frac{R}{12}\phi_a\phi_c$, that is, $R\phi_a\phi_c = R\phi\phi_a\phi_c$ and since $d\phi$ never vanishes on V, $R = \phi R$ on V. If R does not vanish at some $p \in V$, R does not vanish in some open neighbourhood $W \subset V$ of p and so $\phi = 1$ on W. Hence $d\phi$ vanishes on W and this is a contradiction to the definition of V. It follows that $R = 0$ and so $Ricc = \tilde{Ricc}$, that is, $R_{ab} = \psi\phi_a\phi_b$, on V. A substitution into (4.42) and a contraction with ϕ^d shows that $R_{abcd}\phi^d = 0$ on V from which the Ricci identities give $\phi_{a;bc} = \phi_{a;cb}$ and similarly $\phi_{a|bc} = \phi_{a|cb}$. Then (4.47) gives

$\phi_{a;c}\phi_b = \phi_{a;b}\phi_c$ on V. This last equation shows that $\phi_{a;b} = \alpha\phi_a\phi_b = \phi_a(\alpha\phi_b)$ for some smooth, real-valued function α on V and one more application of the Ricci identity on ϕ^a reveals that $\alpha\phi_a$ is locally a gradient on V and so in some open neighbourhood W' of any $p \in V$ there exists a smooth function ρ such that $\alpha\phi_a = \rho_{,a}$. It then follows that $\chi \equiv e^{-\rho}d\phi$ is a nowhere-zero, parallel, null $1-$form on W' and of the form du for some smooth function u on (some possibly reduced) W'. Thus ϕ and ρ are functions of u. Now $R = 0$ on W' and (4.43) shows that $R_{ab;c} = R_{cb;a}$ and hence that $R_{ab} = \gamma(u)\chi_a\chi_b$ on W' for some function $\gamma(u)$. Thus, on some coordinate domain of the open subset W' one has $C = R = 0$ together with a nowhere-zero, parallel, null vector field represented by the $1-$form du. Any point p of the open dense subset of V where $Riem$ and $Riem'$ do not vanish admits a connected neighbourhood on which the conditions of Walker's non-simple K_n^* spaces are satisfied [81] and on which one may choose coordinates u, v, x, y with u as above such that the metric g takes the form

$$ds^2 = H(u, x, y)du^2 + 2dudv + dx^2 + dy^2 \qquad (4.49)$$

and where the conformally flat condition allows the coordinates to be chosen so that that $H(u, x, y) = \delta(u)(x^2 + y^2)$ for some smooth function δ. Such local manifolds as these are well-known from general relativity theory and are the *(conformally flat) plane waves*. The metric g' is also such a plane wave as is easily seen from the above calculations. The Ricci tensor is of Segre type $\{(211)\}$ with zero eigenvalue and represents what is known as a *null fluid* in Einstein's theory. One thus has the following theorem.

Theorem 4.7 *Let M be a 4-dimensional manifold admitting smooth Lorentz metrics g and g'. Suppose that g and g' have the same sectional curvature function at each $p \in M$ and which is not a constant function at each point of some open dense subset of M. Then one may decompose M as above according to $M = V \cup intY \cup K$ where $intK = \emptyset$, $g = g'$ on $intY$ and where V, if not empty, is an open submanifold of M on which g and g' are conformally related, conformally flat plane waves.*

For the given metric g in (4.49) the metric $g' = \phi g$ will satisfy the above conditions on the sectional curvature if and only if the function ϕ satisfies certain conditions which have been given and solved in [63]. Thus such metrics g' different from g always exist.

It is remarked that, by using a similar proof to that in chapter 3, it can be shown that if F and $\overset{*}{F}$ represent *any* dual (orthogonal) pair of space-like/timelike 2-spaces at $p \in M$ their sectional curvatures $\sigma_p(F)$ and $\sigma_p(\overset{*}{F})$ are equal if and only if the Einstein space condition holds at p [64].

4.8   The Ricci Flat (Vacuum) Case

In this section the extra condition that the Ricci tensor *Ricc* is identically zero on M will be imposed. This is usually referred to as the Ricci-flat condition but in the important case of a 4-dimensional manifold admitting a Lorentz metric and its use in Einstein's general relativity theory it will be here be called the *vacuum* condition. It is noted that with this restriction, the curvature and Weyl conformal tensors are equal on M.

The first result is easily derived from the last theorem by noting that if one imposes the non-flat condition on M, that is, the condition that *Riem* does not vanish over any non-empty open subset of M, then neither does the Weyl tensor. Thus if U is the open dense subset of M on which *Riem* and C are nowhere zero the sectional curvature is nowhere a constant function on U; otherwise, for $p \in U$, one would have the constant curvature condition at p and hence, since $Ricc(p) = 0$, the contradiction that $Riem(p) = 0$. Then lemma 4.8 shows that $\phi = 1$ on U and hence on M, and one has

Theorem 4.8 *Let (M, g) be a 4-dimensional manifold admitting a smooth Lorentz metric g and with the vacuum and non-flat conditions holding on M. If g' is any other smooth Lorentz metric on M with the same sectional curvature function on M as g then $g' = g$. Thus, in this case, the sectional curvature uniquely determines the metric and its Levi-Civita connection and is, in this sense, in one-to-one correspondence with non-flat, vacuum metrics on M.*

At this point it is necessary to introduce the concept of a *pp-wave* metric. This term arose, and was described in detail, in [64] as a consequence of the attempt to introduce a solution to Einstein's vacuum field equations which described a source-free idealised solution representing pure gravitational waves. Such a solution can be defined by taking (M, g) to satisfy the vacuum and non-flat conditions and, in addition, to admit a global, smooth, nowhere zero, parallel bivector field. From these assumptions it may be shown that this bivector is necessarily null, the Petrov type is **N** and that locally, (M, g) admits a non-vanishing, parallel, null vector field. From these results it turns out that one may write down the metric g in a local coordinate system u, v, x, y as in (4.49) above with $l_a = u_{,a}$ a parallel null 1−form and with $\partial^2 H / \partial x^2 + \partial^2 H / \partial y^2 = 0$ (to achieve the vacuum condition). One can then, if required, add extra conditions of symmetry to get more specialised solutions of the form (4.49). Important special cases are the *(vacuum) plane waves* and in this case the above local coordinates may be chosen so that (4.49) holds with $H = a(u)(x^2 - y^2) + c(u)xy$. [It is here remarked that one may extend the concept of a pp-wave to that of a *generalised pp-wave* and hence to a *generalised plane wave*, where the Ricci tensor is permitted to take the form of a null fluid mentioned earlier. Such metrics are, in general, of Petrov type

N but can be conformally flat without being flat, and the (conformally flat) plane waves of theorem 4.7 are examples of such generalised plane waves. Such a non-vacuum, type **N** plane wave can be shown to be locally conformally related to a type **N**, vacuum plane wave (see [65] where an extended summary of such properties may be found)]

One may now add the following theorem of Brinkmann [39] (see also [13]) which can be proved using the above techniques. Suppose that g and g' are each smooth, non-flat, conformally related, Lorentz, vacuum metrics on the 4-dimensional manifold M. So $g' = \phi g$ on M for a smooth function $\phi : M \to \mathbb{R}$ and the tensor type $(1,3)$ Weyl tensors of g and g' are equal. Let V be the open subset of M on which the $1-$form $d\phi$ is non-zero and let U be the open, dense subset of M on which the curvature tensors of g and g', which are necessarily equal since g and g' are conformally related and vacuum, are non-zero. Finally let $W = U \cap V$. Then by following similar arguments as for the previous theorem [13] one has the following result which is more interesting than the one in the positive definite case given in the last chapter.

Theorem 4.9 *Let (M, g) be a 4-dimensional manifold admitting smooth, conformally related, vacuum, non-flat metrics g and g' of Lorentz signature with $g' = \phi g$ for some smooth nowhere-zero function $\phi : M \to \mathbb{R}$. Then in the above notation, there is a disjoint decomposition $M = W \cup int(M \setminus V) \cup Z$ where Z is a closed subset of M with $intZ = \emptyset$ and where ϕ is constant on each component of $int(M \setminus V)$ and where each point of W admits a coordinate neighbourhood on which each of g and g' are pp-waves.*

Finally, for the important class of vacuum space-times in general relativity one may establish an analogue of theorem 4.5 for such metrics and which involves the Petrov types. Consider the subset $V \subset M$ defined as the subset of all $p \in M$ at which the equation $R^a{}_{bcd}k^d = 0$ has a non-trivial solution for $k \in T_pM$. If the vacuum condition holds on M the tensors $Riem$ and C are equal on M and so equivalent definitions are that (i) V is the subset of points of M at which the Petrov type is **N** or **O**, and (ii) V is the subset of points of M at which the curvature rank is at most 2. It follows that if $M \setminus V \neq \emptyset$ then at points of $M \setminus V$ the curvature rank (which equals the Weyl rank and is hence even) is at least 4 and hence of curvature class A. The next theorem now follows.

Theorem 4.10 *Let M be a space-time with smooth, vacuum Lorentz metric g such that the subset V above has empty interior. Then the curvature tensor uniquely determines g up to a constant conformal factor.*

Finally one has

Theorem 4.11 *Let (M, g) be a 4-dimensional manifold with smooth Lorentz metric g, which is not flat and satisfies the vacuum condition. Then the holonomy group of (M, g) is R_8, R_{14} or R_{15} [13]. If (M, g) is a proper Einstein space the holonomy group of (M, g) is R_7, R_{14} or R_{15}. [114, 13]*

Proof When (M, g) is vacuum, one has $Ricc = 0$ and hence $Riem = C$ and so the curvature map has the dual invariant property since f_C does. Thus the range space of the curvature map is even-dimensional (and dual invariant) at each $p \in M$ and by the Ambrose-Singer theorem (chapter 2) so is the holonomy algebra ϕ. Thus ϕ is of type R_7, R_8, R_{14} or R_{15}. That it cannot be R_7 follows from the restrictions $R^c{}_{acb} = 0$ and $R_{a[bcd]} = 0$ on $Riem$ and from lemma 3.3. If (M, g) is a proper Einstein space one again has the dual invariant property and similarly obtains the above four possibilities for the holonomy. Then the R_8 type is eliminated because $R \neq 0$. $\square$

Chapter 5

Four-Dimensional Manifolds of Neutral Signature

5.1 Neutral Tangent Space Geometry

In this chapter a study will be made of a 4-dimensional manifold M which admits a metric g of *neutral signature* $(+,+,-,-)$ (and hence a Sylvester basis at any $p \in M$ gives the component form $\text{diag}(1,1,-1,-1)$ for $g(p)$). As usual M is assumed smooth, connected, Hausdorff and second countable and g is assumed smooth. At $p \in M$, one of two choices of basis for $T_p M$ will usually be made. The first basis, called *orthonormal*, consists of $x, y, s, t \in T_p M$ satisfying the metric relations $x \cdot x = y \cdot y = -s \cdot s = -t \cdot t = 1$ with all other inner products between basis members zero, and the second is an associated *null* basis l, n, L, N obtained from the first basis according to $\sqrt{2}l = x + t$, $\sqrt{2}n = x - t$, $\sqrt{2}L = y + s$, $\sqrt{2}N = y - s$, so that $l \cdot n = L \cdot N = 1$. Thus for a null basis one sees that l, n, L, N are null vectors and that all other inner products between basis members, apart from the two given, are zero. So s and t are orthogonal timelike vectors and l and n are orthogonal null vectors as are L and N. Conversely, given independent *null* vectors $l, n, L, N \in T_p M$ satisfying $l \cdot n = L \cdot N = 1$ with all other inner products between them equal to zero, one may construct an orthonormal basis according to the scheme $\sqrt{2}x = l+n$, $\sqrt{2}t = l-n$, $\sqrt{2}y = L+N$, $\sqrt{2}s = L-N$. A pair of bases x, y, s, t and l, n, L, N related as above will be referred to as *corresponding* or *associated* bases. They lead to completeness relations $g_{ab} = x_a x_b + y_a y_b - s_a s_b - t_a t_b = l_a n_b + n_a l_b + L_a N_b + N_a L_b$. Conversely, as detailed previously, either of these relations implies the independence of the basis members and obvious contractions with these basis members reveal the inner products specified between them. Other bases for $T_p M$ are often useful, for example, the "hybrid" bases l, n, y, s and x, t, L, N.

The collection of 2−spaces of $T_p M$ is quite different in this case as compared to the Lorentz case and will be described here. A 2−space $U \subset T_p M$ at p is called *spacelike* if each non-zero member of U is spacelike, or each non-zero member of U is timelike (or, equivalently, if U contains no null members). Thus $x \wedge y$ and $s \wedge t$ are examples of spacelike 2−spaces. A 2−space $U \subset M$ is called *timelike* if it contains exactly two distinct, null directions (referred to

DOI: 10.1201/ 9781003023166-5

as the *principal null directions (pnds)* of U). A timelike 2−space also contains spacelike and timelike vectors. Thus $l \wedge n = t \wedge x$ and $L \wedge N = s \wedge y$ are examples of timelike 2−spaces and $l \pm n$ and $L \pm N$ are, in each case, spacelike and timelike members of them. A 2−space $U \subset M$ is called *null* if it contains exactly one null direction (referred to as its *principal null direction (pnd)*). The other members of a null 2−space are either *all* spacelike (for example, $l \wedge y$) or *all* timelike (for example, $l \wedge s$) and in each case the null member is orthogonal to all other members of U (otherwise extra null members would arise). If two members of a null 2−space are orthogonal, at least one of them is null. To see these results let $U = l \wedge v$ be null at p for $l, v \in T_p M$ and with l null. Then $|v| \neq 0$ by definition of a null 2−space and if $l \cdot v \neq 0$ there exists a null member $l + av \in U$ ($0 \neq a \in \mathbb{R}$) independent of l. This contradiction shows that $l \cdot v = 0$ for each $v \in U$. Then if $w \in U$ is independent of l, $w = v + bl$ ($b \in \mathbb{R}$), $|w| = |v|$ and thus all such w are either spacelike or timelike. Finally, if $cv + dl$ and $ev + fl$ ($c, d, e, f \in \mathbb{R}$) are independent orthogonal members of U, $ce = 0$ and so one of them is proportional to l.

If a general 2−space U contains three or more distinct null vectors then any one of these can be written as a non-trivial linear combination of the other two forcing these two null vectors to be orthogonal. Thus U is spanned by two orthogonal null vectors and as a consequence any non-zero member of U is null. This gives the last possibility for U. A 2−space $U \subset M$ is called *totally null* if each non-zero member of U is null (and hence any two members of U are orthogonal). Thus $l \wedge L$, $l \wedge N$, $n \wedge L$ and $n \wedge N$ are examples of totally null 2−spaces. This completes the classification of 2−spaces at any $p \in M$. For a 4-dimensional manifold, totally null 2−spaces can exist only in neutral signature since they require non-proportional, orthogonal null vectors.

Let W be a 3−space of $T_p M$. Then if $0 \neq k \in T_p M$ spans the normal to W, k is orthogonal to each member of W and the direction determined by k is unique. Then W is called spacelike (respectively, timelike or null) if k is spacelike (respectively timelike or null). This completes the classification of 3−spaces at any $p \in M$ and it is remarked that this labelling convention for 3−spaces is different from that used in the Lorentz case. Thus, in the above tetrads $\mathrm{Sp}(x, s, t)$ is spacelike, $\mathrm{Sp}(x, y, t) = \mathrm{Sp}(l, n, y)$ is timelike and $\mathrm{Sp}(l, L, N)$ is null with their respective normals spanned by y, s and l. Only in the null case is $k \in W$. For a spacelike 2−space the metric induced on it is positive definite whereas for a spacelike or a timelike 3−space or a timelike 2−space it is Lorentz. There are no metrics induced on null or totally null 2−spaces or null 3−spaces.

If U is a spacelike (respectively, timelike) 2−space its orthogonal complement $U^\perp$ is spacelike (respectively, timelike). This differs from the Lorentz situation and easily follows. In either case the span $\mathrm{Sp}(U \cup U^\perp)$ equals $T_p M$. If U is a null 2−space a basis l, n, y, s may be chosen for $T_p M$ so that U is of the form $l \wedge y$ or $l \wedge s$ and then $U^\perp$ is also null and of the form $l \wedge s$ or $l \wedge y$, respectively. In either case $\mathrm{Sp}(U \cup U^\perp)$ is a null 3−space with normal l. If U is a totally null 2−space, $U^\perp = U$ and is thus also totally null. So the orthogonal operator on 2−spaces preserves the type (spacelike, timelike, null or totally null) of that 2−space. In terms of the bivector metric a *simple* bivector F is *spacelike* (respectively, *timelike*, *null* or *totally null*) if its blade is spacelike (respectively, timelike, null or totally null) and then F is spacelike if $|F| > 0$, timelike if $|F| < 0$ and either null or totally null if $|F| = 0$.

The collection of all spacelike members S of T_pM is an open, regular, 4−submanifold of T_pM, as is the collection of all timelike members T. Each of S and T is locally path-connected and is hence connected if and only if it is path-connected (see chapter 1). So let $u, v \in S$ and choose an orthonormal basis x, y, s, t at p such that $u = (\epsilon, 0, 0, 0)$ and $v = (a, b, c, d)$ with $\epsilon, a, b, c, d \in \mathbb{R}$ satisfying $a^2 + b^2 > c^2 + d^2$ and $\epsilon > 0$. Now build obvious smooth paths $(\epsilon, 0, 0, 0) \to (a, b, 0, 0)$ and then $(a, b, 0, 0) \to (a, b, c, d)$ each of which passes only through spacelike vectors. Thus any $v \in S$ may be path-connected to $u \in S$ and the result follows. A similar argument shows that T is connected. However, $S \cup T$ is not connected since any path from a member of S to a member of T must pass through a null or zero member of T_pM. Now choose an orthonormal basis x, y, s, t for T_pM so that for $k \in T_pM$, $k = ax + by + cs + dt$ for $a, b, c, d \in \mathbb{R}$ and, if $k \in N$ where N is the set of all null members of T_pM, one has $a^2 + b^2 = c^2 + d^2$ and with a, b, c, d not all zero. Now define the smooth map $f : (T_pM \setminus \{0\}) \to \mathbb{R}$ given in the above orthonormal basis by $f : (a, b, c, d) \to a^2 + b^2 - c^2 - d^2$. Then $N = f^{-1}(0)$ and f has a non-zero gradient at each point of N. Thus N is a 3−dimensional regular submanifold of T_pM (chapter 2). Further, N is connected and this will be established by showing that it is path connected, that is, by showing that there exists $l \in N$ and a smooth path from l to any other member of N which passes only through null vectors. Choose any $l \in N$ and let $k \in N$ be independent of l. If $l \cdot k = 0$ one may choose a smooth path from l to k through the totally null 2−space $l \wedge k$ (since the latter, as a 2−space of T_pM, has topology homeomorphic to $\mathbb{R}^2$) whilst if $l \cdot k \neq 0$, $l \wedge k$ is timelike and one may choose two independent null vectors r and s in the timelike 2−space $(l \wedge k)^{\perp}$ (so that $l \cdot r = l \cdot s = k \cdot r = k \cdot s = 0$) and path-connect k to r, and r to l by smooth paths passing only through null vectors in the totally null 2−spaces $k \wedge r$ and $r \wedge l$. Thus N is connected. In fact, if $k \in N$ one may write k, as was done in the above orthonormal basis, with components (a, b, c, d) not all zero so that k uniquely determines the points $(a, b) \in \mathbb{R}^2 \setminus \{0\}$ and $(c, d) \in \mathbb{R}^2 \setminus \{0\}$ and the positive real number $a^2 + b^2 = c^2 + d^2$. Going to the usual equivalence relationship of proportionality on N one easily sees that the collection of all null *directions* at p is also connected since it is the continuous image of N under the corresponding natural projection map (Chapter 1).

Clearly $N \cup \{0\}$ is the boundary of S and of T and, since $T_pM \setminus \{0\} = S \cup T \cup N$ is open and connected with S and T open and S, T and N disjoint, N is not open and it is not closed either (since it does not contain its boundary point 0).

5.2   Algebra and Geometry of Bivectors

This section discusses the algebra and geometry of bivectors in neutral signature. Some of these results were obtained in collaboration with Zhixang Wang [66, 67] (see also [68]). As before the *type* of a simple bivector is the type of its blade. One may form duals just as before and then the dual $\overset{*}{F}$ of a simple bivector F is simple and *of the same type* as F and the blades of F and $\overset{*}{F}$ are orthogonal. One may arrange that the following results hold for the bases x, y, s, t and l, n, L, N of T_pM. First note that, in the notation established earlier, $t \wedge x = l \wedge n$ and $L \wedge N = s \wedge y$. Then

$$(x \wedge y)^* = s \wedge t, \qquad (x \wedge t)^* = s \wedge y, \qquad (x \wedge s)^* = y \wedge t. \qquad (5.1)$$

One may construct a basis for Λ_pM of the form $x \wedge y$, $s \wedge t$, $x \wedge s$, $y \wedge s$, $x \wedge t$ and $y \wedge t$ and which is orthonormal (up to a factor 2) with respect to the bivector metric P. It follows that the signature of P is $(+, +, -, -, -, -)$. As in the positive definite case (but unlike the Lorentz case) a bivector F and its dual need not be independent members of Λ_pM. In fact, if $\overset{*}{F} = \lambda F$ for $\lambda \in \mathbb{R}$, the result $\overset{**}{F} = F$ for this signature shows that $\lambda = \pm 1$. So label the ± 1 eigenspaces of the linear duality map at p by the subspaces $\overset{+}{S}_p$ and $\overset{-}{S}_p$ of Λ_pM given by $\overset{+}{S}_p = \{F \in \Lambda_pM : \overset{*}{F} = F\}$ and $\overset{-}{S}_p = \{F \in \Lambda_pM : \overset{*}{F} = -F\}$ and then define, as before, $\widetilde{S}_p = \overset{+}{S}_p \cup \overset{-}{S}_p$. Thus F and $\overset{*}{F}$ are independent if and only if $F \in \Lambda_pM \setminus \widetilde{S}_p$. One has $\overset{+}{S}_p \cap \overset{-}{S}_p = \{0\}$ and may write any $F \in \Lambda_pM$ uniquely as $F = \overset{+}{F} + \overset{-}{F}$ for $\overset{+}{F} \in \overset{+}{S}_p$ and $\overset{-}{F} \in \overset{-}{S}_p$. Thus $\Lambda_pM = \overset{+}{S}_p + \overset{-}{S}_p$. It is easily checked from (5.1) that $l \wedge N = \frac{1}{2}((x \wedge y + s \wedge t) - (x \wedge s + y \wedge t))$ is a member of $\overset{+}{S}_p$, as is $n \wedge L$. Similarly one shows that $l \wedge L, n \wedge N \in \overset{-}{S}_p$. Also $l \wedge n - L \wedge N \in \overset{+}{S}_p$ and $l \wedge n + L \wedge N \in \overset{-}{S}_p$. Since the blade of a totally null bivector equals the blade of its orthogonal complement it follows that any totally null bivector is a member of $\widetilde{S}_p$. Then defining

$$F_1 \equiv \frac{1}{2}(l \wedge n - L \wedge N), \qquad F_2 \equiv \frac{1}{\sqrt{2}}(l \wedge N), \qquad F_3 \equiv \frac{1}{\sqrt{2}}(n \wedge L),$$

$$G_1 \equiv \frac{1}{2}(l \wedge n + L \wedge N), \qquad G_2 \equiv \frac{1}{\sqrt{2}}(l \wedge L), \qquad G_3 \equiv \frac{1}{\sqrt{2}}(n \wedge N), \quad (5.2)$$

one sees that F_1, F_2, F_3 are independent members of $\overset{+}{S}_p$ and that G_1, G_2, G_3 are independent members of $\overset{-}{S}_p$. Thus $\overset{+}{S}_p = \mathrm{Sp}(F_1, F_2, F_3)$ and

$\bar{S}_p = \mathrm{Sp}(G_1, G_2, G_3)$ so that each of $\overset{+}{S}_p$ and $\bar{S}_p$ is 3–dimensional. Also one has

$$|F_1| = |G_1| = -1, \qquad |F_2| = |F_3| = |G_2| = |G_2| = 0,$$
$$F_1 \cdot F_2 = F_1 \cdot F_3 = G_1 \cdot G_2 = G_1 \cdot G_3 = 0, \qquad F_2 \cdot F_3 = G_2 \cdot G_3 = 1 \quad (5.3)$$

and a straightforward calculation shows that if $F \in \overset{+}{S}_p$ and $G \in \bar{S}_p$, necessarily $F \cdot G = 0$. It follows from this that the bivector metric on $\Lambda_p M$, which, as shown above has signature $(+, +, -, -, -, -)$, restricts to Lorentz metrics with signatures $(-, -, +)$ on each of $\overset{+}{S}_p$ and $\bar{S}_p$. Thus one may view $\overset{+}{S}_p$ as a 3–dimensional Lorentz space with F_1 "timelike" and F_2 and F_3 "null" vectors (or, alternatively, $F_2 \pm F_3$ as "spacelike" and "timelike" vectors), and similarly for $\bar{S}_p$. Further it is easily shown that

$$[F_1, F_2] = F_2, \qquad [F_1, F_3] = -F_3, \qquad [F_2, F_3] = -F_1,$$
$$[G_1, G_2] = G_2, \qquad [G_1, G_3] = -G_3, \qquad [G_2, G_3] = -G_1 \qquad (5.4)$$

and if $F \in \overset{+}{S}_p$ and $G \in \bar{S}_p$, one easily computes that $[F, G] = 0$. To get a more familiar picture of this consider the vector space $\mathbb{R}^3$ with Lorentz metric of signature $(-, -, +)$ and let l', n', x' be a basis for $\mathbb{R}^3$ so that, using a dot also for this inner product, l' and n' are null and $l' \cdot n' = 1, x' \cdot x' = -1$, $l' \cdot x' = n' \cdot x' = 0$. Then construct a bivector basis $l' \wedge n', l' \wedge x', x' \wedge n'$ for the associated bivector space. Consider the association between this bivector basis and the basis F_1, F_2, F_3 for $\overset{+}{S}_p$ given by $F_1 \leftrightarrow l' \wedge n', F_2 \leftrightarrow l' \wedge x'$ $F_3 \leftrightarrow x' \wedge n'$. Computing brackets for bivectors in $\mathbb{R}^3$ in the usual way one finds $[(l' \wedge n'), (l' \wedge x')] = l' \wedge x', [(l' \wedge n'), (x' \wedge n')] = -x' \wedge n'$ and $[(l' \wedge x'), (x' \wedge n')] = -l' \wedge n'$ and so the above association is a Lie algebra isomorphism between $\overset{+}{S}_p$ and $o(1, 2)$. Thus $\overset{+}{S}_p$ (and similarly $\bar{S}_p$) are Lie isomorphic to $o(1, 2)$. So

$$\Lambda_p M = o(1, 2) + o(1, 2) \qquad (5.5)$$

and the natural projections $\pi_1 : \Lambda_p M \to \overset{+}{S}_p$ and $\pi_2 : \Lambda_p M \to \bar{S}_p$ are Lie algebra homomorphisms.

At this point one can collect together several further results on the geometry of bivectors for this signature including, for convenience, those described above. The results of lemma 3.4 hold independently of signature (with the proviso that for this chapter one requires $\epsilon = 1$ in part *(iii)* of this lemma).

Lemma 5.1 *(i) If $F \in \overset{+}{S}_p$ and $G \in \bar{S}_p$ then $F \cdot G = 0$ and $[F, G] = 0$. For any $F \in \Lambda_p M$ $|F| = |\overset{*}{F}|$.*

(ii) If $E \in \widetilde{S}_p$ the statements (a) E is simple, (b) E is totally null and (c) $|E| = 0$ are equivalent.

(iii) *If* F, G *are independent and totally null and each is in* $\overset{+}{S}_p$ *or each is in* $\overset{-}{S}_p$ *their blades intersect trivially. Otherwise their blades intersect in a unique null direction at* p.

(iv) *If* $H = F_1$ *or* $H = G_1$, $H_{ac}H^c{}_b = \frac{1}{4}g_{ab}$ *whilst if* $H = \frac{1}{2}(x \wedge y \pm s \wedge t)$, $|H| = 1$ *and* $H_{ac}H^c{}_b = -\frac{1}{4}g_{ab}$.]

(v) *If* $F \in \Lambda_p M$ *satisfies* $F_{ab}k^b = \overset{*}{F}_{ab}k^b = 0$ *for some* $0 \neq k \in T_p M$ *then* k *is necessarily null. These conditions are equivalent to* F *being either null or totally null. If* k *is unique up to a scaling,* F *is null. Otherwise, there are infinitely many solutions for* k *no two of which are proportional and* F *is totally null.* F *is null if and only if* $F = A + B$ *for totally null members* $A \in \overset{+}{S}_p$ *and* $B \in \overset{-}{S}_p$.

Proof

(i) This was given earlier.

(ii) If $E \in \widetilde{S}_p$ is simple, $E = \pm \overset{*}{E}$ and $E = p \wedge q$ for $p, q \in T_p M$. But then the blades of E and $\overset{*}{E}$ are orthogonal and so E is totally null. If E is totally null (and hence in $\widetilde{S}_p$) then clearly $|E| = 0$ and finally, if $E \in \widetilde{S}_p$ with $|E| = 0$, $E \cdot \overset{*}{E} = \pm|E| = 0$ and so (lemma 3.1), E is simple.

(iii) If $F, G \in \overset{+}{S}_p$ are totally null they are simple and if their blades intersect non-trivially, $F = p \wedge q$ and $G = p \wedge r$ for $p, q, r \in T_p M$ null vectors with $p \cdot q = p \cdot r = 0$. Choosing a null basis l, n, L, N with $p = l$, it follows that q may be chosen (up to a real scaling) to be either L or N (and similarly for r) and then $F = l \wedge N$, $G = l \wedge L$ or vice versa. But then one of them is in $\overset{+}{S}_p$ and the other in $\overset{-}{S}_p$ and a contradiction follows. Thus their blades intersect only trivially. If $F \in \overset{+}{S}_p$ and $G \in \overset{-}{S}_p$ are totally null with $F = p \wedge q$ and $G = r \wedge s$ for $p, q, r, s \in T_p M$ all null and $p \cdot q = r \cdot s = 0$ suppose that p, q, r, s are independent. The condition $[F, G] = 0$ then gives $p \cdot r = p \cdot s = q \cdot r = q \cdot s = 0$ and a contradiction since then any of them is orthogonal to each member of the basis p, q, r, s and hence to each member of $T_p M$. Thus p, q, r, s are dependent but cannot span a 2−space of $T_p M$ since F and G are independent in $\Lambda_p M$. Hence their blades intersect in a single null direction. Taking this direction to be spanned by l one gets (see above) $F = l \wedge N \in \overset{+}{S}_p$ and $G = l \wedge L \in \overset{-}{S}_p$, or vice versa.

(iv) The proof of this is a simple calculation.

The proof of (v) is straightforward since the given conditions force F and $\overset{*}{F}$ to be simple with k in each of their blades, and hence null. It is noted here that, in the final part, the blades of A and B intersect non-trivially (in the principal null direction of F from part (iii)). $\qquad\qquad\qquad\square$

It follows (part (ii)) that totally null bivectors in $\widetilde{S}_p$ constitute exactly the collection of simple members of $\widetilde{S}_p$ and then using the fact that $\overset{+}{S}_p$ with the induced metric from P is $3-$dimensional and of Lorentz signature $(-,-,+)$, that the null "cone" of this geometry is the totality of totally null members of $\overset{+}{S}_p$ and hence that two independent such members cannot be orthogonal. Similarly, if $A \cdot B = 0$ for bivectors $A, B \in \overset{+}{S}_p$ with B totally null, $|A| \leq 0$. Similar comments apply to $\overset{-}{S}_p$. If F and G are independent and totally null and if $\dot{F} \cdot G = 0$ then $F \in \overset{+}{S}_p$ and $G \in \overset{-}{S}_p$, or vice versa. Regarding part (iii) suppose that F, G are independent, totally null members of $\overset{+}{S}_p$ and P, Q are independent, totally null members of $\overset{-}{S}_p$. Then, using part (iii), one may determine, up to scaling, four null vectors $p, q, r, s \in T_pM$ by the respective symbolic representations $F \cap P$, $F \cap Q$, $G \cap P$ and $G \cap Q$. Now if p and q are proportional the blades of P and Q intersect non-trivially and this is a contradiction. Thus p and q are independent and similarly one sees that the pairs (p, r), (q, s) and (r, s) are pair-wise independent and that one may write $F = p \wedge q$, $G = r \wedge s$, $P = p \wedge r$ and $Q = q \wedge s$ with $p \cdot q = r \cdot s = p \cdot r = q \cdot s = 0$. It now follows that p, q, r, s are independent. To see this suppose not and from the above given independent pairs deduce that either q is a linear combination of p, r and s or that p is a linear combination of q, r and s. In the first case $F = p \wedge q$ is a linear combination of P and $R \equiv p \wedge s$ and it follows that $q \cdot r = 0$ and this gives the contradiction that q is proportional to p. The other possibility is similar and so p, q, r and s are independent. The two other remaining inner products $p \cdot s$ and $q \cdot r$ must each be non-zero otherwise, say if $p \cdot s = 0$, p would be orthogonal to each member of the basis p, q, r, s and a contradiction arises. It also follows that, starting with independent, totally null members F, G of $\overset{+}{S}_p$ one may choose a null basis l, n, L, N for T_pM such that $F = \alpha l \wedge N$ and $G = \beta n \wedge L$ $(0 \neq \alpha, \beta \in \mathbb{R})$ and if a bivector $A \in \overset{+}{S}_p$ satisfies $A \cdot F = A \cdot G = 0$, A is a multiple of F_1 in (5.2). Similar comments apply to $\overset{-}{S}_p$. It follows that if U, V are totally null 2$-$spaces satisfying $U \cap V = \{0\}$ then one may choose a null basis l, n, L, N with $U = l \wedge N$ and $V = n \wedge L$, or $U = l \wedge L$ and $V = n \wedge N$ depending on whether their respective bivectors are both in $\overset{+}{S}_p$ or both in $\overset{-}{S}_p$. Of course, given a null vector $l \in T_pM$, there exists exactly two independent totally null bivectors whose blades contain l and in a null basis l, n, L, N at p they are $l \wedge N \in \overset{+}{S}_p$ and $l \wedge L \in \overset{-}{S}_p$ (such a bivector must be of the form $l \wedge P$ for a null P in $\mathrm{Sp}(L, N)$). It is easily seen that this result is independent of the basis chosen because if l', n', L', N' is another such null basis with $l' = Al$ for $0 \neq A \in \mathbb{R}$ then $L' = al + bn + cL + dN$ with $b = 0 = cd$ and so $l \wedge L'$ is a multiple of $l \wedge N$ or $l \wedge L$, and similarly for N'.

Lemma 5.2 (i) *For* $F, G \in \Lambda_p M$, $[F, \overset{*}{G}] = [\overset{*}{F}, G] = [F, G]^*$ *and* $[\overset{*}{F}, \overset{*}{G}] = [F, G]$.

(ii) *The only 2-dimensional subalgebra of* $\overset{+}{S}_p$ *is, in the notation (5.2) and up to isomorphism, of the form* $\mathrm{Sp}(F_1, F_2)$ *and it is non-Abelian. This subalgebra is uniquely determined by its totally null member* F_2 *and is denoted, somewhat loosely, by* $\overset{+}{B}_p$ *(or similarly by* $\overset{-}{B}_p$*). Hence for any two independent members* $F, G \in \overset{+}{S}_p$, $[F, G] \neq 0$, *and similarly for* $\overset{-}{S}_p$.

(iii) *Suppose* $0 \neq A \in \Lambda_p M \setminus \widetilde{S}_p$. *Then for* $X \in \Lambda_p M$, $[X, A] = 0 \Rightarrow X$ *is a linear combination of* A *and* $\overset{*}{A}$. *If* $A \in \overset{+}{S}_p$ *and* $[X, A] = 0$ *then* $X = \lambda A + Q$ *where* $\lambda \in \mathbb{R}$ *and* $Q \in \overset{-}{S}_p$ *and similarly for* $A \in \overset{-}{S}_p$.

(iv) *If* $F, G, H \in \overset{+}{S}_p$, *or* $F, G, H \in \overset{-}{S}_p$, *the relations* $\|[F, G]\| = |F||G| - (F \cdot G)^2$ *and* $[F, [G, H]] = (F \cdot H)G - (F \cdot G)H$ *hold.*

(v) *If* f *is a Lie algebra isomorphism between any of* $\overset{+}{S}_p$ *and* $\overset{-}{S}_p$ *to any of* $\overset{+}{S}_p$ *and* $\overset{-}{S}_p$ *then for* $F \in \overset{+}{S}_p$ *(or* $\overset{-}{S}_p$*),* $|F| = |f(F)|$ *(cf.[67]).*

Proof The proof of (i) is as in lemma 3.5. One writes $F = P + Q$ for $P \in \overset{+}{S}_p$ and $Q \in \overset{-}{S}_p$ and similarly for G and recalls lemma 5.1 and the fact that $\overset{+}{S}_p$ and $\overset{-}{S}_p$ are subalgebras. For (ii) one notes that if A, B span a *non-Abelian* 2-dimensional subalgebra of $\overset{+}{S}_p$, $[A, B] = aA + bB \neq 0$ $(a, b \in \mathbb{R})$. If $a \neq 0 \neq b$ one may redefine A as $aA + bB$ to get for the subalgebra $[A, B] = aA$. This is, of course trivially the case if exactly one of a and b vanishes. Since $[A, B]$ is orthogonal to A and B with respect to the bivector metric P (lemma 3.4) one finds $|A| = 0$ and $A \cdot B = 0$, that is, A is totally null and B is orthogonal to it, hence $|B| < 0$. This, and other similar results are, as mentioned earlier, consequences of the fact that the induced bivector inner product on $\overset{+}{S}_p$ and $\overset{-}{S}_p$ is 3–dimensional Lorentz with signature $(-, -, +)$. Thus, in the basis (5.2) and using (5.3) and (5.4) one may take $A = F_2$ and B some linear combination of F_1 and F_2 to get $\mathrm{Sp}(A, B) = \mathrm{Sp}(F_1, F_2)$ and, with a judicious choice of the above linear combination, $[F_1, F_2] = F_2$. Given the totally null F_2 the 2-dimensional subspace of $\overset{+}{S}_p$ orthogonal to F_2 and which constitutes the subalgebra is uniquely determined by F_2. For the Abelian case note that one may assume that one of $|A|$ and $|B|$ is ≤ 0 because, if $|A| > 0 < |B|$ one could find a linear combination C of them with $|C| \leq 0$ (because the signature is $(-, -, +)$). Then, say, Sp(A, B) =Sp(A, C) and choosing $C = F_1$ (if $|C| < 0$) or $C = F_2$ (if $|C| = 0$) a contradiction to the assumed Abelian subalgebra condition for any such A is obtained from (5.4). It follows that the non-Abelian subalgebra above is the only possible 2-dimensional subalgebra of $\overset{+}{S}_p$ (and similarly for $\overset{-}{S}_p$). The proof of (iii) is as in lemma 3.5(iii) but requires the extra information that $\overset{+}{S}_p$ and $\overset{-}{S}_p$ have no 2-dimensional *Abelian* subalgebras. The proof of (iv) is similar to the proof in lemma 3.5(vi); one writes out F, G and H in terms of F_1, F_2 and F_3 and computes. For part (v) choose the above basis F_1, F_2, F_3 for $\overset{+}{S}_p$ so that, for, say, an isomorphism $f : \overset{+}{S}_p \to \overset{-}{S}_p$, $f(F_1) \equiv F_1'$, $f(F_2) \equiv F_2'$, $f(F_3) \equiv F_3'$ is a basis for $\overset{-}{S}_p$. That f is a Lie algebra isomorphism shows, from (5.4), that $[F_1', F_2'] = F_2'$, $[F_1', F_3'] = -F_3'$ and $[F_2', F_3'] = -F_1'$. Thus from lemma 3.4(ii), $|F_2'| = |F_3'| = 0$ and $F_1' \cdot F_2' = F_1' \cdot F_3' = 0$, that is, F_2' and F_2 are totally null and independent members of $\overset{-}{S}_p$. Thus from remarks above one may choose a null basis l, n, L, N at p such that $F_2' = \alpha l \wedge L$, $F_3' = \beta n \wedge N$, $F_1' = \gamma(l \wedge n + L \wedge N)$ $(0 \neq \alpha, \beta, \gamma \in \mathbb{R})$ and so $F_2' \cdot F_3' = 2\alpha\beta$ and $|F_1'| = -4\gamma^2$. Then the equations $[F_2', F_3'] = -F_1'$ and $[F_1', F_2'] = F_2'$ give $\alpha\beta = \gamma = \frac{1}{2}$ and so $|F_1'| = 1$ and $F_2' \cdot F_3' = 1$. It now follows that $|f(F)| = |F|$ for each $F \in \overset{+}{S}_p$. $\qquad\square$

5.3 Classification of Symmetric Second Order Tensors

This section will deal with the classification of symmetric tensors in neutral signature. Such a classification scheme was discussed in [49] and commented on in [79] but with few details given. In this section a direct approach will be described in full and based on [73]. One has a 4-dimensional manifold M with a metric g of neutral signature and $S \neq 0$ is a second order, symmetric tensor at $p \in M$ with associated linear map f on T_pM given for $k \in T_pM$ by $k^a \to S^a{}_b k^b$. The term "complex" when applied to eigenvalues and eigenvectors will mean "complex and non-real" and attention is drawn to the convention about writing Segre symbols given in chapter 1. As in the Lorentz case of chapter 4, f always admits an invariant 2−space but now an extra type of 2−space (the totally null type) is possible. Lemma 4.1 is now replaced by the following series of results and which are proved by simple algebra just as for a similar result in chapter 4.

Lemma 5.3 *Let f be as above. Then f admits an invariant 2−space U and its orthogonal complement $U^\perp$ is also invariant (chapter 3). If $U \subset T_pM$ is an* *invariant 2−space for f, then if U is spacelike it contains two orthogonal (real) eigenvectors of f which are either both spacelike or both timelike. If U is null the principal null direction of U is an eigendirection of f. There may or may not be another real eigendirection of f in U. If U is timelike with principal null directions spanned by l and n then either l and n are eigendirections of f with equal eigenvalues, or f admits two independent real, orthogonal, non-null eigendirections with distinct eigenvalues or f admits a conjugate pair of complex eigendirections (which may, after an appropriate choice of l and n, be taken as l ± in) or exactly one of l and n is the only eigendirection of f. If U is totally null then either there are exactly one or exactly two independent null eigenvectors in U, and in the latter case with either equal or distinct eigenvalues, or a complex conjugate pair of complex null eigenvectors arise for f. In the latter case, one may choose real null vectors l and L so that these complex eigenvectors are l ± iL and U = l ∧ L.*

Lemma 5.4   *(i) Suppose f admits a conjugate pair of complex eigenvectors $x \pm iy$ for $x, y \in T_pM$ with respective eigenvalues $a \pm ib$ $(a, b \in \mathbb{R}$ with $b \neq 0)$. The invariant 2−space spanned by x, y is totally null (respectively, timelike) if one (and hence both of) $x \pm iy$ is null (respectively, not null).*

(ii) Any real or complex eigenvector of f corresponding to a non-simple elementary divisor is null. (Hence any real or complex non-null eigenvector is associated with a simple elementary divisor.) If f admits a real or complex null eigenvector either the corresponding eigenvalue is degenerate or the associated elementary divisor is non-simple.

(iii) If f admits a conjugate pair of complex, null eigenvectors $x \pm iy$ with respective eigenvalues $a \pm ib$ with $b \neq 0$ then f admits no real eigenvectors and no other complex eigenvector with eigenvalue different from $a \pm ib$, that is, $a \pm ib$ are the only eigenvalues of f and are degenerate or are associated with a non-simple elementary divisor of order 2.

(iv) If, for $x, y, p, q \in T_pM$, $x \pm iy$ and $p \pm iq$ are conjugate pairs of complex eigenvectors of f with distinct, respective eigenvalues $a \pm ib$ and $c \pm id$ (so that $a + ib \neq c \pm id$ and $bd \neq 0$) their associated invariant 2−spaces $x \wedge y$ and $p \wedge q$ are timelike and orthogonal and intersect only in $\{0\}$.

Proof

(i) since $a + ib \neq a - ib$, $(x+iy) \cdot (x-iy) = 0$ and so $x \cdot x + y \cdot y = 0$. Then $x \pm iy$ is null implies that $x \cdot x - y \cdot y = x \cdot y = 0$ and so $x \cdot x = y \cdot y = x \cdot y = 0$. Thus $x \wedge y$ is totally null and, conversely, if $x \wedge y$ is totally null, clearly $x \pm iy$ are null. If $x \pm iy$ are non-null one still has $x \cdot x + y \cdot y = 0$ and at least one of $x \cdot x - y \cdot y \neq 0$ and $x \cdot y \neq 0$ holds. Whichever is the case $x \wedge y$ is not totally null and, in fact, is timelike because of the existence of the complex eigenvectors $x \pm iy$ (lemma 5.3).

(*ii*) If f admits a (real or complex) eigenvector k associated with a non-simple elementary divisor, k need not be real (as in the Lorentz case—see lemma 4.1) but with this proviso, the proof given in this last lemma shows that k is null. Now suppose f admits a (real or complex) *null* eigenvector k, $f(k) = \alpha k$ ($\alpha \in \mathbb{C}$) such that α is associated with a simple elementary divisor *and* is not degenerate. Suppose u is another (real or complex) eigenvector of f with eigenvalue $\gamma \in \mathbb{C}$. Then, since α is not degenerate, $\alpha \neq \gamma$ and so $k \cdot u = 0$. Thus k is orthogonal to every eigenvector of f (including itself) and it follows that the collection of all such eigenvectors cannot form a basis for $T_p M$. Thus some such eigenvector u, independent of k and with eigenvalue $\alpha \neq \gamma$ (hence $k \cdot u = 0$), must be associated with a *non-simple* elementary divisor and so there exist members $u, r, s, \dots$ of a Jordan basis such that (chapter 1) $f(u) = \gamma u$, $f(r) = \gamma r + u$, $f(s) = \gamma s + r, \dots$ Then by the symmetry of S, $k \cdot f(r) = r \cdot f(k)$, $k \cdot f(s) = s \cdot f(k), \dots$, and it follows that $r \cdot k = s \cdot k = \dots = 0$ and so k is orthogonal to each of the independent vectors $u, r, s, \dots$. It is now clear that k is orthogonal to each member of a full Jordan basis for f. This contradiction completes the proof.

(*iii*) Any real eigenvector k of f has a real eigenvalue and is thus different from $a \pm ib$ since $b \neq 0$. Hence $k \cdot (x \pm iy) = 0$, that is, $k \cdot x = k \cdot y = 0$. This means that $k \in (x \wedge y)^{\perp} = x \wedge y$, since $x \wedge y$ is totally null (from part (*i*) since $x \pm iy$ is null), contradicting the independence of k and $x \pm iy$. If $p \pm iq$ is another conjugate pair of complex eigenvectors with eigenvalues $c \pm id$ ($d \neq 0$) and with $a + ib \neq c \pm id$, $(x \pm iy) \cdot (p \pm iq) = 0$ and so $x \cdot p = x \cdot q = y \cdot p = y \cdot q = 0$. Since $x \wedge y$ is totally null, from part (*i*) this implies that $p, q \in x \wedge y$ and so $p \pm iq$ are easily checked to be complex linear combinations of $x \pm iy$ contradicting independence. The proof now follows.

(*iv*) One has $(x \pm y) \cdot (p \pm iq) = 0$ and so, as before, $x \cdot p = x \cdot q = y \cdot p = y \cdot q = 0$. Thus $x \wedge y$ and $p \wedge q$ are orthogonal complements. With $U = x \wedge y$, if $U = U^{\perp}$, U is totally null and $x \pm y$ are (complex) linear combinations of $p \pm iq$, contradicting the inequality of the eigenvalues. If $U \cap U^{\perp}$ is 1-dimensional (that is, if U is null) a real eigenvector results lying in U and $U^{\perp}$ and a contradiction arises. Thus U and $U^{\perp}$ intersect only in $\{0\}$ and are each necessarily timelike (lemma 5.3). □

It is remarked that several results in this lemma (suitably reworded) can be generalised [73]. Also the notation used in describing Segre symbols is that positive integer entries represent real eigenvalues except for the case $\{22\}$ (where complex eigenvalues arise) and this fact is indicated after the symbol. Again, conjugate pairs of complex eigenvalues arising from simple elementary divisors are denoted by the pair entry $\{z\bar{z}\}$.

Theorem 5.1 *Let M be a 4-dimensional manifold with a smooth metric g of neutral signature and S a non-zero, second order, symmetric tensor at* *p ∈ M with associated linear map f on T_pM. The Jordan/Segre type for f is one of the following: $\{1111\}$, $\{z\bar{z}11\}$, $\{z\bar{z}w\bar{w}\}$, $\{211\}$, $\{2z\bar{z}\}$, $\{22\}$ (with real eigenvalues), $\{22\}$ (with complex eigenvalues), $\{31\}$ or $\{4\}$, together with their possible degeneracies, each of which can occur.*

Proof First suppose that f admits a real, non-null eigenvector k which then has a simple elementary divisor from lemma 5.4(*ii*). The 3−dimensional subspace $U \equiv (\mathrm{Sp}(k))^\perp$ orthogonal to k is then invariant and has an induced Lorentz metric from g. Lemma 4.2 then applies to the action of f on U and together with the eigenvector k gives possible Segre types $\{1111\}$, $\{z\bar{z}11\}$, $\{211\}$ and $\{31\}$, or their degeneracies, for f.

Next suppose that the only eigenvectors of f are either complex, or real and null. If a complex null eigenvector exists for f lemma 5.4(*ii*) shows that its eigenvalue is either degenerate or is associated with a non-simple elementary divisor. Thus the possible Segre types are $\{22\}$ (complex eigenvalues) or $\{(z z)(\bar{z}\bar{z})\}$. If a complex non-null eigenvector exists for f its real and imaginary parts give rise to a timelike invariant 2−space U for f (lemma 5.4(*i*)) whose orthogonal complement $W^\perp$ is also invariant and timelike and with $U \cap U^\perp = \{0\}$. The invariant 2−space $U^\perp$ may admit either a single independent real null eigenvector, or two independent real eigenvectors, or a complex conjugate pair of complex eigenvectors. However, the second of these possibilities leads, from lemma 5.3, either directly to a real non-null eigenvector for f or to an independent pair of null eigenvectors for f with equal eigenvalues and hence to non-null eigenvectors in the consequent eigenspace, and a contradiction. So in this case either $U^\perp$ contains a real null eigenvector associated with a non-simple elementary divisor from lemma 5.4(*ii*) (and necessarily of order two) or $U^\perp$ gives rise to a conjugate pair of complex eigenvectors and so the possible Segre types for f are $\{2z\bar{z}\}$ and $\{z\bar{z}w\bar{w}\}$ or the degeneracy of the last type.

Finally suppose all the eigenvectors of f are real and null. If there are at least three independent such vectors, say p, q, r then they cannot be mutually orthogonal because if $W \equiv \mathrm{Sp}(p, q, r)$ then $\dim W = 3$ (and hence $\dim W^\perp = 1$). But p, q, r are independent members of $W^\perp$ and a contradiction follows. Then if, say, $p \cdot q \neq 0$ and $f(p) = \alpha p$, $f(q) = \beta q$, $(\alpha, \beta \in \mathbb{R})$, the condition $p \cdot f(q) = q \cdot f(p)$ shows that $\alpha = \beta$ and hence the contradiction that there exist real, non-null eigenvectors in $p \wedge q$. So either there is only one independent real null eigenvector (and hence the Segre type is either $\{4\}$) or there are exactly two such vectors. In the latter case the only possibilities are $\{22\}$ with real eigenvalues, and $\{31\}$. In the last of these cases the two real null eigenvectors must have equal eigenvalues by lemma 5.4(*ii*), (since the second null eigenvector corresponds to a simple elementary divisor) and must be orthogonal otherwise extra independent non-null eigenvectors would arise in the (timelike) eigenspace which they span. Thus this eigenspace is totally null. One is led to the degenerate type $\{(31)\}$ with eigenvalue α but, in fact, this type cannot exist. To see this let p, q, r, s be a Jordan basis for f so that $f(p) = \alpha p$, $f(q) = \alpha q + p$, $f(r) = \alpha r + q$ and $f(s) = \alpha s$ with p, s null and

$p \cdot s = 0$. The symmetry of S gives $s \cdot f(r) = r \cdot f(s)$ and $p \cdot f(r) = r \cdot f(p)$ and so $q \cdot s = 0$ and $q \cdot p = 0$ which reveal that q is in the totally null blade $p \wedge s$ contradicting the fact that p, q, r, s is a (Jordan) basis. It follows that the only possibility in this case is $\{22\}$ (or its degeneracy) with real eigenvalues. This completes the proof. □

One can now find canonical forms for each (non-trivial) type in terms of an orthonormal basis x, y, s, t or a null basis l, n, L, N at p or a hybrid basis, for example, y, s, l, n. Each Segre symbol given is taken to include each of its degeneracies and all Greek letters refer to real numbers.

For type $\{1111\}$ one can choose an orthonormal basis x, y, s, t each member of which is an eigenvector of f with corresponding eigenvalues $\alpha, \beta, \gamma, \delta$ and a canonical form

$$S_{ab} = \alpha x_a x_b + \beta y_a y_b - \gamma s_a s_b - \delta t_a t_b. \tag{5.6}$$

For type $\{z\bar{z}11\}$ the complex eigenvalues are associated with a timelike $2-$space, say $l \wedge n$ (lemma 5.4(i)) and can be chosen as $l \pm in$ and eigenvalues $\gamma \pm i\delta$ ($\delta \neq 0$). The real eigenvectors lie in the orthogonal complement $y \wedge s$ with eigenvalues α and β. Thus one achieves

$$S_{ab} = \alpha y_a y_b - \beta s_a s_b + \gamma(l_a n_b + n_a l_b) + \delta(l_a l_b - n_a n_b). \tag{5.7}$$

For type $\{z\bar{z}w\bar{w}\}$ with distinct conjugate pairs of complex eigenvalues one chooses the orthogonal, invariant, timelike $2-$spaces as $l \wedge n$ and $L \wedge N$ and eigenvectors $l \pm in$ and $L \pm iN$ with associated eigenvalues $\alpha \pm i\beta$ and $\gamma \pm i\delta$ with $\beta\delta \neq 0$. In the basis l, n, L, N one gets

$$S_{ab} = \alpha(l_a n_b + n_a l_b) + \beta(l_a l_b - n_a n_b) + \gamma(L_a N_b + N_a L_b) + \delta(L_a L_b - N_a N_b). \tag{5.8}$$

In the event that the complex eigenvalues are equal, say, $\alpha = \gamma$ and $\beta = \delta$ the $\alpha + i\beta$ complex eigenspace spanned by $l + in$ and $L + iN$ contains exactly two complex null eigendirections spanned by $l \mp N + i(n \pm L)$ and this suggests changing the null basis from l, n, L, N to l', n', L', N' according to $\frac{1}{\sqrt{2}}n = l' - N'$, $\frac{1}{\sqrt{2}}l = n' - L'$, $\frac{1}{\sqrt{2}}L = l' + N'$ and $\frac{1}{\sqrt{2}}N = n' + L'$, so that $l' = \frac{1}{2\sqrt{2}}(n + L)$, $n' = \frac{1}{2\sqrt{2}}(l + N)$, $L' = \frac{1}{2\sqrt{2}}(N - l)$ and $N' = \frac{1}{2\sqrt{2}}(L - n)$ and then l', n', L', N' is a null basis and using the appropriate completeness relation the expression (5.8) becomes, after dropping primes,

$$S_{ab} = \alpha g_{ab} + \beta(l_a N_b + N_a l_b - n_a L_b - L_a n_b). \tag{5.9}$$

In this new basis the $\alpha + i\beta$ eigenspace is spanned by $l + iL$ and $n - iN$ and the $\alpha - i\beta$ eigenspace is spanned by $l - iL$ and $n + iN$. The Segre type is $\{(zz)(\bar{z}\bar{z})\}$.

For the Segre type $\{211\}$ there is a real null eigenvector l with eigenvalue α corresponding to the non-simple elementary divisor of order 2 and two other eigenvectors with simple elementary divisors which are orthogonal to l (see the first part of the proof of theorem 5.1). They may be chosen as y and s

with eigenvalues β and γ. One may then choose a (hybrid) tetrad l, n, y, s to get

$$S_{ab} = \alpha(l_a n_b + n_a l_b) + \lambda l_a l_b + \beta y_a y_b - \gamma s_a s_b \qquad (5.10)$$

where, because of the Segre type, $0 \neq \lambda \in \mathbb{R}$ and a scaling of l and a compensatory inverse scaling of n, which cancel out, may be used to set $\lambda = \pm 1$. A Jordan basis, up to a scaling, is l, n, y, s. It is noted here that, unlike the Lorentz case, spacelike and timelike eigenvectors are admitted and that further null eigenvectors L and N are admitted if $\beta = \gamma$.

For Segre type $\{2z\bar{z}\}$, again l is a real null eigenvector with associated non-simple elementary divisor of order 2 and eigenvalue α and a conjugate pair of complex eigenvectors $L \pm iN$ with eigenvalues $\gamma + i\delta$ ($\delta \neq 0$). One then gets

$$S_{ab} = \alpha(l_a n_b + n_a l_b) + \lambda l_a l_b + \gamma(L_a N_b + N_a L_b) + \delta(L_a L_b - N_a N_b) \quad (5.11)$$

and, as before, one may scale l so that $\lambda = \pm 1$.

For Segre type $\{22\}$ with complex eigenvalues one has two independent complex conjugate *null* eigenvectors associated with a totally null invariant $2-$space and which may be chosen as $l \wedge L$ with elementary divisors of order 2 (see lemma 5.4(i) and (ii)) and after a (possible) adjustment of the choice of l and L the complex eigenvectors may be taken as $l \pm iL$ with respective eigenvalues $\alpha \pm i\beta$ ($\beta \neq 0$). One can obtain a canonical form for the tensor components S_{ab} at p by first writing them out as a linear combination of the ten independent symmetrised products of a null basis l, n, L, N at p given by $l_a l_b, n_a n_b, ... l_{(a} n_{b)}, ... L_{a(}N_{b)}$ with l and L as given earlier, using the eigenvector conditions $S_{ab} l^b = \alpha l_a - \beta L_a$ and $S_{ab} L^b = \beta l_a + \alpha L_a$ and the completeness relation for such a basis given in section 5.1. (Of course this could have been used in the previous cases.) One finds

$$S_{ab} = \alpha g_{ab} + \omega(l_a L_b + L_a l_b) + \mu l_a l_b + \nu L_a L_b + \beta(l_a N_b + N_a l_b - n_a L_b - L_a n_b) \tag{5.12}$$

for $\omega, \mu, \nu \in \mathbb{R}$. If $\mu \neq \nu$ in (5.12) a change of null basis given by $l' = l$, $L' = L$, $n' = n + xL$ and $N' = N - xl$ ($x \in \mathbb{R}$) preserves the form (5.12) in the new basis but with the coefficients μ and ν satisfying $\mu = -\nu$ and with $\mu^2 + \nu^2 \neq 0$. Then (5.12) gives $S_{ab}(n^b + iN^b) = (\alpha - i\beta)(n_a + iN_a) + (\mu + i\omega)(l_a - iL_a)$ with $\mu + i\omega \neq 0$ and so in this new basis $l \pm iL$ and $n \pm iN$ constitute, up to scaling, a Jordan basis for f and the Segre type is $\{22\}$ with complex eigenvalues. If $\mu = \nu$ the above basis change can be used to set $\mu = \nu = 0$ in (5.12) and again this Segre type emerges provided $\omega \neq 0$ but if $\omega = 0$, (5.12) reduces to (5.9) and the Segre type is $\{(zz)(\bar{z}\bar{z})\}$. Thus (5.12) is of Segre type $\{22\}$ with complex eigenvalues if $\mu \neq \nu$ or if $\mu = \nu$ and $\omega \neq 0$ and of Segre type $\{(zz)(\bar{z}\bar{z})\}$ if $\mu = \nu$ and $\omega = 0$.

For Segre type $\{22\}$ with real eigenvalues one has orthogonal null eigenvectors l and L (lemma 5.4(ii)), say, with eigenvalues α and β. The orthogonality follows irrespective of the values of α and β since if $l \cdot L \neq 0$ the $2-$space $l \wedge L$ is timelike and its (timelike) orthogonal complement would produce extra (forbidden) eigenvectors. Thus $l \wedge L$ is totally null. If one extends this to a null tetrad l, n, L, N one achieves from the eigenvector conditions

$$S_{ab} = \alpha(l_a n_b + n_a l_b) + \beta(L_a N_b + N_a L_b) + \omega(l_a L_b + L_a l_b) + \mu l_a l_b + \nu L_a L_b. \quad (5.13)$$

If $\alpha \neq \beta$ a basis change given by $l' = l$, $L' = L$, $n' = n + xL$ and $N' = N - xl$, for $x \in \mathbb{R}$, allows one to set $\omega = 0$. In this case one must insist that $\mu \neq 0 \neq \nu$ to avoid n and/or N being eigenvectors and then (5.13) shows that l, n, L, N is, up to scaling, a Jordan basis for f and that the Segre type is $\{22\}$ with real eigenvalues. If $\alpha = \beta$ an application of the completeness relation in (5.13) shows that

$$S_{ab} = \alpha g_{ab} + \omega(l_a L_b + L_a l_b) + \mu l_a l_b + \nu L_a L_b. \quad (5.14)$$

The Jordan matrix for this Segre type forces $S - \alpha g$ to have rank 2 and so $\mu\nu \neq \omega^2$. [It is noticed that if $\mu\nu = \omega^2$ then if $\mu > 0 < \nu$, $S - \alpha g$ takes the form $S_{ab} - \alpha g_{ab} = k_a k_b$ where k is the null (co)vector $\sqrt{\mu}l \pm \sqrt{\nu}L$ and if $\mu < 0 > \nu$, the form $-k_a k_b$ where $k = \sqrt{-\mu}l \mp \sqrt{-\nu}L$. In these cases S has Segre type $\{(211)\}$ and satisfies (5.10) with $\alpha = \beta = \gamma$.] Using the freedom of choice of l and L in the α-eigenspace one may change $l \to l + aL$ and $L \to L + bl$ for $a, b \in \mathbb{R}$ provided $(ab - 1) \neq 0$ and see that, in the new basis, the coefficients μ' of $l_a l_b$, ν' of $L_a L_b$ and ω' of $(l_a L_b + L_a l_b)$ satisfy

$$\mu' = \mu + 2\omega b + \nu b^2, \qquad \nu' = \nu + 2\omega a + \mu a^2, \qquad \omega' = \omega + \mu a + \nu b + ab\omega. \quad (5.15)$$

Now suppose $\omega^2 > \mu\nu$ in the original basis. Then $\omega'^2 > \mu'\nu'$ holds in the new basis since such basis changes preserve the Segre type of S and hence the sign of $\omega^2 - \mu\nu$, or by direct calculation from (5.15). The idea now is to show that one can choose the new basis so that $\mu' = \nu' = 0$. If $\mu = \nu = 0$ then it is done by choosing $a = b = 0$. If $\mu = 0 \neq \nu$ then $\omega \neq 0$ and one may choose a so that $\nu + 2\omega a = 0$ to get $\nu' = 0$ and $b = 0$ to keep $\mu' = 0$ (and $ab \neq 1$). One proceeds similarly if $\nu = 0 \neq \mu$. If $\mu \neq 0 \neq \nu$, the condition $\omega^2 > \mu\nu$ allows the above quadratic expressions for μ' and ν' to be set to zero and solved for a and b, the two solutions obtained for each allowing one to ensure that $ab \neq 1$. Thus $\mu' = \nu' = 0$ (and $\omega \neq 0$). If $\omega^2 < \mu\nu$ then $\mu\nu > 0$ (so that $\mu \neq 0 \neq \nu$) and one can show that ω may be chosen to be zero. To see this note that if $\omega = 0$ then all is well and if $\omega \neq 0$ one chooses $b = 0$ and a such that $\omega' = 0$ in the 3rd expression of (5.15). Thus if $\omega^2 > \mu\nu$ one may achieve (5.14) with $\mu = \nu = 0 \neq \omega$ and l, n, L, N is a Jordan basis for f. If $\omega^2 < \mu\nu$ one gets (5.14) with $\mu \neq 0 \neq \nu$ and $\omega = 0$ and the same Jordan basis. In all cases the Segre type is $\{22\}$ (or $\{(22)\}$) with real eigenvalues.

If the Segre type is $\{31\}$ one has a real null eigenvector l corresponding to the non-simple elementary divisor of order 3 and with eigenvalue α. From the proof of theorem 5.1 the other eigenvector k is non-null and orthogonal to l showing that $l \wedge k$ is a null 2-space. Choose a hybrid basis l, n, y, s at p with $k = s$ and let the eigenvalue associated with s be β. This information gives, at p

$$S_{ab} = \mu l_a l_b + \alpha(l_a n_b + n_a l_b) + A y_a y_b + B(l_a y_b + y_a l_b) - \beta s_a s_b \qquad (5.16)$$

for $A, B, \mu \in \mathbb{R}$. If $B = 0$, y becomes an extra eigenvector so one must take $B \neq 0$. If $A \neq \alpha$, $l + \frac{A - \alpha}{B} y$ becomes an extra eigenvector so one must take $A = \alpha$. Scalings of l and n can be used to make $B = 1$ and a basis change $l' = l$, $s' = s$, $y' = y + al$, $n' = n - ay - \frac{a^2}{2} l$ can be used to set the coefficient of $l_a l_b$ to zero. Use of the completeness relation then gives

$$S_{ab} = \alpha g_{ab} + (\alpha - \beta) s_a s_b + (l_a y_b + y_a l_b). \qquad (5.17)$$

No further eigenvectors can occur since now l, y, n, s give, up to a scaling, a Jordan basis. One could have chosen the eigenvector $k = y$ which has the effect of switching y and s in (5.17) and some straightforward changing of signs.

Finally suppose the Segre type is $\{4\}$ with the single, necessarily null, independent eigenvector l and real eigenvalue α. One may then choose a Jordan basis containing $l, k \in T_p M$ such that the associated map f satisfies $f(l) = \alpha l$ and $f(k) = \alpha k + l$. Then $l \wedge k$ is an invariant 2−space containing only one eigendirection and is hence timelike, null or totally null. If $l \wedge k$ is null one gets $(l \wedge k)^\perp = l \wedge k'$ for $k' \in T_p M$ with l, k, k' independent and the 2−space $l \wedge k'$ invariant and containing only one independent eigendirection. From this last fact it follows that $f(k') = \alpha k' + al$ for $0 \neq a \in \mathbb{R}$, the appearance of the eigenvalue α here being necessary to avoid extra independent eigenvectors in $l \wedge k'$. One may then scale k' so that $f(k') = \alpha k' + l$ and which gives the contradiction that $k - k'$ is an eigenvector of f independent of l. If $l \wedge k$ is timelike its orthogonal complement leads to extra independent eigenvectors and another contradiction. Thus $l \wedge k$ is totally null and one can choose a null basis l, n, L, N with $k = N$. Then $f(l) = \alpha l$, $f(N) = \alpha N + l$ and $f(L) = aL + bN + cl + dn$ for $a, b, c, d \in \mathbb{R}$. Using the fact that S is symmetric, one has $L \cdot f(l) = l \cdot f(L)$ one gets $0 = d$ and so $f(L) = aL + bN + cl$. Similarly the condition $L \cdot f(N) = N \cdot f(L)$ gives $\alpha = a$. Thus $f(L) = \alpha L + bN + cl$. If $b = 0$ then either $c = 0$ (which gives an extra, forbidden, independent eigenvector L) or $c \neq 0$ in which case $N - c^{-1} L$ is an extra independent eigenvector. Thus $b \neq 0$. In this null basis one then has

$$S_{ab} = \alpha g_{ab} + \mu l_a l_b + b N_a N_b + c(l_a N_b + N_a l_b) + (l_a L_b + L_a l_b). \qquad (5.18)$$

One can now perform the basis change $l \to l$, $L \to L$, $n \to n - \lambda L$ and $N \to N + \lambda l$, $(\lambda \in \mathbb{R})$, and choose the real number λ to set $c = 0$, followed by the basis change $l \to l$, $N \to N$, $L \to L + \lambda' l$ and $n \to n - \lambda' N$, $(\lambda' \in \mathbb{R})$, to set $\mu = 0$. Finally a scaling of N together with compensatory scalings in L and l may be used to set $b = \pm 1$. The final result is

$$S_{ab} = \alpha g_{ab} + b N_a N_b + l_a L_b + L_a l_b \qquad (5.19)$$

with $b = \pm 1$ and with l, N, L, n constituting, up to a scaling, a Jordan basis.

This completes the argument for obtaining the canonical forms for S (or f). It is clear, by simply equating eigenvalues, where allowable, that each degeneracy of each Segre type is possible and that they are more complicated than those in the case of Lorentz signature. □

It is noted that for this signature, unlike the other two, a real eigenvector need not exist (types $\{z\bar{z}w\bar{w}\}$ and $\{22\}$ with complex eigenvalues). It is also observed that $\{31\}$ admits an invariant null 2−space containing only one eigendirection and one containing two independent eigendirections. Similarly, one of the type $\{(11)11\}$ cases admits a timelike invariant 2−space containing two independent null eigenvectors and one admitting an independent pair of spacelike/timelike eigenvectors whilst type $\{211\}$ admits a timelike invariant 2−space admitting a single (real, null) eigendirection and type $\{2z\bar{z}\}$ admits a timelike invariant 2−space containing a conjugate pair of complex eigenvectors (cf. the Lorentz case—and lemma 5.3).

5.4 Classification of Bivectors

In this section a complete classification scheme for second order, skew-symmetric tensors (bivectors) for neutral signature will be described. Thus for a non-zero bivector F one considers the associated linear map f on T_pM given, for $k \in T_pM$, by $k^a \to F^a{}_b k^b$. Thus $f(k) \cdot k = 0$ for each $k \in T_pM$. As a reminder (see chapter 3) it is recalled that if k is a *non-null* (real or complex) eigenvector of f the corresponding eigenvalue is zero and that if k and k' are eigenvectors of f with eigenvalues α and β then either $\alpha = -\beta$ or $k \cdot k' = 0$. Also if U is an invariant subspace for f so is its orthogonal complement $U^\perp$ and a 2-dimensional invariant 2−space for f always exists, as in the previous section. If f admits a spacelike invariant 2−space U spanned by orthogonal spacelike vectors x and y, then either f acts trivially on U or $f(x) = ax + by$, $f(y) = cx + dy$ $(a, b, c, d \in \mathbb{R})$ and the skew-symmetry of F shows that $a = d = 0$ and $b = -c$. Thus $f(x) = by$ and $f(y) = -bx$ $(0 \neq b \in \mathbb{R})$ and so $x \pm iy$ are complex eigenvectors for f with eigenvalues $\mp ib$. If U is null with pnd l then l is a real eigenvector for f and U may or may not contain another eigenvector. If U is timelike with pnds spanned by null vectors l and n with $l \cdot n = 1$ then $f(l) = al + bn$ and $f(n) = cl + dn$ and, as above, $b = c = 0$ and $a = -d \neq 0$. Thus l and n are real eigenvectors for f whose eigenvalues differ only in sign. If U is totally null there are either one or two independent real eigenvectors for f in U or a conjugate pair of complex eigenvectors for f whose real and imaginary parts span U. All these possibilities will be seen to occur. It is also remarked for future purposes that if $p, q, r, s \in T_pM$ are such that $p \wedge q = r \wedge s$ then $p \pm iq$ are each complex linear combinations of $r + is$ and $r - is$.

If F is *simple* it may be written in terms of some basis, as shown earlier, (up to a real scaling) as $F = x \wedge y$ (F spacelike), $F = l \wedge n$ (F timelike),

$F = l \wedge y$ (F null) and $F = l \wedge N$ (F totally null). In the spacelike case a conjugate pair of eigenvectors arise as a consequence of the 2−space $x \wedge y$ whilst in the timelike case one has $f(l) = l$ and $f(n) = -n$. In the null case, and in the above basis, one gets $f(l) = 0$, $f(y) = l$, $f(n) = -y$ and $f(s) = 0$ whilst in the totally null case one has $f(l) = 0$, $f(L) = l$, $f(N) = 0$ and $f(n) = -N$. Thus, spotting the obvious Jordan bases in each case, the Segre types are, respectively, $\{z\bar{z}(11)\}$, $\{11(11)\}$, $\{(31)\}$ and $\{(22)\}$ with eigenvalues zero in the last two cases. The null case reveals that a null invariant 2−space may contain exactly one or exactly two independent eigenvectors. The totally null case reveals a totally null invariant 2−space with two independent real eigenvectors.

If F is a *non-simple* bivector *each of its eigenvalues is non-zero* (by non-degeneracy) and hence *each of its eigenvectors is null*. Suppose F is non-simple with complex null eigenvectors $p \pm iq$ and associated (non-zero) eigenvalues $\alpha \pm i\beta$ for $p, q \in T_pM$ and $\alpha, \beta \in \mathbb{R}$ with $\beta \neq 0$. Then since $p \pm iq$ are null, $|p| = |q|$ and $p \cdot q = 0$ and, in addition, from a remark at the start of this section, either $(p+iq) \cdot (p-iq) = 0$ or $\alpha = 0$. Thus either $|p| + |q| = 0$ or $\alpha = 0$ which combines with the previous restriction to give either $|p| = |q| = 0 = p \cdot q$ or $|p| = |q|$, $p \cdot q = 0$ and $\alpha = 0$. It follows that $p \wedge q$ is either spacelike or totally null and if spacelike, $\alpha = 0$, and so the eigenvalues are $\pm i\beta$. Then if F also admits a real (necessarily null) eigenvector k with eigenvalue $0 \neq \gamma \in \mathbb{R}$, $\gamma \neq \alpha \pm i\beta$ and so $k \cdot p = k \cdot q = 0$. If $p \wedge q$ is spacelike so also is $(p \wedge q)^{\perp}$ and $k \in (p \wedge q)^{\perp}$ which contradicts the fact that k is null. If $p \wedge q$ is totally null then $k \in p \wedge q$ and is hence a (complex) linear combination of $p \pm iq$, contradicting the independence of k and $p \pm iq$. *Thus the eigenvalues of a non-simple bivector are either all real or all complex.* [This last result fails for simple bivectors as was seen earlier.] One can say more here. If F is non-simple with all eigenvalues complex then with the exception of the degenerate case of Segre type $\{(zz)(\bar{z}\bar{z})\}$ the invariant spaces arising from its (complex) eigenvectors are *either all spacelike or all totally null*. To see this let $p \pm iq$ and $r \pm is$ be independent complex eigenvectors with associated invariant 2−spaces $p \wedge q$ (totally null) and $r \wedge s$ (spacelike) and respective eigenvalues $\alpha \pm i\beta$ and $\pm i\gamma$ ($p, q, r, s \in T_pM$ and $\alpha, \beta, \gamma \in \mathbb{R}$) and, from the excluded degenerate case, $\pm i\gamma \neq \alpha \pm i\beta$. Then $(p+iq) \cdot (r \pm is) = 0$ and so $p \cdot r = p \cdot s = q \cdot r = q \cdot s = 0$ which gives the contradiction $r, s \in p \wedge q$. That the degenerate case must be excluded here will be seen later.

So suppose F is non-simple and suppose the map f associated with F has all eigenvalues complex with $p \pm iq$ being an eigenvector with eigenvalues $a \pm ib$ for $p, q \in T_pM$ and $a, b \in \mathbb{R}$, $b \neq 0$. The subspace $U = p \wedge q$ associated with $p \pm iq$ is then invariant, as is $U^{\perp}$. Now $p \pm iq$ must be null and U (as shown above) must be spacelike or totally null. Suppose U is spacelike, so that $U^{\perp}$ is also spacelike. Choosing a tetrad with $p = x$ and $q = y$ one has, from remarks above, $a = 0$, $f(x) = -by$ and $f(y) = bx$. Similarly, applying this method to $U^{\perp} = s \wedge t$ for an orthonormal basis x, y, s, t gives $f(t) = -ds$ and $f(s) = dt$ ($0 \neq d \in \mathbb{R}$) revealing that $s \pm it$ are (complex) eigenvectors with eigenvalues

$\mp id$. The Segre type is $\{z\bar{z}w\bar{w}\}$ with F given by

$$F_{ab} = b(x_a y_b - y_a x_b) + d(s_a t_b - t_a s_b)[= b(x \wedge y) + d(s \wedge t)]. \qquad (5.20)$$

Thus if $b \neq \pm d$ each (complex) eigenvector gives rise to a spacelike invariant 2–space and it is seen that these are the only invariant 2–spaces for f.

There is a degenerate case when $b = \pm d$. In the case $b = d$ (5.20) gives

$$F_{ab} = b((x_a y_b - y_a x_b) + (s_a t_b - t_a s_b))[= b((x \wedge y) + (s \wedge t))] \qquad (5.21)$$

and the Segre type $\{(zz)(\bar{z}\bar{z})\}$ arises. Similar comments apply if $b = -d$ and it is noted that these degenerate cases lie in $\overset{+}{S}_p$ and $\overset{-}{S}_p$, respectively. In this case the ib–eigenspace is spanned by $x + iy$ and $s - it$ and contains the eigenvector $(x+iy)+(s-it)$. This eigenvector gives rise to a totally null invariant 2–space $(x+s) \wedge (y-t)$. These are in addition to the spacelike ones arising from $x + iy$ and $s - it$ and justify an earlier remark.

Now suppose that U is totally null with associated complex eigenvectors $l \pm iL$ with complex eigenvalues $a \pm ib$ and $b \neq 0$. Then $f(l) = al - bL$, $f(L) = bl + aL$ and so, extending to a null basis l, n, L, N for T_pM the above expressions for $f(l)$ and $f(L)$ lead (from the skew-symmetry of F using relations like $n \cdot f(n) = \dots = 0$ and $n \cdot f(L) = -L \cdot f(n)\dots$) to $f(N) = \rho l + bn - aN$ and $f(n) = -\rho L - an - bN$ for $\rho \in \mathbb{R}$. Then in the corresponding bivector basis $l \wedge n, l \wedge L, l \wedge N, n \wedge L, n \wedge N$ and $L \wedge N$ at p one has

$$F = \alpha(l \wedge n) + \beta(l \wedge L) + \gamma(l \wedge N) + \delta(n \wedge L) + \mu(n \wedge N) + \nu(L \wedge N) \quad (5.22)$$

for $\alpha, \dots, \delta \in \mathbb{R}$ and thus $\alpha = \nu = a$, $\beta = \rho$, $\gamma = \delta = b$ and $\mu = 0$. This gives

$$F = a(l \wedge n + L \wedge N) + b(l \wedge N + n \wedge L) + \rho(l \wedge L). \qquad (5.23)$$

If $a \neq 0$ a basis change $l' = l$, $L' = L$, $n' = n + \kappa L$ and $N' = N - \kappa l$ with $\kappa = -\frac{\rho}{2a}$ will (dropping primes) remove the final term in $l \wedge L$ in (5.23) (thus correcting a typo in [73]). In this case $(n \pm iN)$ are easily checked to be eigenvectors of F with eigenvalues $-a \pm ib$ (and with associated invariant 2–spaces totally null) and F is diagonalisable over $\mathbb{C}$ with Segre type $\{z\bar{z}w\bar{w}\}$ with no degeneracies possible since $a \neq 0$. [This case gives an example of a totally null invariant 2–space with a conjugate pair of complex eigenvectors.] In the case $a = \rho = 0$ in (5.23) one gets the degenerate case (5.21) since, as is easily checked from the relationships between the basis members, $l \wedge N + n \wedge L = x \wedge y + s \wedge t$, and the degenerate Segre type $\{(zz)(\bar{z}\bar{z})\}$ arises.

Next suppose $a = 0 \neq \rho$ in (5.23) so that $l \pm iL$ are eigenvectors with eigenvalues $\pm ib$. Recalling the above remarks about this degenerate case and noting that the bivector in (5.23) is now *not* in $\overset{+}{S}_p$ or $\overset{-}{S}_p$, it is clear that any other independent eigenvectors for F are (complex and) null and it can be checked that the associated eigenvalues are $\pm ib$ but then a contradiction is achieved so that the resulting Segre type is $\{22\}$ with complex eigenvalues.

However, a more direct approach is to show, from (5.23) that $f(N) = bn + \rho l$ and $f(n) = -bN - \rho L$ and hence that $f(N \pm in) = \rho(l \mp iL) \mp ib(N \pm in)$ and so $l \pm iL$ and $N \pm in$ give, up to a scaling, a Jordan basis for f, revealing it to have Segre type $\{22\}$ with complex eigenvalues (and no degeneracy).

Finally suppose that all eigenvectors of f are real (and necessarily null). Then there exists a null $l \in T_pM$ such that $f(l) = \lambda l$ for $0 \neq \lambda \in \mathbb{R}$. Choosing a null basis containing l one has $\alpha = \lambda$ and $\mu = \delta = 0$ in (5.22) and then, since F is non-simple, $\nu \neq 0$ otherwise one would have the simple condition $F = l \wedge p$ for some $p \in T_pM$. Suppose that $\beta \neq 0 \neq \gamma$ and define

$$\kappa = \frac{\nu - \lambda}{\gamma}, \qquad \kappa' = \frac{-\nu - \lambda}{\beta}. \tag{5.24}$$

It follows that $\kappa = \kappa' = 0$ is impossible since it is equivalent to $\nu = \lambda = 0$. So suppose that $\kappa \neq 0 \neq \kappa'$ and which gives $\lambda \neq \pm\nu$. Then $L' \equiv l + \kappa L$ and $N' \equiv l + \kappa'N$ are also independent real null eigenvectors for f with respective eigenvalues ν and $-\nu$ and $L' \cdot N' = \kappa\kappa'$. The condition $F^c{}_c = 0$ shows that another real (null) eigenvector, say n' exists with eigenvalue $-\lambda \neq 0$ and so f is diagonalisable over $\mathbb{R}$. Then l, L', N', n' is a new basis consisting of null eigenvectors and since $\lambda \neq \pm\nu$ and $\lambda \neq 0$, $n' \cdot L' = n' \cdot N' = 0$ (and so $l \cdot n' \neq 0$). Thus f has Segre type $\{1111\}$ with no degeneracies and, after scalings to make $l, L,' N,' n'$ a null basis and then, dropping primes, one has

$$F_{ab} = \lambda(l_a n_b - n_a l_b) + \nu(L_a N_b - N_a L_b) \tag{5.25}$$

which is (5.22) with $\beta = \gamma = \delta = \mu = 0$. Retaining the conditions $\beta \neq 0 \neq \gamma$ suppose that $\kappa = 0 \neq \kappa'$, which is equivalent to $\lambda = \nu$. In (5.22) one has $\alpha = \lambda$, $\mu = \delta = 0$ and a basis change $l'' = l$, $L'' = L$, $n'' = n - \frac{\beta}{2\lambda}L$ and $N'' = N + \frac{\beta}{2\lambda}l$ can be used to set $\beta = 0$ in (5.22). Dropping primes one gets

$$F_{ab} = \lambda(l_a n_b - n_a l_b + L_a N_b - N_a L_b) + \gamma(l_a N_b - N_a l_b) \tag{5.26}$$

and hence $f(l) = \lambda l$, $f(N) = -\lambda N$, $f(L) = \lambda L + \gamma l$ and $f(n) = -\lambda n - \gamma N$. Thus, since $\gamma \neq 0$, l, L, N, n form a Jordan basis and the Segre type is $\{22\}$ with real eigenvalues and no degeneracy. [In this case one has an example of totally null invariant 2−spaces with either exactly one or exactly two independent real eigenvectors.] Similar comments apply if $\kappa' = 0 \neq \kappa$. Thus if exactly one of κ and κ' is zero, one of β and γ may be set to zero. If $\beta = \gamma = 0$ one gets (5.25) and Segre type $\{1111\}$ and if, additionally, $\lambda = \pm\nu$, the Segre type is $\{(11)(11)\}$. Thus, in summary, one has the following theorem.

Theorem 5.2 *Let M be a 4-dimensional manifold with smooth metric g of neutral signature and let F be a non-zero bivector at $p \in M$. If F is simple its Segre type is $\{z\bar{z}(11)\}$ (blade of F spacelike), $\{11(11)\}$ (timelike), $\{(31)\}$ (null) and $\{(22)\}$ (totally null) with eigenvalues zero in the last two cases. If F is non-simple its eigenvalues are either all real or all complex and the possible Segre types are $\{z\bar{z}w\bar{w}\}$ ((5.20) with $0 \neq b \neq \pm d \neq 0$, or (5.23) with*

$a \neq 0 = \rho$), $\{(zz)(\bar{z}\bar{z})\}$ ((5.21), or (5.23) with $a = \rho = 0 \neq b$), $\{22\}$ with complex eigenvalues ((5.23) with $a = 0 \neq \rho$ and $b \neq 0$), $\{1111\}$ ((5.25) with $\lambda \neq \pm\nu$), $\{(11)(11)\}$ ((5.25) with $\lambda = \pm\nu$) and $\{22\}$ with real eigenvalues ((5.26) with $\lambda \neq 0 \neq \gamma$). No further degeneracies are possible.

As consequences one can see that if F is non-simple then $F \in \widetilde{S_p}$ if and only if its Segre type has an eigenvalue degeneracy. Also, a consideration of the above canonical forms shows (again) that if $F \in \overset{+}{S_p}$, and $|F| < 0$ there exists a basis l, n, L, N at p such that F is a multiple of $l \wedge n - L \wedge N$ with all eigenvalues real, whilst if $|F| > 0$ there exists a basis x, y, s, t at p such that F is a multiple of $x \wedge y + s \wedge t$ with all eigenvalues complex and if $|F| = 0$, F is totally null and there exists a basis l, n, L, N at p such that F is a multiple of $l \wedge N$ (with all eigenvalues zero) with similar obvious comments if $F \in \overset{-}{S_p}$ (cf lemma 5.5 in the next section). Incidentally the above canonical forms show that any bivector F may be written as a sum of two simple bivectors in neutral signature (this result being obvious for the other two signatures).

It is remarked here that in the case of a non-simple bivector F with $|F| > 0$ and $F \notin \widetilde{S_p}$ and hence Segre type $\{z\bar{z}w\bar{w}\}$ ((5.20) with $b \neq \pm d$) the orthogonal pair of spacelike 2$-$spaces $x \wedge y$ and $s \wedge t$ are clearly uniquely determined by F and referred to as the *canonical blade pair for F*. However, in the degenerate case when F is non-simple with $|F| > 0$ and $F \in \widetilde{S_p}$ and of Segre type $\{(zz)(\bar{z}\bar{z})\}$ given by (5.20) with $d = \pm b$ the blade pair $x \wedge y$ and $s \wedge t$ is not uniquely determined in the sense that there exists infinitely many orthonormal bases x', y', s', t' with $F(= b(x \wedge y + s \wedge t)) = b(x' \wedge y' + s' \wedge t') \in \overset{+}{S_p}$ (and similarly for $\overset{-}{S_p}$) but with the 2$-$space $x \wedge y$ (respectively, $s \wedge t$) distinct from $x' \wedge y'$ (respectively, $s' \wedge t'$) (see lemma 5.10(vi) in the next section and cf. section 3.4). In this case each such pair $(x \wedge y, s \wedge t)$, $(x' \wedge y', s' \wedge t')$ is called a *canonical blade pair for F*. It is easily checked that if F and G are non-simple members of $\overset{+}{S_p}$ (or $\overset{-}{S_p}$) with $|F| > 0 < |G|$, then if F and G have canonical blade pairs with a common member then these pairs are equal as (non-ordered) pairs and F and G are proportional. The proof is similar to that in lemma 3.5. Similarly if F is non-simple with $F \notin \widetilde{S_p}$ and $|F| < 0$ and of Segre type $\{1111\}$ ((5.25) with $\lambda \neq \pm\nu$) the orthogonal pair of timelike 2$-$spaces $l \wedge n$ and $L \wedge N$ are uniquely determined by F and called the *canonical blade pair for F*. Again this fails in the degenerate case (Segre type $\{(11)(11)\}$ and (5.25) with $\lambda = \pm\nu$) in the sense that there exists infinitely many null bases l', n', L', N' with $F(= \lambda(l \wedge n - L \wedge N)) = \lambda(l' \wedge n' - L' \wedge N') \in \overset{+}{S_p}$ but with the 2$-$spaces $l \wedge n$ (respectively, $L \wedge N$) distinct from $l' \wedge n'$ (respectively, $L' \wedge N'$) (and similarly for $\overset{-}{S_p}$). Such pairs of orthogonal timelike 2$-$spaces are also called *canonical blade pairs for F*. Again if F and G are non-simple members of $\overset{+}{S_p}$ (or $\overset{-}{S_p}$) with $|F| < 0 > |G|$ and have canonical blade pairs with a common member these pairs agree and F and G are proportional. One way of viewing this lack of uniqueness in the degenerate case can be gleaned from lemma 5.5(i) below by, in the notation of that lemma, fixing F and changing G. That infinitely many such blade pairs exist can be inferred from section 3.4 by, for example, in the $|F| > 0$ case, concentrating on the complex 2-dimensional i−eigenspaces admitted by $F = x \wedge y + s \wedge t$ and $F' = x' \wedge y' + s' \wedge t'$ given by $\mathrm{Sp}(x + iy, s - it)$ and $\mathrm{Sp}(x' + iy', s' - it')$ and which must agree if $F = F'$. The case when $|F| < 0$ is similar.

5.5　Lie Algebra o(2,2)

This 6−dimensional Lie algebra was considered, and its subalgebras listed, in [74]. Here a different approach will be used to cast the list of subalgebras into a form consistent with the approach taken in this book and is based on the work in [66, 67]. Before embarking on this procedure another lemma is required which was given in [66].

Lemma 5.5 *Let $F \in \overset{+}{S}_p$ and $G \in \overset{-}{S}_p$ and let x, y, s, t and l, n, L, N be, respectively, orthonormal and null corresponding bases at $p \in M$.*

(i) *If $|F| > 0 < |G|$ one may choose the above bases such that F is a multiple of $x \wedge y + s \wedge t$ and G is a multiple of $x \wedge y - s \wedge t$. If $|F| < 0 > |G|$ one may choose the bases such that F is a multiple of $l \wedge n - L \wedge N$ (and hence of $x \wedge t - y \wedge s$) and G a multiple of $l \wedge n + L \wedge N$ (and hence of $x \wedge t + y \wedge s$).*

(ii) *If $|F| > 0$ (respectively < 0) and $|G| < 0$ (respectively > 0) one may choose these bases such that F and G are multiples of $x \wedge y + s \wedge t$ and $l \wedge n + L \wedge N$ (respectively, of $l \wedge n - L \wedge N$ and $x \wedge y - s \wedge t$).*

(iii) *If $|F| > 0$ (respectively $= 0$) and $|G| = 0$ (respectively > 0) one may choose these bases such that F and G are multiples of $x \wedge y + s \wedge t$ and $l \wedge L$ (respectively, $l \wedge N$ and $x \wedge y - s \wedge t$).*

(iv) *If $|F| < 0$ (respectively $= 0$) and $|G| = 0$ (respectively < 0) one may choose the null basis l, n, L, N such that F and G are multiples of $l \wedge n - L \wedge N$ and $l \wedge L$ (respectively, $l \wedge N$ and $l \wedge n + L \wedge N$).*

(v) *If $|F| = |G| = 0$ there exists a null basis such that F and G are multiples of $l \wedge N$ and $l \wedge L$, respectively.*

(vi) *If $\overset{+}{B}_p$ (respectively, $\overset{-}{B}_p$) is a 2-dimensional subalgebra of $F \in \overset{+}{S}_p$ (respectively, $F \in \overset{+}{S}_p$) one may choose a null basis such that $\overset{+}{B}_p = Sp(l \wedge N, l \wedge n - L \wedge N)$ and $\overset{-}{B}_p = Sp(l \wedge L, l \wedge n + L \wedge N)$.*

(vii) Let $A \in \overset{+}{S}_p$ be a Lie subalgebra of $\overset{+}{S}_p$ of dimension 1 or 2. Then there is no Lie algebra homomorphism from $\overset{+}{S}_p$ onto A (and similarly for $\overset{-}{S}_p$).

Proof For part (i), suppose $|F| > 0 < |G|$. Then since $F \in \overset{+}{S}_p$ and $G \in \overset{-}{S}_p$ the bivector $aF + bG$ is simple if and only $P((aF + bG), (a\overset{*}{F} + b\overset{*}{G})) = a^2|F| - b^2|G| = 0$ (lemma 3.1). Thus, although neither F nor G is simple, $H \equiv F + \kappa G$ and $\overset{*}{H} \equiv F - \kappa G$ are simple for $\kappa = (\frac{|F|}{|G|})^{\frac{1}{2}} \in \mathbb{R}$. But then $|H| > 0 < |\overset{*}{H}|$ and so H and $\overset{*}{H}$ are each spacelike with orthogonal blades and hence a basis x, y, s, t may be chosen so that H and $\overset{*}{H}$ are (the same) multiples of $x \wedge y$ and $s \wedge t$, respectively, and the result follows. The proof when $|F| < 0 > |G|$ is similar.

For part (ii) one notes first that if $|F| > 0 > |G|$ then in some null basis l, n, L, N, noting that the desired result is independent of any scalings applied to F and G, one may write $G = l \wedge n + L \wedge N$ (from theorem 5.2—see the remarks following this theorem). Letting f_G denote the linear isomorphism associated with G (as in the last section and noting that it is an isomorphism since G is non-simple) one finds $f_G(l) = l$, $f_G(n) = -n$, $f_G(L) = L$ and $f_G(N) = -N$ and so $l \wedge L$ and $n \wedge N$ are totally null eigenspaces of f_G with non-zero eigenvalues differing only in sign. Since $F \in \overset{+}{S}_p$ and $G \in \overset{-}{S}_p$, $[F, G] = 0$ and so, on contracting this equation successively with l, n, L and N one easily sees that $l \wedge L$ and $n \wedge N$ are invariant 2−spaces of the linear isomorphism f_F arising from F on each of which f_F acts as an isomorphism. Now define $L' \equiv f_F(l)$ which is necessarily null (since it is in the f_F-invariant space $l \wedge L$) and is not proportional to l (since, from the work of the last section, f_F has no real eigenvectors). Now N spans the unique null direction in $n \wedge N$ which is orthogonal to l and so, since f_F acts as an isomorphism on $n \wedge N$, there exists a unique null direction spanned by $n' \in n \wedge N$ such that the null vector $N' \equiv f_F(n')$ is proportional to N (and hence orthogonal to l) and not proportional to n'. Thus l, L', n', N' are independent, null members of T_pM and which satisfy $l \cdot L' = l \cdot N' = n' \cdot N' = 0$ and also $L' \cdot n' = f_F(l) \cdot n' = -f_F(n') \cdot l = -N' \cdot l = 0$. Now none of the members of the basis l, L', n', N' can be orthogonal to all members of this basis and so $l \cdot n'$ and $L' \cdot N'$ are non-zero. Using the above freedom in scaling on n' one can choose $l \cdot n' = 1$. Now define the basis x, y, s, t at p by $\sqrt{2}x \equiv l + n'$, $\sqrt{2}y \equiv L' + N'$, $\sqrt{2}s \equiv L' - N'$ and $\sqrt{2}t \equiv l - n'$, the members of which are easily seen to be mutually orthogonal. Now, from part (i) of this lemma and lemma 5.1(iv) one may scale F so that $F^a{}_c F^c{}_b = -\delta^a_b$ and so $f_F \circ f_F$ is minus the identity map on T_pM. Thus one has $f_F(x) = \frac{1}{\sqrt{2}} f_F(l + n') = \frac{1}{\sqrt{2}}(L' + N') = y$, $f_F(y) = -x$, $f_F(t) = s$ and $f_F(s) = -t$. This shows that $x \pm iy$ and $s \pm it$ are complex, null eigenvectors of f_F with eigenvalues $\pm i$ and $\mp i$, respectively, and hence that $x \cdot x = y \cdot y$ and $s \cdot s = t \cdot t$. It follows that $l \cdot n' = L' \cdot N' (= 1)$ and then that x, y, s, t is an orthonormal basis at p. In this basis F is a multiple of $x \wedge y + s \wedge t$, as required. The other parts follow similarly.

For part (iii) note that since $|F| > 0$ and $|G| = 0$ one may choose an orthonormal basis x, y, s, t for T_pM in which, after a scaling of F, $F = x \wedge y + s \wedge t$ and $G = p \wedge q$ for null, orthogonal $p, q \in T_pM$. Since $[F, G] = 0$, obvious contractions of this last equation with p and q show that the $0-$eigenspace $p \wedge q$ for G is invariant for F and hence, since f_F has no real eigenvectors, and, after a possible redefinition of p and q within the $2-$space $p \wedge q$, f_F has complex eigenvectors $p \pm iq$ which must lie in the $\pm i$ eigenspaces of f_F. Thus $p + iq = x + iy + (a + ib)(s - it) = (x + as + bt) + i(y + bs - at)$ $(a, b \in \mathbb{R})$ and the fact that p and q are null reveals that $a^2 + b^2 = 1$. Now set $S = bs - at$ and $T = as + bt$ to see that $S \cdot S = T \cdot T = -1$ and $S \cdot T = 0$, that x, y, S, T is an orthonormal basis with $S \wedge T = s \wedge t$ and that F is proportional to $x \wedge y + S \wedge T$. Then $G = p \wedge q = (x + T) \wedge (y + S)$ is proportional to $l \wedge L$ where $\sqrt{2}l = x + T$ and $\sqrt{2}L = y + S$. The last part is similar and this completes the proof of (iii).

For part (iv) suppose $|F| < 0 = |G|$ and select a basis in which F is a multiple of $l \wedge n - L \wedge N$ and write $G = p \wedge q$ for $p, q \in T_pM$ which are null and orthogonal. As before, since p and q span the 0-eigenspace of f_G, they span a totally null invariant $2-$space for f_F. Now the only eigenspaces for F are $l \wedge N$ and $n \wedge L$ and they are in $\overset{+}{S}_p$ and since $G \in \overset{-}{S}_p$ the invariant $2-$space $p \wedge q$ intersects them in two independent eigendirections for F (lemma 5.1(iii)). Since $p \wedge q$ is not an eigenspace for F these latter (null) eigenvectors for F have distinct real eigenvalues ± 1. Calling these latter eigenvectors p' and q' one has $f_F(p') = p'$ and $f_F(q') = -q'$. Choosing a tetrad l, n, L, N with $l = p'$ and $L = q'$ one finds from (5.22) that $F = (l \wedge n - L \wedge N) + \beta l \wedge L$, $(\beta \in \mathbb{R})$ and G is a multiple of $l \wedge L$. Since $F \in \overset{+}{S}_p$, $\beta = 0$ and the result follows.

For part (v), using lemma 5.1(iii), the blades of F and G intersect in a null direction spanned by, say, the null vector l. Then, up to multiples, $F = l \wedge p$ and $G = l \wedge q$ for null vectors p, q each orthogonal to l. By extending l to a basis l, n, L, N, p and q lie in $\mathrm{Sp}(l, L, N)$ and recalling that $F \in \overset{+}{S}_p$ and $G \in \overset{-}{S}_p$ part (v) follows. For part (vi) one has $\overset{+}{B}_p = \mathrm{Sp}(P, R)$ and $\overset{-}{B}_p = \mathrm{Sp}(Q, S)$ for $P, R \in \overset{+}{S}_p$ and $Q, S \in \overset{-}{S}_p$ and with P, Q totally null and R, S satisfying $|R| < 0 > |S|$ and with $P \cdot R = Q \cdot S = 0$ (lemma 5.2(ii)). The previous part shows that one may choose a tetrad l, n, L, N with P proportional to $l \wedge N$ and Q proportional to $l \wedge L$. But $P \cdot R = Q \cdot S = 0$ means that R must be a linear combination of $l \wedge N$ and $l \wedge n - L \wedge N$ and S a linear combination of $l \wedge L$ and $l \wedge n + L \wedge N$. The result now follows.

Part (vii) is a simple example of a more general result concerning simple Lie algebras. Suppose $f : \overset{+}{S}_p \to A$ is such a map and that $\dim A = 2$. Then from lemma 5.2(ii) one may choose a basis F_1, F_2, F_3 for $\overset{+}{S}_p$ as in (5.2)-(5.4) so that $A = \mathrm{Sp}(F_1, F_2)$. Writing $f(F_i) = a_iF_1 + b_iF_2$ $(1 \le i \le 3, a_i, b_i \in \mathbb{R})$ one gets $f(F_1) = -f([F_2, F_3]) = -[f(F_2), f(F_3)]$ etc and easily finds that each of $f(F_i)$ is a multiple of F_2 contradicting the fact that f has rank 2. If $\dim A = 1$ one similarly finds the contradiction that f is the trivial map. $\qquad\square$

It is remarked here that this lemma can be used to give an alternative proof of theorem 5.2. One writes a general bivector F at $p \in M$ (uniquely) as $F = \overset{+}{F} + \overset{-}{F}$ with $\overset{+}{F} \in \overset{+}{S}_p$ and $\overset{-}{F} \in \overset{-}{S}_p$. Then one pairs off the cases when $|\overset{+}{F}|$ is $> 0, = 0, < 0$, with the cases when $|\overset{-}{F}|$ is $> 0, = 0, < 0$, to get the appropriate canonical forms and Segre types. Thus writing the case $|\overset{+}{F}| > 0 < |\overset{-}{F}|$ as $(+, +)$ one appeals to lemma 5.5(*i*) to get (5.20), for $(+, -)$ lemma 5.5(*ii*) leads to (5.23) with $a \neq 0 \neq b$, $\rho = 0$, and for $(-, -)$, lemma 5.5(*i*) leads to (5.25). The case $(+, 0)$ and lemma 5.5(*iii*) lead to (5.23) with $a = 0 \neq b$, the case $(0, -)$ and lemma 5.5(*iv*) lead to (5.26) and the case $(0, 0)$ and lemma 5.5(*v*) lead to F being simple and null. The simple spacelike and timelike possibilities for F are special cases of (5.20) and (5.25) whilst the totally null case for F is a special case of $(0, 0)$.

Let M be a 4-dimensional manifold admitting a metric g of neutral signature $(+, +, -, -)$, let $p \in M$ and let $\mathcal{N}$ denote the collection of all linear maps $f : T_p M \to T_p M$ which preserve the metric $g(p)$, that is, $g(p)(f(u), f(v)) = g(p)(u, v)$ where $g(p)$ is the metric at $p \in M$. Then, as in the cases of the other signatures, $\mathcal{N}$ is a 6−dimensional Lie group called here the *neutral group* on $T_p M$. Each member of $\mathcal{N}$ is an isomorphism on $T_p M$ and one may choose coordinates at p so that $g(p)$ takes the Sylvester form $\omega \equiv \mathrm{diag}(1, 1, -1, -1)$. A matrix A representing such a transformation satisfies $A\omega A^T = \omega$ and hence $\det A = \pm 1$. The group $\mathcal{N}$ may be split into four components. Only the (connected) identity component $\mathcal{N}_0$ of $\mathcal{N}$ will be needed here. The Lie algebra of $\mathcal{N}_0$, labelled $o(2, 2)$, is represented as the bivector algebra $\Lambda_p M = \overset{+}{S}_p + \overset{-}{S}_p$ under matrix commutation as described earlier. In this section $\overset{+}{S}$, $\overset{-}{S}$, $\widetilde{S}$ and Λ are used, for convenience, to denote $\overset{+}{S}_p$, $\overset{-}{S}_p$, $\widetilde{S}_p$ and $\Lambda_p M$, respectively, with the first two identified with the subalgebras $\overset{+}{S} + \{0\}$ and $\{0\} + \overset{-}{S}$ of Λ and $\overset{+}{B}$ and $\overset{-}{B}$ respectively representing $\overset{+}{B}_p + \{0\}$ and $\{0\} + \overset{-}{B}_p$ (lemma 5.2). It is also convenient to make some definitions before proceeding and which are, for a subalgebra V of $o(2, 2)$, $\overset{+}{V} \equiv V \cap \overset{+}{S}$ and $\overset{-}{V} \equiv V \cap \overset{-}{S}$ with $\overset{+}{V}$ a subalgebra of V and $\overset{+}{S}$, and similarly for $\overset{-}{V}$. The natural projections $\pi_1 : \Lambda \to \overset{+}{S}$ and $\pi_2 : \Lambda \to \overset{-}{S}$ are also needed and are easily checked to be Lie algebra homomorphisms. It is useful here to note that, from (5.4), the set of commutators of $\overset{+}{S}$ (respectively, $\overset{-}{S}$) spans $\overset{+}{S}$ (respectively, $\overset{-}{S}$). Since, for $v \in V$ one may always write (uniquely) $v = \overset{+}{v} + \overset{-}{v}$ for $\overset{+}{v} \in \overset{+}{S}$ and $\overset{-}{v} \in \overset{-}{S}$, it follows that $\overset{+}{V} + \overset{-}{V} \subset V \subset \pi_1(V) + \pi_2(V)$ where $\pi_1(V)$ and $\pi_2(V)$

are subalgebras of $\overset{+}{S}$ and $\overset{-}{S}$ (and $\overset{+}{V} \cap \overset{-}{V} = \{0\}$). One also recalls the dimension formula (chapter 1) that if W is a vector space with subspaces A and B then $\dim \mathrm{Sp}(A,B) = \dim A + \dim B - \dim(A \cap B)$. If V is a subalgebra of $o(2,2)$ then $\dim V = n$ where $0 \le n \le 6$ and it will be seen that all dimensional possibilities for V can occur and if $\dim V = 6$, V coincides with $o(2,2)$. The idea now is to find a convenient classification of the subalgebras of $o(2,2)$. Each subalgebra will be labelled by an integer, which is its dimension, followed by a letter which distinguishes the different types of subalgebra of that dimension. The approach given here is based on joint work between the author and Dr Zhixiang Wang [66, 67] with the notation taken largely from [66].

It is convenient to deal with the 1-dimensional cases first. So suppose $\dim V = 1$. Here $V = \mathrm{Sp}(F)$ for some non-zero bivector F. The different types are then listed according to the different algebraic (Segre) types of F using theorem 5.2. These types are then, first for F simple, $1a$ (for F timelike, Segre type $\{11(11)\}$), $1b$ (F spacelike, $\{z\bar z(11)\}$), $1c$ (F null, $\{(31)\}$), and $1d$ (F totally null, $\{(22)\}$) with eigenvalues zero in the last two cases. For F non-simple and using the identity $l \wedge N + n \wedge L = x \wedge y + s \wedge t \in \overset{+}{S}$ the types are: $1e$ ($F = \alpha(x \wedge y) + \beta(s \wedge t) \notin \widetilde{S}$, $\{z\bar z w \bar w\}$), $1f$ ($F = \alpha(l \wedge n) + \beta(L \wedge N) \notin \widetilde{S}$, $\{1111\}$), $1g$ ($F = \alpha(l \wedge N + n \wedge L) + \beta(l \wedge n + L \wedge N) = \alpha(x \wedge y + s \wedge t) + \beta(l \wedge n + L \wedge N) \notin \widetilde{S}$, $\{z\bar z w \bar w\}$), $1h$ ($F = \alpha(l \wedge N + n \wedge L) + \beta(l \wedge L) = \alpha(x \wedge y + s \wedge t) + \beta(l \wedge L) \notin \widetilde{S}$, $\{22\}$ with complex eigenvalues), $1j$ ($F = x \wedge y + s \wedge t \in \overset{+}{S}$, $\{(zz)(\bar z \bar z)\}$), $1k$ ($F = l \wedge n - L \wedge N \in \overset{+}{S}$, $\{(11)(11)\}$), $1l$ ($F = \alpha(l \wedge n - L \wedge N) + \beta(l \wedge L) \notin \widetilde{S}$, $\{22\}$ with real eigenvalues). In this list $\alpha, \beta \in \mathbb{R}$ and, in addition for subalgebras $1e$ and $1f$, one requires $0 \ne \alpha \ne \pm \beta \ne 0$ whereas for subalgebras $1g$, $1h$ and $1l$, $\alpha\beta \ne 0$. It is noted that the types $1d$, $1j$ and $1k$ are the only types which lie in $\overset{+}{S}$ (or with the obvious changes, in $\overset{-}{S}$). A pattern becomes clear from the 1-dimensional subalgebras spanned by a non-simple bivector F not in $\overset{+}{S}$ nor $\overset{-}{S}$ in that they are a sum of non-zero bivectors $\overset{+}{F} \in \overset{+}{S}$ and $\overset{-}{F} \in \overset{-}{S}$ with the sign pairs $(|\overset{+}{F}|, |\overset{-}{F}|)$ being $(+,+)$, $(-,-)$, $(+,-)$, $(+,0)$ and $(-,0)$ for types $1e$, $1f$, $1g$, $1h$ and $1l$, respectively. (Note that the case $(0,0)$ is simple.) Lemma 5.5 is useful here.

Suppose $\dim V = 5$. An above dimension formula gives $\dim(V + \overset{+}{S}) = 5 + 3 - \dim(\overset{+}{V})$ and so $\dim \overset{+}{V} \ge 2$ and similarly $\dim \overset{-}{V} \ge 2$. Now $\overset{+}{V} + \overset{-}{V} \subset V$ from above so that $\dim \overset{+}{V} = \dim \overset{-}{V} = 3$ gives a contradiction and $\dim \overset{+}{V} = 3$, $\dim \overset{-}{V} = 2$ leads, from lemma 5.2(ii), to the possibility that $V = \overset{+}{S} + \overset{-}{B}$ (and similarly to the possibility that $V = \overset{+}{B} + \overset{-}{S}$). Otherwise, $\dim \overset{+}{V} = \dim \overset{-}{V} = 2$ and $\overset{+}{V} + \overset{-}{V} \subset V$ but $V \ne \overset{+}{V} + \overset{-}{V}$. Thus $\overset{+}{V} = \overset{+}{B}$ and $\overset{-}{V} = \overset{-}{B}$ and since $V \subset \pi_1(V) + \pi_2(V)$ one sees that either $\overset{+}{V}(= \overset{+}{B}) \ne \pi_1(V)$ or $\overset{-}{V}(= \overset{-}{B}) \ne \pi_2(V)$.

In the former case one sees that $\overset{+}{B}$ is a proper subalgebra of $\pi_1(V)$ and hence that $\pi_1(V) = \overset{+}{S}$. So let $h = (\overset{+}{h}, \overset{-}{h}) \in V$ with $h \notin \overset{+}{V} + \overset{-}{V}$, $\overset{+}{h} \in \overset{+}{S} \setminus \overset{+}{B}$, $\overset{-}{h} \in \pi_2(V)$. Then the collection of all Lie brackets of h with members of $\overset{+}{V} + \{0\} = \overset{+}{B} + \{0\}$ is in V and consists of members of the form $[\overset{+}{h}, b]$, $b \in \overset{+}{B}$ each of which is in $\overset{+}{V} = \overset{+}{B}$. Hence $[\overset{+}{h}, b] \in \overset{+}{V}$ for each $\overset{+}{h} \in \overset{+}{S} \setminus \overset{+}{B}$ and (5.4) reveals that this is a contradiction. It follows that V is the product $V = \overset{+}{B} + \overset{-}{S}$ (or $\overset{+}{S} + \overset{-}{B}$) and is the only possibility for $\dim V = 5$. This 5-dimensional subalgebra is labelled *type 5*.

Suppose $\dim V = 4$. If V is not a product of subalgebras of $\overset{+}{S}$ and $\overset{-}{S}$ and since $V \subset \pi_1(V) + \pi_2(V)$, it follows that V is a *proper* subalgebra of $\pi_1(V) + \pi_2(V)$ and so $\dim(\pi_1(V) + \pi_2(V)) \geq 5$. It follows that one of $\pi_1(V)$ and $\pi_2(V)$ is 3−dimensional, say $\dim \pi_1(V) = 3$, and that $\dim \pi_2(V) \geq 2$. Applying the dimension formula used earlier to the subspaces $V, \overset{+}{S}$ and $V + \overset{+}{S}$ gives $\dim \overset{+}{V} \geq 1$ and similarly $\dim \overset{-}{V} \geq 1$ and since $\overset{+}{V} + \overset{-}{V} \subset V$ but $V \neq \overset{+}{V} + \overset{-}{V}$, (and V is not a product), gives $\dim(\overset{+}{V} + \overset{-}{V}) \leq 3$ and so $1 \leq \dim \overset{+}{V}$, $\dim \overset{-}{V} \leq 2$. So let $h = (\overset{+}{h}, \overset{-}{h}) \in V$ with $\overset{+}{h} \in \pi_1(V) \setminus \overset{+}{V} = \overset{+}{S} \setminus \overset{+}{V}$ and $\overset{-}{h} \in \pi_2(V)$ so that the collection of Lie brackets of h with each member of V lies in V. Then $[\overset{+}{h}, b] \in \overset{+}{V}$ for each $b \in \overset{+}{V}$ and each $\overset{+}{h} \in \overset{+}{S} \setminus \overset{+}{V}$ and this is a contradiction whether $\dim \overset{+}{V}$ is 1 or 2. It follows that V is a product of subalgebras of $\overset{+}{S}$ and $\overset{-}{S}$ and is thus either of the form $\overset{+}{S} + K$ where K is a 1-dimensional subalgebra of $\overset{-}{S}$ (types 4a, 4b and 4d) or of the form $\overset{+}{B} + \overset{-}{B}$ (type 4c), up to isomorphism and a switching between $\overset{+}{S}$ and $\overset{-}{S}$. They are collected together in Table 5.1 where lemma 5.5(*vi*) is used.

Now suppose that $\dim V = 3$. If $\pi_1(V) = \overset{+}{S}$ and $\pi_2(V) = \overset{-}{S}$ the projection maps are Lie algebra isomorphisms $V \to \overset{+}{S}$ and $V \to \overset{-}{S}$ and hence $\pi_2 \circ \pi_1^{-1}$ is a Lie algebra isomorphism $\overset{+}{S} \to \overset{-}{S}$ which, by lemma 5.2(*v*), preserves the bivector metric. So if $v \in V$, $v = v_1 + v_2$ with $v_1 = \pi_1(v) \in \overset{+}{S}$ and $v_2 = \pi_2(v) \in \overset{-}{S}$, one has $|v_1| = |v_2|$ and so $(v_1 + v_2) \cdot (v_1 + v_2)^* = (v_1 + v_2) \cdot (v_1 - v_2) = 0$ and so $v = v_1 + v_2$ is simple. It follows that each non-zero member of V is simple and one may use lemma 3.2 to see that either the blades of the members of V have a common direction k or that the blades of the duals of the members of V have a common direction k'. In the first case it is easily checked that one may choose a basis for V spanned by $A = k \wedge p$, $B = k \wedge q$ and $C = k \wedge r$ with k, p, q, r independent vectors and, in addition, if k is not null, $k \cdot p = k \cdot q = k \cdot r = 0$. It then follows that $[A, B] \notin V$ and so in this case, since V is a subalgebra, k must be null. Choosing a null basis l, n, L, N with $k = l$

TABLE 5.1: Lie subalgebras for $(+, +, -, -)$.

Type	Dimension	Basis
1a	1	$l \wedge n$
1b	1	$x \wedge y$
1c	1	$l \wedge y$ or $l \wedge s$
1d	1	$l \wedge L$
2a	2	$l \wedge n - L \wedge N, l \wedge N (= \overset{+}{B})$
2b	2	$l \wedge n, L \wedge N$
2c	2	$l \wedge n - L \wedge N, l \wedge L + n \wedge N$
2d	2	$l \wedge n - L \wedge N, l \wedge L$
2e	2	$x \wedge y, s \wedge t$
2f	2	$l \wedge N + n \wedge L, l \wedge L$
2g	2	$l \wedge N, l \wedge L$
2h	2	$l \wedge N, \alpha(l \wedge n) + \beta(L \wedge N)$
2j	2	$l \wedge N, \alpha(l \wedge n - L \wedge N) + \beta(l \wedge L)$
2k	2	$l \wedge y, l \wedge n$ or $l \wedge s, l \wedge n$
2l	2	$l \wedge N, \alpha(l \wedge n - L \wedge N) + \beta(l \wedge L + n \wedge N)$
3a	3	$l \wedge n, l \wedge N, L \wedge N$
3b	3	$l \wedge n - L \wedge N, l \wedge N, l \wedge L$
3c	3	$x \wedge y, x \wedge t, y \wedge t$ or $x \wedge s, x \wedge t, s \wedge t$
3d	3	$l \wedge N, l \wedge L, \alpha(l \wedge n) + \beta(L \wedge N)$
3e	3	$\overset{+}{S}$
3f	3	$\overset{+}{B}, l \wedge L + n \wedge N$
4a	4	$\overset{+}{S}, l \wedge n + L \wedge N$
4b	4	$\overset{+}{S}, l \wedge L + n \wedge N$
4c	4	$\overset{+}{B}, \overset{-}{B}$ or $l \wedge L, l \wedge N, l \wedge n, L \wedge N$
4d	4	$\overset{+}{S}, l \wedge L$
5	5	$\overset{+}{S}, \overset{-}{B}$
6	6	$o(2, 2)$

each 2−space containing l must intersect $\mathrm{Sp}(n, L, N)$ in a unique direction and so $V = \mathrm{Sp}(l \wedge n, l \wedge L, l \wedge N)$. But this is impossible since then $l \wedge L$ and $l \wedge N$ would then span a 2-dimensional Abelian subalgebra of V which has been seen to be impossible since V is isomorphic to the subalgebra $\overset{+}{S}$ (lemma 5.2(ii)). It follows that the duals of the members of V have a common direction, say k and hence that k is orthogonal to each member of V. If k is null then with the choice $k = l$ and null basis l, n, L, N one gets $V = \mathrm{Sp}(l \wedge L, l \wedge N, L \wedge N)$ and a similar argument to that above gives a contradiction. Thus k is not null and orthonormal bases may be chosen so that $V = \mathrm{Sp}(x \wedge y, x \wedge t, y \wedge t)$ or $V = \mathrm{Sp}(x \wedge s, x \wedge t, s \wedge t)$. These are isomorphic and labelled type 3c. So suppose that $\pi_1(V) = \overset{+}{S}$ and $\pi_2(V) \neq \overset{-}{S}$. If $\pi_2(V)$ is trivial, $V = \overset{+}{S}$ and this is labelled type 3e. If $\pi_2(V)$ is of dimension 2 or 1 the projection map $\pi_1 : V \to \pi_2(V)$ is a Lie algebra homomorphism onto a 1− or 2-dimensional subalgebra of $\overset{-}{S}$ and this is impossible from lemma 5.5(vi).

Finally suppose that $\pi_1(V) \neq \overset{+}{S}$ and $\pi_2(V) \neq \overset{-}{S}$. Then V is a subset of $\pi_1(V) + \pi_2(V)$ and it follows that either $\dim\pi_1(V) = 2$ and $\dim\pi_2(V) = 1$, or vice versa, or $\dim\pi_1(V) =\dim\pi_2(V) = 2$. The first case(s) shows that (up to isomorphism) V is the product of $\overset{+}{B}$ and a 1-dimensional subalgebra of $\overset{-}{S}$. Thus, if the 1-dimensional subalgebra is spanned by $F \in \overset{-}{S}$, $|F| < 0$, one may appeal to lemma 5.5(iv) and choose a basis l, n, L, N such that the null member of $\overset{+}{B}$ and the bivector in $\overset{-}{S}$ are $l \wedge N$ and $l \wedge n + L \wedge N$, respectively. But then the general theory of section 5.2 shows that the bivector $l \wedge n - L \wedge N$ can be used to make $\overset{+}{B} =\mathrm{Sp}(l \wedge N, l \wedge n - L \wedge N)$ and then, taking obvious linear combinations, one gets $V =\mathrm{Sp}(l \wedge N, l \wedge n, L \wedge N)$ (sometimes written $\mathrm{Sp}(\overset{+}{B}, l \wedge n + L \wedge N)$). This is labelled type 3a. If the 1-dimensional subalgebra in $\overset{-}{S}$ is totally null a similar argument using lemma 5.5 gives $V =\mathrm{Sp}(l \wedge N, l \wedge L, l \wedge n - L \wedge N)$ (or $V =\mathrm{Sp}(\overset{+}{B}, l \wedge L)$ and this is labelled 3b. If the bivector $F \in \overset{-}{S}$ satisfies $|F| > 0$ one gets in a mixed basis or a null basis $V =\mathrm{Sp}(\overset{+}{B}, x \wedge y - s \wedge t)$ or $V =\mathrm{Sp}(\overset{+}{B}, l \wedge L + n \wedge N)$, from lemma 5.5($iii$). This is labelled 3$f$. In the second case when $\dim\pi_1(V) =\dim\pi_2(V) = 2$ one has $\pi_1(V) = \overset{+}{B}$ and $\pi_2(V) = \overset{-}{B}$ and also $V \subset \overset{+}{B} + \overset{-}{B}$ and hence $V + \overset{+}{B} \subset \overset{+}{B} + \overset{-}{B}$. Thus use of the dimension formula gives $4 \geq \dim\mathrm{Sp}(\overset{+}{B}, V)=3 + 2-\dim(V \cap \overset{+}{B})$ and so $\dim(V \cap \overset{+}{B}) \geq 1$, and similarly, $\dim(V \cap \overset{-}{B}) \geq 1$. So one may write $V =\mathrm{Sp}(u, v, w)$ with $u = (a, 0)$, $v = (0, b)$ and $w = (c, d)$ for $a, c \in \overset{+}{B}$ and $b, d \in \overset{-}{B}$. If one of c, d is trivial one repeats the above products of types 3a, 3b and 3f. So choose $c \neq 0 \neq d$. Then a, c are independent members of $\overset{+}{B}$ otherwise one could, by taking linear combinations of u and w reduce the situation to one of the previous product types. Thus a, c and similarly, b, d are independent and $[a, c] \neq 0 \neq [b, d]$ (lemma 5.2(ii)). Finally, since $a, c \in \overset{+}{B}$ and $b, d \in \overset{-}{B}$, one can, by taking linear combinations of u and w, take $u = (a, 0)$ with a totally null, and similarly take $v = (0, b)$ with b totally null. Then (lemma 5.5(v)) one may choose a null basis l, n, L, N with $a = l \wedge N$, $b = l \wedge L$, $c \in \mathrm{Sp}(l \wedge N, l \wedge n - L \wedge N)$ and $d \in \mathrm{Sp}(l \wedge L, l \wedge n + L \wedge N)$ and after a final taking of linear combinations of u, v, w, one achieves $V =\mathrm{Sp}(l \wedge N, l \wedge L, \alpha(l \wedge n) + \beta(L \wedge N))$ for $\alpha, \beta \in \mathbb{R}$ with $\beta \neq \pm\alpha$, the latter restriction being to avoid $w \in \widetilde{S}$ (and hence V being a product). This type is labelled 3d.

Now suppose that $\dim V = 2$. Since $V \subset \pi_1(V) + \pi_2(V)$ it follows that $\pi_1(V)$ and $\pi_2(V)$ have dimension at most 2. If one of these is trivial then V is isomorphic to $\overset{+}{B}$ (or $\overset{-}{B}$) and this type is labelled 2a. If each has dimension 1, $V = \pi_1(V) + \pi_2(V)$ and there are six types depending on whether $\pi_1(V)$ is spanned by $F \in \overset{+}{S}$ with $|F| > 0$, $|F| = 0$ or $|F| < 0$—(and similarly for $\pi_2(V)$). These six types (after use of lemma 5.5) are labelled 2b,...,2g. If, however, $\pi_1(V)$ and $\pi_2(V)$ are each 2-dimensional (and hence isomorphic to $\overset{+}{B}$ and $\overset{-}{B}$, respectively) then V is isomorphic to $\overset{+}{B}$ under the map π_1 (or $\overset{-}{B}$ under π_2) and is a subalgebra of $\overset{+}{B} + \overset{-}{B}$. Then the map $\pi_2 \circ \pi_1^{-1}$ is a Lie algebra isomorphism $\overset{+}{B} \to \overset{-}{B}$ which preserves the bivector metric (lemma 5.2(v)). Thus, as proved in the $\dim V = 3$ case, each non-zero member of V is a simple bivector. But $\dim V = 2$ and so from lemma 3.2, there exists a vector $k \neq 0$ which is contained in each of the blades of the members of V. Thus one may write $V = \text{Sp}(A, B)$ for bivectors A, B and, since V is isomorphic to $\overset{+}{B}$, one may take $[A, B] = A$ with $A = k \wedge p$ and $B = k \wedge q$ for non-zero vectors p, q (with p, q, k independent) from which it follows by expanding the Lie bracket relation that k is null and $k \cdot p = 0 \neq k \cdot q$. It follows that A is either a null bivector (if p is not null) or a totally null bivector (if p is null). If p is null then $A \in \tilde{S}$ and so one achieves the contradiction that one of $\pi_1(V)$ and $\pi_2(V)$ is 1-dimensional and it follows that p is not null and B is a null bivector. If q is not null one may replace q by a linear combination of k, q which is null and labelling this vector n, taking $k \equiv l$ and choosing a hybrid basis l, n, s, y in which $p \equiv y$ (or s) one gets $V = \text{Sp}(l \wedge n, l \wedge y)$ (or $V = \text{Sp}(l \wedge n, l \wedge s)$) and this type is labelled 2k. Finally suppose that $\dim \pi_1(V) = 2$ and $\dim \pi_2(V) = 1$ (the reverse case is similar) so that V is isomorphic to $\pi_1(V)$, that is, to $\overset{+}{B}$ and so V is a subalgebra of $\overset{+}{B} + \pi_2(V)$. Thus V and $\overset{+}{B} + \{0\}$ are subspaces of $\overset{+}{B} + \pi_2(V)$ and the dimension formula gives $\dim(V \cap \overset{+}{B}) = 1$ and so $V \cap \overset{+}{B} = V \cap \overset{+}{S} = \overset{+}{V} = \text{Sp}(v)$ for some $0 \neq v \in \overset{+}{S}$. Now choose $0 \neq u \in V$ which is independent of $(v, 0)$ and with $u = \overset{+}{u} + \overset{-}{u}$ with $\overset{+}{u} \in \overset{+}{S}$ independent of v and $0 \neq \overset{-}{u} \in \overset{-}{S}$. It follows that $[u, v] \in V$. But $[u, v] = [\overset{+}{u}, v] \in \overset{+}{S}$ and so (lemma 5.2(ii)), $[\overset{+}{u}, v]$ is a non-zero member of $\overset{+}{V}$ and hence $[u, v]$ is some non-zero multiple of $(v, 0)$. Recalling the Lie bracket restrictions on the subalgebras $\overset{+}{S}$ and $\overset{-}{S}$ and the fact that V is isomorphic to $\overset{+}{B}$ it is seen that v must be a totally null bivector. This shows that V is spanned by $(v, 0)$ and a bivector $w \notin \tilde{S}$ to avoid contradicting $\dim \overset{+}{V} = 1$ or reproducing one of the above Lie algebra products for V. It follows that $w = c + d$ with $c \in \pi_1(V)$ and $d \in \pi_2(V)$ and that $|c| < 0$ since, together with v, it forms a subalgebra isomorphic to $\overset{+}{B}$. Thus $V = \text{Sp}(v, c + d)$ and one appeals to lemma 5.5 to get the possibilities for V in the following way. One chooses a null basis for the pair $v \in \overset{+}{S}$ and $d \in \overset{-}{S}$, the choice of v then determining the choice of c up to multiples of v, which may be ignored. Thus if $|d| > 0$ a null basis l, n, L, N, with associated basis x, y, s, t, may be chosen so that $v = l \wedge N$ (and hence $c = l \wedge n - L \wedge N$) and $d = x \wedge y - s \wedge t$. Then $V = \mathrm{Sp}(l \wedge N, \alpha(l \wedge n - L \wedge N) + \beta(x \wedge y - s \wedge t))$ for non-zero $\alpha, \beta \in \mathbb{R}$. This type is labelled $2l$. The others are similarly computed and are, for $(|d| < 0)$, $V = \mathrm{Sp}(l \wedge N, \alpha(l \wedge n) + \beta(L \wedge N))$, labelled $2h$, and, for d totally null $(|d| = 0)$ $V = \mathrm{Sp}(l \wedge N, \alpha(l \wedge n - L \wedge N) + \beta l \wedge L)$, labelled $2j$. In type $2h$ one requires $\alpha \neq \pm\beta$ (to avoid reproducing one of the earlier product algebras) and in type $2j$ one needs $\alpha \neq 0 \neq \beta$. This completes the classification of the subalgebras of $o(2,2)$ and they are listed in Table 5.1 and which is an extension of the work in [66]. [The author is to blame for the somewhat idiosyncratic labelling.] A number of these cases cannot arise for the holonomy algebra of a metric connection (see below).

Apart from the trivial subalgebra and the full subalgebra $o(2,2)$ there are eleven types of 1-dimensional subalgebras, $1a,...,1l$, eleven 2-dimensional subalgebras, $2a,...,2l$, six 3-dimensional subalgebras, $3a,...,3f$, four 4-dimensional subalgebras, $4a,...,4d$, and one 5-dimensional subalgebra, 5, a total of thirty three proper subalgebras. Any two subalgebras with a different label are non-isomorphic but it is not claimed that two subalgebras under the same label are isomorphic.

A full list of subalgebras of $o(2,2)$ was given in [74] (in a different labelling and format and using different techniques to those given above) where a brief history of the subject can also be found. In this reference there are, apart from the trivial subalgebra and $o(2,2)$, thirty one proper subalgebras labelled $A_1 - A_{31}$ and $A_{32} = o(2,2)$. The difference in the total numbers of subalgebras lies in the absorbing of the types here labelled $1a$ and $1b$ inside other types in [74]. The retaining of $1a$ and $1b$ as separate cases together with the general bivector approach followed here are more convenient for the present needs. The link between this labelling is implicitly given in [67]. In Table 5.1 some types contain real parameters α and β restricted, as described earlier and repeated in more detail here. For the 1-dimensional subalgebra types and in cases $1e$ and $1f$, one has $0 \neq \alpha \neq \pm\beta \neq 0$ since the cases $\alpha = 0$ and $\beta = 0$ repeat types $1a$ and $1b$, whilst $\alpha = \pm\beta$, repeat types $1j$ and $1k$. In types $1g$, $1h$ and $1l$, $\alpha \neq 0 \neq \beta$, otherwise one repeats types $1j$, $1k$ or $1d$. In the 2-dimensional types $2j$ and $2l$, one requires $\alpha \neq 0 \neq \beta$ to avoid repeating types $2g$, $2a$ and $2f$. In type $2h$ one needs $\alpha \neq \pm\beta$ to avoid repeating types $2a$ and $2d$ but $\alpha = 0$ or $\beta = 0$ (not both) is allowed. In the 3-dimensional type $3d$, $\alpha = 0$ or $\beta = 0$ (not both) is allowed but one requires $\alpha \neq \pm\beta$ to avoid repeating type $3b$. For type $3f$ it is noted that $l \wedge L + n \wedge N = x \wedge y - s \wedge t$ and that in type $2l$, $\alpha(l \wedge n - L \wedge N) + \beta(x \wedge y - s \wedge t)$ is non-simple for each $\alpha, \beta \in \mathbb{R}, \alpha^2 + \beta^2 \neq 0$.

The labelling here for subalgebras of dimension ≥ 2 correspond in a one-to-one way with the types labelled $A_{10} - A_{20}$ (2-dimensional cases), $A_{21} - A_{26}$ (3–dimensional cases), $A_{27} - A_{30}$ (4-dimensional cases), A_{31} (5-dimensional case) and A_{32} (6–dimensional case) in [74]. Now for the situation where, for (M, g), g has neutral signature, previous work shows that its holonomy group Φ is a Lie subgroup of $\mathcal{N}$ and so the holonomy algebra ϕ is a subalgebra of $o(2, 2)$. It is then remarked that the cases A_{12} $(2l)$, A_{21} $(3e)$, A_{24} $(3f)$ and A_{29} $(4d)$ cannot occur as the holonomy group of (M, g) since the connection ∇ is metric whilst it seems that A_{13} $(2j)$ possibly cannot occur for the same reason ([74] and references given therein). As mentioned before (section 3.7), the 1-dimensional subalgebras can only represent a metric connection if they are spanned by a simple bivector. Thus, in the above notation, the only relevant 1-dimensional subalgebras here are $1a$, $1b$, $1c$ and $1d$ and for this reason and also reasons of typographical convenience, these 1-dimensional non-simple cases are omitted from Table 5.1.

5.6   Curvature Tensor

As in the positive definite and Lorentz cases one can construct the linear curvature map f on bivectors and use it to classify $Riem(p)$ at $p \in M$ into the five curvature classes A, B, C, D, O as detailed in chapter 3. Regarding this map as arising from the symmetric matrix R_{AB} it is not necessarily diagonalisable (over $\mathbb{R}$ or $\mathbb{C}$) since the bivector metric has signature $(+, +, -, -, -, -)$. In class D $\dim rgf(p) = 1$ with $rgf(p)$ spanned by a *simple* bivector (see chapters 3 or 4) and this type may be subdivided into subclasses where this bivector is spacelike, timelike, null or totally null. Always one has $rgf(p) = \overline{rgf(p)}$ (the smallest subalgebra containing $rgf(p)$) and this subalgebra is of type $1a$, $1b$, $1c$ or $1d$. In class B one has $rgf(p) = \mathrm{Sp}(P, Q)$ for independent bivectors P, Q with no common annihilator and satisfying $[P, Q] = 0$. Writing $P = \overset{+}{P} + \overset{-}{P}$ for $\overset{+}{P} \in \overset{+}{S}$ and $\overset{-}{P} \in \overset{-}{S}$, and similarly for Q, the condition $[P, Q] = 0$ implies $[\overset{+}{P}, \overset{+}{Q}] = [\overset{-}{P}, \overset{-}{Q}] = 0$. Then an almost identical proof to that in chapter 3, but recalling that $\overset{+}{S}_p$ and $\overset{-}{S}_p$ now have no 2-dimensional *Abelian* subalgebras, shows that $rgf(p)$ can be spanned by two independent bivectors with no common annihilator, one in $\overset{+}{S}_p$ and one in $\overset{-}{S}_p$ (and their bracket is then necessarily zero). Thus $rgf(p) = \overline{rgf(p)}$ and this subalgebra type is easily checked to be either $2b$, $2c$, $2d$, $2e$ or $2f$, the other 2-dimensional subalgebras either being non-Abelian or having a common annihilator. For class C, $rgf(p)$ is such that its members have a *unique* independent common annihilator $0 \neq k \in T_pM$. Thus all members of $rgf(p)$ are simple and from lemma 3.2, $\dim rgf(p) \leq 3$

and if it is 3, $rgf(p) = \overline{rgf(p)}$ and it is a subalgebra. If $rgf(p)$ is a subalgebra it follows that it is of type 2g, 2h (with $\alpha\beta = 0$), 2k, 3c or 3d (with $\alpha = 0$). The curvature class O at p means that $Riem(p) = 0$ and if, at p, the class is none of B, C, D or O it is of class A. In addition, allowing, as before, any of the letters $A, ..., O$ to denote precisely the subset of points of M where the curvature class is equal to that letter one can achieve the results of theorem 3.1. The existence of a non-trivial solution $k \in T_pM$ to the equation $R^a{}_{bcd}k^d = 0$ is again confined to $p \in C \cup D \cup O$.

5.7   Weyl Conformal Tensor I

Now consider the Weyl conformal tensor C and its algebraic classification. A detailed account of this was given in [68] and later, during the preparation of this latter paper, it was pointed out to this author that an earlier but less detailed discussion of this problem has been given, both in the language of spinors in [75, 76] and in tensorial notation in [77, 78, 80]. Here the discussion will follow [68] with the analysis largely following the approach used in chapter 4 for the Petrov classification of this tensor in the case of Lorentz signature. In this approach and following the initial derivation of the algebraic types, Bel-type criteria will be developed as was done in the Lorentz case. It is believed that the tensor approach followed here is simpler and more amenable to differential geometers and is especially useful in calculations. The case of neutral signature is a little more complicated than that in the Lorentz (Petrov) case but exhibits a rich structure and will be arranged in such a way as to establish clearly the close relationship between the Lorentz and neutral signature cases.

The Weyl conformal tensor is the type $(1, 3)$ tensor denoted by C and with components given by $C^a{}_{bcd}$. Also useful will be the type $(0, 4)$ Weyl tensor with components $C_{abcd} \equiv g_{ae}C^e{}_{bcd}$ which has the index symmetries given in chapter 3. For neutral signature one has $^*C = C^*$ and $^*C^* = C$, the double dual for this signature being the identity map, and one can again introduce the linear map f_C from Λ_pM to the vector space of type $(1, 1)$ tensors at p given by $f_C : F^{ab} \to C^a{}_{bcd}F^{cd}$ called the *Weyl map* at p and whose rank is referred to as the *Weyl rank* at p. Again, since g is given, one may introduce the related map (also denoted by f_C) given by $F^{ab} \to C_{abcd}F^{cd}$ and then in an obvious shorthand way (using the identifications arising from the metric g) as $f_C : F \to CF$. Then $(f_C F)^* = (CF)^* = (^*C)F = (C^*)F = C\overset{*}{F}$ and so f_C maps the subspaces $\overset{+}{S}_p$ and $\overset{-}{S}_p$ of Λ_pM into themselves, that is, they are invariant subspaces of f_C. It also follows that if $F \in rgf_C(p)$ then, at p, $F = CG$ for some $G \in \Lambda_pM$ and then $\overset{*}{F} =^* CG = C^*G = C\overset{*}{G}$ which shows that $\overset{*}{F}$ (which may not be independent of F) is also in $rgf_C(p)$, that is, $rgf(p)$ is "dual invariant". One can decompose the type $(0,4)$ Weyl tensor as

$$C = \overset{+}{W} + \overset{-}{W}, \qquad \overset{+}{W} = \frac{1}{2}(C + {}^*C), \qquad \overset{-}{W} = \frac{1}{2}(C - {}^*C), \qquad (5.27)$$

where the type $(0,4)$ tensors $\overset{+}{W}$ and $\overset{-}{W}$, which are uniquely determined by C, are the *self dual* and *anti-self dual* parts of C, respectively, and satisfy $\overset{+}{W}{}^* = {}^*\overset{+}{W} = \overset{+}{W}$ and $\overset{-}{W}{}^* = {}^*\overset{-}{W} = -\overset{-}{W}$. It also follows (chapter 3) that $\overset{+}{W}$ and $\overset{-}{W}$ are skew-symmetric in their first two and also in their last two indices, that $\overset{+}{W}_{abcd} = \overset{+}{W}_{cdab}$, that $\overset{+}{W}_{a[bcd]} = 0$ and that $\overset{+}{W}{}^c{}_{acb} = 0$ and similarly for $\overset{-}{W}$. The tensors $\overset{+}{W}$ and $\overset{-}{W}$ give rise, in an obvious way, to linear maps $\overset{+}{f_C}$ and $\overset{-}{f_C}$ on $\Lambda_p M$ constructed from them as f_C was from C with $\overset{+}{f_C}$ restricting to a linear map $\overset{+}{S_p} \to \overset{+}{S_p}$ and to the trivial map on $\overset{-}{S_p}$. This follows since $(\overset{+}{f_C}(F))^* = (\overset{+}{W}F)^* = {}^*\overset{+}{W}F = \overset{+}{W}\overset{*}{F}$ and so $F \in \overset{+}{S_p} \Rightarrow \overset{+}{f_C}(F) \in \overset{+}{S_p}$ whereas if $F \in \overset{-}{S_p}$, $\overset{+}{f_C}(F) = \overset{+}{W}F = \overset{+}{W}{}^*F = \overset{+}{W}\overset{*}{F} = -\overset{+}{W}F$. Similarly, $\overset{-}{f_C}$ restricts to a (linear) map $\overset{-}{S_p} \to \overset{-}{S_p}$ and to the trivial map on $\overset{+}{S_p}$. Also one has $f_C = \overset{+}{f_C} + \overset{-}{f_C}$. To achieve a classification of C one classifies the independent maps $\overset{+}{f_C}$ and $\overset{-}{f_C}$.

Considering the map $\overset{+}{f_C}$ as a linear map on the $3-$dimensional real vector space $\overset{+}{S_p}$ to itself, the latter having Lorentz signature, one may appeal to the work in chapter 4 (lemmas 4.1 and 4.2) and the symmetry $\overset{+}{W}_{abcd} = \overset{+}{W}_{cdab}$ to get a classification of $\overset{+}{f_C}$, and similarly for the (independent) map $\overset{-}{f_C}$ on $\overset{-}{S_p}$. The eigenvectors of $\overset{+}{f_C}$ are then the eigenbivectors of $\overset{+}{W}(p)$ in $\overset{+}{S_p}$, and similarly for $\overset{-}{f_C}$ and $\overset{-}{W}(p)$ in $\overset{-}{S_p}$. For this the bases F_1, F_2, F_3 for $\overset{+}{S_p}$ and G_1, G_2, G_3 for $\overset{-}{S_p}$ given earlier in (5.2) play exactly the role required. To see this one makes the identification $F_1 \leftrightarrow x$, $F_2 \leftrightarrow l$ and $F_3 \leftrightarrow n$ where l, n, x constitute the basis used in lemma 4.2, noting that the signature used in this lemma was $(-,+,+)$ whereas for $\overset{+}{S_p}$ it is $(-,-,+)$ (see (5.3)). A similar identification may be made for $\overset{-}{S_p}$. Considering the map $\overset{+}{f_C}$ (the discussion of $\overset{-}{f_C}$ is similar) one sees that there are four general Jordan/Segre forms for this map which are $\{111\}$ (diagonalisable over $\mathbb{R}$ with three real eigenvalues), $\{1z\bar{z}\}$ (diagonal over $\mathbb{C}$ with one real eigenvalue and a pair of complex conjugate eigenvalues), $\{21\}$ (one independent, totally null eigenbivector with real eigenvalue and associated elementary divisor of order 2 and one independent, eigenbivector with negative square with respect to the bivector metric P and with real eigenvalue and associated elementary divisor simple) and $\{3\}$ (with a single, necessarily totally null, independent eigenbivector with real eigenvalue and associated elementary divisor of order 3). The basis F_1, F_2, F_3 above will be used but, for ease of notation, will be temporarily redefined according to $F \equiv F_1$, $G \equiv F_2$ and $H \equiv F_3$. Thus one has for these canonical forms at $p \in M$ using (4.1)–(4.3)

$$\overset{+}{W}_{abcd} = -\rho_3 F_{ab}F_{cd} + \rho_1(G_{ab}H_{cd} + H_{ab}G_{cd}) + \rho_2(G_{ab}G_{cd} \pm H_{ab}H_{cd}), \quad (5.28)$$

$$\overset{+}{W}_{abcd} = \rho_1(G_{ab}H_{cd} + H_{ab}G_{cd}) + \lambda G_{ab}G_{cd} - \rho_2 F_{ab}F_{cd}, \quad (5.29)$$

$$\overset{+}{W}_{abcd} = \rho_1(G_{ab}H_{cd} + H_{ab}G_{cd}) + \mu(G_{ab}F_{cd} + F_{ab}G_{cd}) - \rho_1 F_{ab}F_{cd}, \quad (5.30)$$

where $\rho_1, \rho_2, \rho_3 \in \mathbb{R}$, where the $+$ sign (respectively the $-$sign) in (5.28) gives type $\{111\}$ (respectively, type $\{1z\bar{z}\}$) and where μ and λ are non-zero real numbers which may be chosen, after basis rescalings, as $\lambda = \pm 1$ and $\mu = 1$. (Change the null basis l, n, L, N to $\alpha l, \alpha^{-1}n, L, N$ with $0 \neq \alpha \in \mathbb{R}$ so that $F \to F, G \to \alpha G, H \to \alpha^{-1}H$ and choose $\alpha^2\lambda = \pm 1$ in (5.29) and $\mu\alpha = 1$ in (5.30)). In (5.28) the eigen(bi)vector/eigenvalue pairs are F (ρ_3) and $G \pm H$ ($\rho_1 \pm \rho_2$) if the plus sign is chosen and F (ρ_3) and $G \pm iH$ ($\rho_1 \pm i\rho_2$) ($\rho_2 \neq 0$) if the minus sign is chosen, whilst in (5.29) (Segre type $\{21\}$) they are F (ρ_2) and G (ρ_1) and in (5.30) (Segre type $\{3\}$) it is G (ρ_1). It is remarked that the eigenvalue associated with a totally null eigenbivector is always real.

After the basis rescalings mentioned above and noting that

$$G^c{}_aG_{cb} = H^c{}_aH_{cb} = G^c{}_aF_{cb} + F^c{}_aG_{cb} = H^c{}_aF_{cb} + F^c{}_aH_{cb} = 0,$$
$$2(G^c{}_aH_{cb} + H^c{}_aG_{cb}) = -4F^c{}_aF_{cb} = g_{ab}, \quad (5.31)$$

one can apply the tracefree condition $\overset{+}{W}{}^c{}_{acb} = 0$ ($\Leftrightarrow \overset{*+}{W}_{a[bcd]} = 0 \Leftrightarrow \overset{+}{W}_{a[bcd]} = 0$) to these expressions which gives $\rho_3 = -2\rho_1$ in (5.28) (for either sign), $\rho_2 = -2\rho_1$ in (5.29) and $\rho_1 = 0$ in (5.30). Then one gets for these three canonical forms, respectively,

$$\overset{+}{W}_{abcd} = \rho_1(G_{ab}H_{cd} + H_{ab}G_{cd} + 2F_{ab}F_{cd}) + \rho_2(G_{ab}G_{cd} \pm H_{ab}H_{cd}), \quad (5.32)$$

$$\overset{+}{W}_{abcd} = \rho_1(G_{ab}H_{cd} + H_{ab}G_{cd} + 2F_{ab}F_{cd}) \pm G_{ab}G_{cd}, \quad (5.33)$$

$$\overset{+}{W}_{abcd} = (G_{ab}F_{cd} + F_{ab}G_{cd}). \quad (5.34)$$

At this point it is convenient to label these algebraic types by analogy with the Petrov types in the Lorentz case. Thus if $\overset{+}{f}_C$ has three *distinct* (real or complex) eigenvalues at p, as in (5.32), with either the $+$ sign and eigenvalues $-2\rho_1$ and $\rho_1 \pm \rho_2$ ($0 \neq \pm\rho_2 \neq 3\rho_1$) or the $-$ sign and eigenvalues $-2\rho_1$ and $\rho_1 \pm i\rho_2$ ($\rho_2 \neq 0$), its type will be referred to as type **I** (and this type can be subdivided into type $\mathbf{I_\mathbb{R}}$ if these distinct eigenvalues are all real and $\mathbf{I_C}$ if two of them are complex). If $\overset{+}{f}_C$ is as in (5.33) with distinct eigenvalues $-2\rho_1$ and

ρ_1, ($\rho_1 \neq 0$) the type will be labelled type **II** and for (5.34) (with eigenvalues necessarily zero) the type will be labelled as type **III**. There is a degeneracy in the type $\mathbf{I}_{\mathbb{R}}$ case when two eigenvalues are equal. This can only occur if all eigenvalues are real but can occur, geometrically, in two distinct ways. Thus in (5.32) if the Segre type is $\{(11)1\}$ consider the 2-dimensional eigenspace in $\overset{+}{S}_p$ which results. If this has an induced Lorentz signature one may choose the eigenspace to be spanned by $G + H$ and $G - H$, giving $\rho_2 = 0$, and this type will be labelled $\mathbf{D}_1$ whilst if it is Euclidean (negative definite in this case) the eigenspace may be chosen to be spanned by F and $G - H$, giving $3\rho_1 = \rho_2 \neq 0$ and this type will be labelled $\mathbf{D}_2$. Thus for type $\mathbf{D}_1$ one has an eigenbivector F with eigenvalue $-2\rho_1$ and a ρ_1-eigenspace which is Lorentz and spanned by G and H whilst for type $\mathbf{D}_2$ one has an eigenbivector $G + H$ with eigenvalue $4\rho_1$ and a $(-2\rho_1)-$eigenspace which is Euclidean and spanned by F and $G - H$. The degenerate case $\mathbf{D}_2$ cannot admit any totally null eigenbivectors since it contains only eigenbivectors Q with $|Q| \neq 0$. There is also a degeneracy of type **II** (Segre type $\{(21)\}$) which arises when $\rho_1 = 0$ and this will be labelled type **N**. Thus these degenerate types are (with $\rho_1 \neq 0$)

$$\overset{+}{W}_{abcd} = \rho_1(G_{ab}H_{cd} + H_{ab}G_{cd} + 2F_{ab}F_{cd}), \qquad (type\ \mathbf{D}_1) \qquad (5.35)$$

$$\overset{+}{W}_{abcd} = \rho_1(G_{ab}H_{cd} + H_{ab}G_{cd} + 2F_{ab}F_{cd}) + 3\rho_1(G_{ab}G_{cd} + H_{ab}H_{cd}), \qquad (type\ \mathbf{D}_2) \qquad (5.36)$$

$$\overset{+}{W}_{abcd} = \pm G_{ab}G_{cd}. \qquad (type\ \mathbf{N}) \qquad (5.37)$$

A little simplification of these expressions could be achieved by using the *bivector completeness relation* (restricted to $\overset{+}{S}_p$) given by $\overset{+}{P}_{abcd} \equiv (G_{ab}H_{cd} + H_{ab}G_{cd} - F_{ab}F_{cd})$. Thus $\overset{+}{P}_{abcd}G^{cd} = G_{ab}$, etc. (A similar one is available for $\overset{-}{S}_p$). If one adds the type **O** for the case when $\overset{+}{W}(p) = 0$ one has a complete algebraic classification of $\overset{+}{W}(p)$ (and similarly for $\overset{-}{W}(p)$). A simple calculation (or a comparison with the results of lemma 4.2) shows that $\overset{+}{W}(p)$ admits no totally null eigenbivectors if and only if it is of type **I** or type $\mathbf{D}_2$, that $\overset{+}{W}(p)$ admits exactly two independent totally null eigenbivectors if and only if it is of type $\mathbf{D}_1$ (and then their eigenvalues are equal, non-zero and real) and that $\overset{+}{W}(p)$ admits exactly one independent totally null eigenbivector if and only if it is of type **II** (eigenvalue non-zero) or type **III** or **N** (eigenvalue zero). These results will be slightly augmented in lemma 5.6 below.

It is convenient here to consider Bel-type classifications (section 4.5) of $\overset{+}{W}(p)$ and $\overset{-}{W}(p)$ which can later be used to give a similar classification for $C(p)$. First one writes out the general form for $\overset{+}{W}(p)$ in terms of the basis F, G, H for $\overset{+}{S}_p$ in a null basis l, n, L, N at p and with the tracefree condition applied

$$\overset{+}{W}_{abcd} = A(G_{ab}H_{cd} + H_{ab}G_{cd} + 2F_{ab}F_{cd}) + BG_{ab}G_{cd} + CH_{ab}H_{cd}$$
$$+ D(G_{ab}F_{cd} + F_{ab}G_{cd}) + E(H_{ab}F_{cd} + F_{ab}H_{cd}), \quad (5.38)$$

for $A, B, C, D, E \in \mathbb{R}$.

Lemma 5.6 *Let $p \in M$ and suppose $\overset{+}{W}(p) \neq 0$.*

(i) *There exists $0 \neq k \in T_pM$ such that $\overset{+}{W}_{abcd}k^d = 0$ if and only if $\overset{+}{W}(p)$ is of type **N**. The vector k is necessarily null and may be any non-zero member of the blade of the (unique up to scaling) totally null eigenbivector of $\overset{+}{W}(p)$, and only these.*

(ii) *Suppose for $\alpha \in \mathbb{R}$*

$$\overset{+}{W}_{abcd}k^b k^d = \alpha k_a k_c \qquad (5.39)$$

*for some $0 \neq k \in T_pM$. Then k is necessarily null and (5.39) holds with the same α for any non-zero member of a certain totally null $2-$space containing k. The totally null bivector with this $2-$space as its blade is in $\overset{+}{S}_p$ and is an eigenbivector of $\overset{+}{W}(p)$ with eigenvalue 2α. Conversely, if $Q \in \overset{+}{S}_p$ is a totally null eigenbivector of $\overset{+}{W}(p)$ with eigenvalue γ, each non-zero (necessarily null) member of the blade of Q satisfies (5.39) with $\alpha = \frac{1}{2}\gamma$. Further if $\alpha \neq 0$ the type of $\overset{+}{W}(p)$ is **II** or $\mathbf{D_1}$ and if $\alpha = 0$ the type of $\overset{+}{W}(p)$ is **III** or **N**. If the type is $\mathbf{D_1}$, two independent totally null eigenbivectors (with equal eigenvalues) arise and the non-zero members of their blades each satisfy (5.39) for the same α.*

(iii) *There exists $0 \neq k \in T_pM$ such that $\overset{+}{W}_{abcd}k^d = J_{ab}k_c$ for a non-zero bivector J if and only if $\overset{+}{W}(p)$ is of type **III**. The bivector J is necessarily a totally null bivector in $\overset{+}{S}_p$ and k is necessarily null and lies in the blade of J. One may then choose a null basis l, n, L, N so that $k = l$ and J a multiple of G. Thus J is uniquely determined up to a scaling and k is any non-zero member of the blade of J.*

Proof (i) Suppose $\overset{+}{W}_{abcd}k^d = 0$. Then $\overset{+}{W}{}^*_{abcd}k^d = 0$ and so $\epsilon^{rscd}\overset{+}{W}_{abrs}k_d = 0$ and hence $\epsilon^{rsdc}\overset{+}{W}_{abrs}k_d = 0$ from which it follows that $\overset{+}{W}_{ab[cd}k_{e]} = 0$ (see a similar proof in the Lorentz case in chapter 4). A contraction of this last equation with k^c shows that k is null. Choosing a null basis l, n, L, N with $l = k$ a contraction of (5.38) with l^d gives $A = C = D = E = 0$

and then (5.37) shows that $\overset{+}{W}(p)$ is of type **N**. The converse is immediate. It is now clear that k could be any non-zero member of the blade of the eigenbivector G of $\overset{+}{W}(p)$, and only these.

(*ii*) If (5.39) holds with $\alpha \neq 0$ a contraction with k^a easily reveals that k is null whereas if $\alpha = 0$ one has $\overset{+}{W}{}^*{}_{abcd}k^b k^d = 0$ and so $(\epsilon_a{}^{bmn}k_b W_{mncd})k^d = 0$ which, after rearranging indices, gives $k^a \overset{+}{W}_{ab[cd}k_{e]} = 0$. A contraction of this equation with k^e shows that either k is null or that $\overset{+}{W}_{abcd}k^d = 0$, the latter revealing also that k is null from part (*i*). [Another proof of the fact that k is null in part (*i*) and also in part (*ii*) with $\alpha = 0$ arises by noting, in an obvious notation, that each leads to the result $P\overset{+}{W}Q = 0$ at p for any simple bivectors P and Q at p whose blade contains k. Then since $\overset{+}{W}(p) = \overset{+}{W}{}^*(p) = {}^*\overset{+}{W}(p)$, one gets $P\overset{+}{W}{}^*Q = P{}^*\overset{+}{W}Q = {}^*P\overset{+}{W}{}^*Q = 0$ at p. Now if k is *not* null one can choose a basis for $\Lambda_p M$ of the form $A_1, A_2, A_3, \overset{*}{A_1}, \overset{*}{A_2}$ and $\overset{*}{A_3}$ where A_1, A_2 and A_3 are independent simple bivectors with k in their blades and then $\overset{+}{W}A_1, ..., \overset{+}{W}\overset{*}{A_3}$ are each orthogonal to each member of this basis and hence zero. Thus one achieves the contradiction that $\overset{+}{W}(p) = 0$ and so k is null. (It is easily checked that no such basis exists if k is null.)]

Thus k is null and a choice of tetrad l, n, L, N with $k = l$ and use of (5.38) shows that $C = E = 0$ and that (5.39) holds for any non-zero member of the blade $l \wedge N$ of $G \in \overset{+}{S}_p$ (note that this blade is an eigenspace of the bivector F) with $A = 2\alpha$. It then easily follows that G is a totally null eigenbivector of $\overset{+}{W}(p)$ with eigenvalue A. If $Q \in \overset{+}{S}_p$ is a totally null eigenbivector of $\overset{+}{W}(p)$ with eigenvalue γ one can choose a null basis l, n, L, N such that $Q = G = l \wedge N$ and then (5.38) gives $C = E = 0$ and $A = \gamma$. Then (5.38) shows that (5.39) holds with $2\alpha = \gamma$ for each non-zero member of the blade of G. The admission by $\overset{+}{W}(p)$ of a totally null eigenbivector means, as shown earlier, that it cannot be of type $\mathbf{D_2}$. Now suppose that $A \neq 0$ ($\Leftrightarrow \alpha \neq 0$) and change the null basis to l', n', L', N' where $l' = l, N' = N, n' = n + \lambda N$ and $L' = L - \lambda l$ ($\lambda \in \mathbb{R}$). Then, since the 2−space $l \wedge N$ is preserved in this new basis, one achieves an expression like (5.38) with $C = E = 0$ but now one may choose λ so that $D = 0$ and so the type of $\overset{+}{W}(p)$ is **II** or $\mathbf{D_1}$. If $A = 0$ ($\Leftrightarrow \alpha = 0$) one, of course, still has $C = E = 0$ and a similar basis change can be used, if $D \neq 0$, to set $B = 0$ and so the type of $\overset{+}{W}(p)$ is **III** or **N**. If the type is $\mathbf{D_1}$ two independent totally null eigenbivectors G and H arise (with equal eigenvalues) and the non-zero members of their blades each satisfy (5.39) with the same α.

For part (*iii*) J is easily seen to necessarily be in $\overset{+}{S}_p$ and the condition $\overset{+}{W}_{a[bcd]} = 0$ shows, using lemma 3.1, that J is simple (and hence totally null) and that k lies in its blade (and is hence null). So one may choose a null basis l, n, L, N with $k = l$ and then a contraction of (5.38) with k^d shows that $A = C = E = 0$ and that $D \neq 0$ and so J is a multiple of G and k may be any member of its blade. It follows that $\overset{+}{W}(p)$ is of type **III** since if $B \neq 0$ a basis change of the type given in the previous part may be used to set it to zero. The converse is immediate. □

A (necessarily null) vector k satisfying (5.39) is said to span a *repeated principal null direction* (a repeated pnd) for $\overset{+}{W}(p)$ (the reason for the term "repeated" will appear later). It is straightforward to check, using the symmetry of the expression $\overset{+}{W}_{abcd} k^b k^d$ in the indices a and c, that an equivalent condition at p to (5.39) is $k_{[e} \overset{+}{W}_{a]bcd} k^b k^c = 0$. Now consider the equations at p for $0 \neq k \in T_p M$ given for $\overset{+}{W}(p) \neq 0$ by

$$k_{[e} \overset{+}{W}_{a]bc[d} k_{f]} k^b k^c = 0, \qquad \overset{+}{W}_{abcd} k^b k^c = k_a q_d + q_a k_d, \qquad (5.40)$$

where q is a $1-$form at p. These equations are easily checked to be equivalent, using the symmetry immediately above, and the special case when q (which is uniquely determined by $\overset{+}{W}$, k and (5.40)) is proportional to k (possibly zero) gives rise to the case when k spans a repeated pnd (see (5.39)). The more general case occurs when q is neither zero nor proportional to k. In this situation the tracefree condition on $\overset{+}{W}$ in the second equation in (5.40) shows that since $q \neq 0$, $k \cdot q = 0$ and then the second equation in (5.40) contracted with k^a reveals that k is necessarily null. So introducing a null basis l, n, L, N with $k = l$ and recalling from the properties of the map $\overset{+}{f}_C$ that $\overset{+}{W}_{abcd} R^{cd} = 0$ where $R = l \wedge L \in \overset{-}{S}_p$ a contraction of the second equation in (5.40) with L^a shows that $q \cdot L = 0$ and so $q \in l \wedge L$ and is hence also null. The vector k in (5.40) is said to span a (*non-repeated*) *principal null direction* (a (*non-repeated*) pnd) for $\overset{+}{W}(p)$ with q as its *associated 1-form* (and any scaling of k results in a similar scaling of q). One will generally refer to repeated and non-repeated pnds simply as pnds.

One can say a little more about pnds and their associated $1-$forms. First construct a null basis l, n, L, N about l at p and use (5.38) quite generally to get

$$\overset{+}{W}_{abcd} l^b l^d = \frac{A}{2} l_a l_c + \frac{C}{2} L_a L_c - \frac{E}{2\sqrt{2}} (l_a L_c + L_a l_c), \qquad (5.41)$$

$$\overset{+}{W}_{abcd} n^b n^d = \frac{A}{2} n_a n_c + \frac{B}{2} N_a N_c + \frac{D}{2\sqrt{2}} (n_a N_c + N_a n_c), \qquad (5.42)$$

$$\overset{+}{W}_{abcd} L^b L^d = \frac{A}{2} L_a L_c + \frac{B}{2} l_a l_c - \frac{D}{2\sqrt{2}} (l_a L_c + L_a l_c), \qquad (5.43)$$

$$\overset{+}{W}_{abcd} N^b N^d = \frac{A}{2} N_a N_c + \frac{C}{2} n_a n_c + \frac{E}{2\sqrt{2}} (n_a N_c + N_a n_c). \qquad (5.44)$$

The condition that l spans a pnd is easily checked, from (5.40), to be equivalent to $C = 0$ which is equivalent to N spanning a pnd. Also l spans a repeated pnd if and only if $C = E = 0$ if and only if N also spans a repeated pnd. If l and hence N span pnds (so that $C = 0$) their associated 1−forms are $q = \frac{A}{4}l - \frac{E}{2\sqrt{2}}L$ and $q' = \frac{A}{4}N + \frac{E}{2\sqrt{2}}n$, respectively, and are null and orthogonal to l, respectively, N, the former lying in the totally null 2−space $l \wedge L$ and the latter lying in the totally null 2−space $n \wedge N$. Since, given l, the null basis above is otherwise arbitrary, it follows that each of the non-zero members of the unique (up to scaling), totally null member of $\overset{+}{S}_p$ containing l span pnds. Further these pnds are either all repeated or all non-repeated (depending on whether $E = 0$ or $E \neq 0$) and for the non-repeated ones it is easily checked that the associated 1−forms are "additive" in the sense that if l and N are non-repeated as above with associated 1−forms q and q' and $\lambda \in \mathbb{R}$, $l + \lambda N$ is non-repeated with associated 1−form $q + \lambda q'$. It is thus seen from this and from lemma 5.6 that, at least for the tensors $\overset{+}{W}(p)$ (and similarly for $\overset{-}{W}(p)$), attention is drawn to such totally null 2−spaces whose non-zero members are repeated pnds (respectively, non-repeated pnds) and which will be called *repeated, principal, (totally null) 2−spaces, (respectively, principal, (totally null) 2−spaces)*, for $\overset{+}{W}(p)$ (and for $\overset{-}{W}(p)$).

Lemma 5.6 showed that a totally null member of $\overset{+}{S}_p$ gives rise to a repeated principal 2−space of $\overset{+}{W}(p)$ if and only if it is an eigenbivector of $\overset{+}{W}(p)$ and this reveals how to compute repeated pnds for $\overset{+}{W}(p)$. There is a similar characterisation of non-repeated principal 2−spaces.

Lemma 5.7 *Any non-repeated principal null direction for $\overset{+}{W}(p)$ lies in a 2−space which is the blade of a simple bivector in $\overset{+}{S}_p$ each non-zero member of which is a non-repeated principal null direction for $\overset{+}{W}(p)$ (a principal 2−space for $\overset{+}{W}(p)$). If $Q \in \overset{+}{S}_p$ is totally null the blade of Q is a (non-repeated) principal 2−space of $\overset{+}{W}(p)$ if and only if $\overset{+}{W}_{abcd}Q^{cd} = aQ_{ab} + Z_{ab}$ ($a \in \mathbb{R}$) where $Z \in \overset{+}{S}_p$ is non-zero, not proportional to Q and $Q \cdot Z = 0$. These equivalent conditions are themselves equivalent to the conditions that Q is not an eigenbivector of $\overset{+}{W}(p)$ and $\overset{+}{W}_{abcd}Q^{ab}Q^{cd} = 0$.*

Proof The first part was given above. For the remainder choose a null basis such that, in the present language, $Q = G$ is a (non-repeated) principal 2−space of $\overset{+}{W}(p)$ and then using (5.38) one sees that $C = 0 \neq E$ and $Z = EF$ so that $0 \neq Z \in \overset{+}{S}_p$. Conversely, if Q is totally null and $\overset{+}{W}_{abcd}Q^{cd} = aQ_{ab} + Z_{ab}$ with $a \in \mathbb{R}$, $0 \neq Z$, $Q \cdot Z = 0$ and Z not proportional to Q, choose $Q = G$ and use (5.38) to get $C = 0 \neq E$ and $Q = G$ is a non-repeated principal 2−space of $\overset{+}{W}(p)$. The last part of the lemma is clear since the choice $Q = G$ in (5.38) is then equivalent to $C = 0 \neq E$. $\qquad\square$

To be able to compute the number of repeated and non-repeated principal 2−spaces for $\overset{+}{W}(p)$ first write out $\overset{+}{W}(p)$ in terms of a null basis l', n', L', N' as in (5.38) but with the corresponding coefficients primed, $A', ..., E'$. Now perform a change of null basis to l, n, L, N which fixes the bivector $l \wedge N$ up to a scaling and hence fixes its blade. Thus $l' = al + bN, N' = cN + dl$ with $a, b, c, d \in \mathbb{R}$ and $ac - bd \neq 0$. The expressions for n' and L' are then found from the fact that l', n', L', N' is a null basis. The result is, after some calculation, and for $\rho \in \mathbb{R}$ and $\lambda = (ac - bd)^{-1}$,

$$l' = al + bN, \qquad N' = cN + dl,$$
$$n' = \lambda(cn - dL) - \rho(cN + dl), \qquad L' = \lambda(aL - bn) + \rho(bN + al). \quad (5.45)$$

One then uses (5.45) to write out the associated bivectors G', F' and H' (written in terms of the basis l', n', L', N' just as F, G, H are in terms of the basis l, n, L, N) in terms of F, G and H. Before proceeding further, it is remarked that the general idea is to show that using the transformations (5.45) (fixing the blade of $G = l \wedge N$) one may map $H = n \wedge L$, up to a scaling, to *any* other totally null bivector $H' = n' \wedge L' \in \overset{+}{S}_p$ except G, that is, it maps the blade of H to any other totally null 2−space at p except the blade of G. It is then clear from (5.45) that the totally null 2−spaces achievable under this transformation is not restricted by seeking $(\lambda^{-1}n') \wedge (\lambda^{-1}L')$, that is, one may simplify (5.45) by setting $\lambda = 1$. Then (5.45) gives

$$G' = G, \qquad F' = F - \sqrt{2}\rho G, \qquad H' = H + \rho^2 G - \sqrt{2}\rho F. \quad (5.46)$$

A substitution of these into (5.38) leads to a comparison of the coefficients $A', ... E'$ with $A, ..., E$ due to this basis change. This calculation gives

$$A = A' + \rho^2 C' - \sqrt{2}\rho E',$$
$$B = B' + 6\rho^2 A' + \rho^4 C' - 2\sqrt{2}\rho D' - 2\sqrt{2}\rho^3 E',$$
$$C = C',$$
$$D = D' - 3\sqrt{2}\rho A' - \sqrt{2}\rho^3 C' + 3\rho^2 E',$$
$$E = E' - \sqrt{2}\rho C'. \quad (5.47)$$

To see that under the above basis transformations the totally null bivector H may be transformed by (5.46) to (a multiple of) *any* other totally null bivector in $\overset{+}{S}_p$ except G note that one may write this latter totally null bivector (up to proportionality) as $H + \beta F + \frac{\beta^2}{2}G$ ($\beta \in \mathbb{R}$) and then the transformation (5.46) with $-\sqrt{2}\rho = \beta$ ((5.45) with $\lambda = 1$ and $-\sqrt{2}\rho = \beta$) is the required one.

The method of finding (repeated and non-repeated) principal 2−spaces for a general $\overset{+}{W}(p)$ involves first writing out $\overset{+}{W}(p)$ in terms of a general null basis

l', n', L', N' and associated coefficients $A', ..., E'$ as in (5.38). One then seeks a transformation of the form (5.45), allied with (5.46), to a new (unprimed) null basis l, n, L, N in which $\overset{+}{W}_{abcd}H^{ab}H^{cd} = 0$, that is, $B = 0$, and which, from lemma 5.7, gives a (repeated or non-repeated) principal 2−space which is the blade of H. Equation (5.47) then shows that $B = 0$ is a polynomial equation for ρ for which one needs *real* solutions and there are at most four of them. Thus $\overset{+}{W}(p)$ admits at most four (repeated or non-repeated) principal 2−spaces. If one specifically requires *repeated* principal 2−spaces one needs real solutions for ρ of $B = D = 0$. If it should happen that G' was a (repeated or non-repeated) principal 2−space for $\overset{+}{W}(p)$, and so $\overset{+}{W}_{abcd}G'^{ab}G'^{cd} = 0$, one has $C' = 0$ and the above polynomial condition $B = 0$ becomes a cubic, as expected. [Thus whether G is a principal 2−space or not one essentially "anchors down" $G = G'$ and uses (5.45) to scan all other totally null 2−spaces to see if any of them are principal. To this collection G is added should it be principal.]

More specifically one first selects an algebraic type for $\overset{+}{W}(p)$ from the above list, written out in the notation of (5.38) but with primes on the associated coefficients and basis bivectors. Thus for type $\mathbf{I_R}$, $A' = \rho_1$, $B' = C' = \rho_2$, $D' = E' = 0$ from (5.32) and, to avoid eigenvalue degeneracies, $3\rho_1 \neq \pm\rho_2 \neq 0$. Since $\overset{+}{W}_{abcd}G'^{ab}G'^{cd} = C' \neq 0$ the 2−space represented by G' is not a (repeated or non-repeated) principal 2−space and the quartic equation $B = 0$ (actually a quadratic equation in ρ^2) requires $9\epsilon^2 - 1 \geq 0$, ($\epsilon = \frac{\rho_1}{\rho_2}$), in order for there to be real solutions for ρ^2. The inequalities $0 \neq \pm\rho_2 \neq 3\rho_1$, to ensure distinct eigenvalues, then give $9\epsilon^2 - 1 \neq 0$ and so one requires $9\epsilon^2 - 1 > 0$ and the real solutions for ρ^2 are distinct. Writing $\mu^2 \equiv 9\epsilon^2 - 1 > 0$ one gets $9\epsilon^2 - \mu^2 > 0$ and so $(3\epsilon + \mu)(3\epsilon - \mu) > 0$. Thus the numbers $3\epsilon \pm \mu$ are distinct and are either both positive or both negative and the condition $B = 0$ gives the solution $\rho^2 = -3\epsilon \pm \mu$. Thus these two (distinct) solutions for ρ^2 are either both positive or both negative. It follows that either there are four distinct solutions for ρ, or there are none. The former case is the one required here and is given by $\rho^2 = -3\epsilon \pm \mu > 0$. In this case, there are four distinct *non-repeated* principal 2−spaces for $\overset{+}{W}(p)$ since with these values for ρ_1 and ρ_2 there can be no real solutions for ρ of $D = 0$ (otherwise one gets $3\epsilon = -\rho^2$ and the contradiction $\mu = 0$). Thus there are no real repeated principal 2−spaces. For type $\mathbf{I_C}$ one has $A' = \rho_1$, $B' = -C' = \rho_2$, $D' = E' = 0$ and a similar argument shows that there are two distinct real and two distinct complex solutions for ρ to the quartic $B = 0$ and none for $D = 0$ and hence two distinct non-repeated principal 2−spaces arise for $\overset{+}{W}(p)$. The other types are handled similarly. For type $\mathbf{N}$ one has $A' = C' = D' = E' = 0$ and $B' = \pm 1$ and, of course, $G' = l' \wedge N'$ is a repeated principal 2−space for $\overset{+}{W}(p)$. Then there are no other solutions of $B = 0$. Thus G' is the unique principal, necessarily repeated

2−space for $\overset{+}{W}(p)$ and it is, in this sense, a quadruply repeated solution of the quartic $B = 0$. For type **III** one has $A' = B' = C' = E' = 0$ and $D' = 1$ and it is known that $G' = l' \wedge N'$ is a repeated principal 2−space for $\overset{+}{W}(p)$. Then the polynomial equation $B = 0$ is a linear equation with root $\rho = 0$ and then $D = D' = 1$. The 2−space resulting from $\rho = 0$ is thus a non-repeated principal 2−space for $\overset{+}{W}(p)$ and is, from (5.45), $n' \wedge L' = n \wedge L$. The solution $l \wedge N$ is thus a triply repeated principal 2−space for $\overset{+}{W}(p)$. For type **II** one has $C' = D' = E' = 0$, $B' = \pm 1$ and $A' = \rho_1 \neq 0$ and it is known that $G' = l' \wedge N'$ is a repeated principal 2−space for $\overset{+}{W}(p)$. The equation $B = 0$ gives the quadratic $0 = \pm 1 + 6\rho_1\rho^2$ and which reveals that the above known solution is doubly repeated. In addition, should two complex solutions arise from $B = 0$, this known solution above is the only (real) one whereas if two real (necessarily distinct) solutions arise from $B = 0$, one gets $D \neq 0$ in each case and these solutions are non-repeated principal 2−spaces for $\overset{+}{W}(p)$ and can be calculated from (5.45). If the type is $\mathbf{D_1}$ one has $A' = \rho_1 \neq 0$ and $B' = C' = D' = E' = 0$ and one already knows that $G = l \wedge N$ is a repeated principal 2−space for $\overset{+}{W}(p)$. The equation $B = 0$ then gives the equation $\rho^2 = 0$ whose repeated solutions $\rho = 0$ each give $D = 0$ and so $n' \wedge L' = n \wedge L$ is also a repeated principal 2−space for $\overset{+}{W}(p)$. Thus there are two doubly repeated principal 2−spaces $l' \wedge N' = l \wedge N$ and $n' \wedge L' = n \wedge L$ for in this case. If the type is $\mathbf{D_2}$ one has $A' = \rho_1$, $B' = C' = 3\rho_1$ ($\rho_1 \neq 0$) and $D' = E' = 0$ and the equation $B = 0$ gives a quartic which actually a quadratic in ρ^2 with two repeated roots $\pm i$ and no real solutions result, as observed earlier. This completes the survey of the algebraic types. In summary, one can, for each (non-zero) type, append a pair (m, n) where m (respectively n,) is the number of repeated (respectively, non-repeated,) *real* principal 2−spaces for that type. These are $\mathbf{I_{\mathbb{R}}}$ ((0,0) or (0,4)), $\mathbf{I_{\mathbb{C}}}$ ((0,2)), **II** ((1,0) or (1,2)), **III** ((1,1)), $\mathbf{D_1}$ ((2,0)), $\mathbf{D_2}$ ((0,0)) and **N** ((1,0)). This argument justifies the use of the term "repeated" in describing principal totally null 2−spaces.

5.8   Weyl Conformal Tensor II

The last section described the classification of $\overset{+}{W}(p)$ and the tensor $\overset{-}{W}(p)$ can similarly be dealt with. This gives an algebraic classification of the Weyl conformal tensor $C(p)$ described by the pair $(\mathbf{A}, \mathbf{B})$ where $\mathbf{A}$ is one of the above algebraic types for $\overset{+}{W}(p)$, $\mathbf{B}$ is the algebraic type for $\overset{-}{W}(p)$ and where, in addition, $\mathbf{A}$ or $\mathbf{B}$ could be the trivial type $\mathbf{o}$. The trivial case for $C(p)$ is thus $(\mathbf{O}, \mathbf{O})$. For such purposes the pairs $(\mathbf{A}, \mathbf{B})$ and $(\mathbf{B}, \mathbf{A})$ will be identified, that is, such pairs are here regarded as unordered. There are Bel-type criteria for $C(p)$ similar to those for $\overset{+}{W}(p)$ and this will be developed now. Before this is done it is remarked that if P and Q are bivectors in $\Lambda_p M$ one may, as described earlier, decompose them uniquely as $P = \overset{+}{P} + \overset{-}{P}$ for $\overset{+}{P} \in \overset{+}{S}_p$ and $\overset{-}{P} \in \overset{-}{S}_p$, and similarly for Q. Now suppose that $C_{abcd}P^{cd} = Q_{ab}$ (written briefly as $CP = Q$) at p. Then it follows from the remarks above that $\overset{+}{W}\overset{+}{P} = \overset{+}{Q}$ and $\overset{-}{W}\overset{-}{P} = \overset{-}{Q}$. It follows that any *real* eigenbivector of $C(p)$ which is *not* in $\widetilde{S}_p$ decomposes into an eigenbivector of $\overset{+}{W}(p)$ in $\overset{+}{S}_p$ and an eigenbivector of $\overset{-}{W}(p)$ in $\overset{-}{S}_p$, each of which is an eigenbivector of $C(p)$ and all of these have the same eigenvalue. [Alternatively, one could take the dual of the equation $CP = \lambda P$ for $\lambda \in \mathbb{R}$ and $P \notin \widetilde{S}_p$ to see that $CP^* = \lambda P^*$ with P and P^* independent. Adding these leads to the same conclusion.] Thus any eigenbivector of $C(p)$ which is *not* in $\widetilde{S}_p$ lies in an eigenspace of $C(p)$ of dimension at least 2 and which intersects both $\overset{+}{S}_p$ and $\overset{-}{S}_p$ and hence, in this sense, one may regard all real eigenbivectors of $C(p)$ as lying in $\widetilde{S}_p$. In particular if $P \notin \widetilde{S}_p$ and $CP = 0$ then $\overset{+}{W}\overset{+}{P} = 0$ and $\overset{-}{W}\overset{-}{P} = 0$ and so each of $\overset{+}{W}$ and $\overset{-}{W}$ has a zero eigenvalue.

Lemma 5.8 (*i*) *Suppose* $C(p) \neq 0$. *There exists* $0 \neq k \in T_p$ *such that* $C_{abcd}k^d = 0$ *if and only if* $C(p)$ *is of type* $(\mathbf{N}, \mathbf{N})$ *or type* $(\mathbf{N}, \mathbf{O})$. *The vector* k *is necessarily null and for the first of these types is unique up to a scaling whilst for the second* k *may be any non-zero member of a certain totally null* $2-$*space at* p.

(*ii*) *Suppose* $C(p) \neq 0$ *and let* Q *be a (real) null eigenbivector of* $C(p)$ *with eigenvalue* $\lambda \in \mathbb{R}$. *Then using the above decomposition of* Q, $\overset{+}{Q}$ *and* $\overset{-}{Q}$ *are totally null eigenbivectors of* $\overset{+}{W}(p)$ *and* $\overset{-}{W}(p)$, *respectively, with eigenvalue* λ. *If* $\lambda \neq 0$, $C(p)$ *has algebraic type* $(\mathbf{A}, \mathbf{B})$ *where* $\mathbf{A}$ *is* $\mathbf{II}$ *or* $\mathbf{D_1}$, *and similarly for* $\mathbf{B}$. *If* $\lambda = 0$, $C(p)$ *has type* $(\mathbf{A}, \mathbf{B})$ *where* $\mathbf{A}$ *is* $\mathbf{III}$, $\mathbf{N}$ *or* $\mathbf{O}$, *and similarly for* $\mathbf{B}$ (*and, of course, type* $(\mathbf{O}, \mathbf{O})$ *is forbidden*).

(*iii*) *Suppose* $C(p) \neq 0$ *and that there exists* $0 \neq k \in T_p M$ *with* $C_{abcd}k^d = Q_{ab}k_c$ *for some bivector* $Q \neq 0$. *Then* Q *is simple,* k *lies in its blade and* k *is necessarily null. If* $Q \notin \widetilde{S}_p$ *it is necessarily null,* $\overset{+}{Q}$ *and* $\overset{-}{Q}$ *are each totally null,* k *is unique up to a scaling and* $C(p)$ *is of type* $(\mathbf{III}, \mathbf{III})$. *If* $Q \in \overset{+}{S}_p$, $\overset{-}{Q} = 0$ *and the type for* $C(p)$ *is either* $(\mathbf{III}, \mathbf{N})$ *with* k *unique up to a scaling, or* $(\mathbf{III}, \mathbf{O})$ *with* k *any non-zero member of the blade of* Q.

Proof For part (*i*) one has $*C_{abcd}k^d = 0$ and so $\overset{+}{W}_{abcd}k^d = 0$ and $\overset{-}{W}_{abcd}k^d = 0$ and the type for $C(p)$ follows from lemma 5.6(*i*) as does the fact that k is null since if $C(p) \neq 0$ one of $\overset{+}{W}(p)$ and $\overset{-}{W}(p)$ is not zero. The same lemma shows that if $\overset{+}{W}(p)$ and $\overset{-}{W}(p)$ are of type **N** they may be written as αPP and βQQ, respectively, with P and Q totally null, $P \in \overset{+}{S}(p)$ and $Q \in \overset{-}{S}(p)$ and $\alpha, \beta \in \mathbb{R}$. Lemma 5.5(v) then shows that one may construct a null basis l', n', L', N' at p so that, for non-zero $\alpha, \beta \in \mathbb{R}$, $C(p) = \alpha G'G' + \beta \bar{H}'\bar{H}'$ with $G' = l' \wedge N'$ and $\bar{H}' = l' \wedge L'$ and with the blades of G' and $\bar{H}'$ intersecting in the unique null direction spanned by $l'(= k)$. A change of null basis to l, n, L, N with $l' = \lambda l$, $n' = \lambda^{-1}n$, $L' = \mu L$ and $N' = \mu^{-1}N$ with $\mu, \lambda \in \mathbb{R}$, $\mu^4 = \frac{|\alpha|}{|\beta|}$ and $\lambda^2 = |\beta|^{-1}\mu^{-2}$ then gives $C_{abcd} = \pm G_{ab}G_{cd} \pm \bar{H}_{ab}\bar{H}_{cd}$ with $G = l \wedge N$ and $\bar{H} = l \wedge L$. For type $(\mathbf{N}, \mathbf{O})$ any non-zero member of the blade of the totally null eigenbivector in lemma 5.6(i) may be identified with k. For part (ii) one has $CQ = \lambda Q$ and hence $\overset{+}{W}\overset{+}{Q} = \lambda \overset{+}{Q}$ and $\overset{-}{W}\overset{-}{Q} = \lambda \overset{-}{Q}$ with $\overset{+}{Q}, \overset{-}{Q}$ totally null and $\overset{+}{Q} \in \overset{+}{S}_p$, $\overset{-}{Q} \in \overset{-}{S}_p$ (see lemma 5.1(v)). The result now follows from lemma 5.6(ii). For part (iii) the given condition together with $C_{a[bcd]} = 0$ shows that Q is simple with k in its blade, then $C_{abcd}k^c k^d = 0$ implies that k is null and finally $C^c{}_{abc} = 0$ implies that $Q_{ab}k^b = 0$. Now $\overset{*}{C}_{abcd}k^d = \overset{*}{Q}_{ab}k_c (\Rightarrow \overset{*}{Q}_{ab}k^b = 0)$ and so $\overset{+}{W}_{abcd}k^d = \overset{+}{Q}_{ab}k_c$ and $\overset{-}{W}_{abcd}k^d = \overset{-}{Q}_{ab}k_c$ with $\overset{+}{Q} = Q + \overset{*}{Q} \in \overset{+}{S}_p$ and $\overset{-}{Q} = Q - \overset{*}{Q} \in \overset{-}{S}_p$. Thus if $Q \notin \widetilde{S}_p$, $\overset{+}{Q} \neq 0 \neq \overset{-}{Q}$, Q is null, $\overset{+}{Q}$ and $\overset{-}{Q}$ totally null and k is unique up to scaling (see lemma 5.1(v)). Then lemma 5.6(iii) shows that the type of $C(p)$ is $(\mathbf{III}, \mathbf{III})$. If, however, $Q \in \overset{+}{S}_p$ with $\overset{-}{Q} = 0$, lemma 5.6(iii) shows that the type of $C(p)$ is either $(\mathbf{III}, \mathbf{N})$ with k unique up to scaling, or $(\mathbf{III}, \mathbf{O})$ with k any non-zero member of the blade of $\overset{+}{Q}$.

It is remarked that, with careful wording, simple converses of parts (ii) and (iii) exist and that the unqualified "and conversely" used in [68] is misleading. □

One may now proceed to the concepts of repeated and non-repeated principal null directions for $C(p)$. Suppose that $C(p) \neq 0$, that $0 \neq k \in T_pM$ and consider the following equations

$$k_{[e}C_{a]bc[d}k_{f]}k^b k^c = 0, \qquad C_{abcd}k^b k^c = k_a q_d + q_a k_d, \qquad (5.48)$$

where q is a non-zero 1–form at p. It is straightforward to check that these two equations are equivalent by a consideration of the second-order *symmetric* tensor $C_{abcd}k^b k^d$ and the index symmetries of C. Applying the tracefree condition $g^{ad}C_{abcd} = 0$ to the second of (5.48) gives $k \cdot q = 0$ and then a contraction of the same equation with k^a reveals that k is necessarily null. A (null) vector k satisfying (5.48) is said to span a *principal null direction* (a *pnd*) of $C(p)$. It is remarked here that if a term $\lambda k_a k_b$ $(\lambda \in \mathbb{R})$ is added to the right hand side of the second equation of (5.48) the original form is retained by replacing q by $q + \frac{\lambda}{2}k$. This is useful in establishing the equivalence of the two equations in (5.48). Now consider the equations

$$C_{abcd}k^b k^d = \alpha k_a k_c, \qquad k_{[e}C_{a]bcd}k^b k^d = 0, \qquad (5.49)$$

for $0 \neq k \in T_p M$ and $\alpha \in \mathbb{R}$. It is again easily checked that these two equations are equivalent. If $\alpha \neq 0$ then a contraction of the first equation with k^a shows that k is necessarily null. However, if $\alpha = 0$ this is not true since if, in a null basis l, n, L, N, $C_{abcd} = G_{ab}G_{cd} - \bar{G}_{ab}\bar{G}_{cd}$ with $G = l \wedge N$ and $\bar{G} = l \wedge L$ (this is of type $(\mathbf{N}, \mathbf{N})$ in the above notation), $C_{abcd}r^b r^d = 0$, where $r = L + N$ is not null. This leads to the following definition; If k satisfies (5.49) *and is null*, k is said to span a *repeated principal null direction* (a *repeated pnd*) of $C(p)$. If k satisfies (5.48) (and is hence null) but not (5.49) it is called a *non-repeated principal null direction* (*non-repeated pnd*) of $C(p)$. This will occur if and only if (5.48) holds with $q \neq 0$ *and* not proportional to k. In general one often abuses notation by referring to repeated and non-repeated principal null directions as principal null directions (pnds) unless problems might arise.

One now seeks a relationship between the non-repeated and the repeated pnds of $\overset{+}{W}(p)$ and those of $C(p)$. This is provided by the following lemma which is also taken to justify the use of the term "repeated" for pnds of $C(p)$.

Lemma 5.9 (*i*) *A (null) vector $l \in T_p M$ spans a repeated pnd for $C(p)$ if and only if it spans a repeated pnd for each of $\overset{+}{W}(p)$ and $\overset{-}{W}(p)$.*

(*ii*) *A (null) vector $l \in T_p M$ spans a non-repeated pnd for $C(p)$ if and only if it spans a pnd for each of $\overset{+}{W}(p)$ and $\overset{-}{W}(p)$ and is non-repeated for at least one of them.*

Proof First let l span a pnd for $C(p)$ so that (5.48) holds with q possibly a multiple of l. Then write $C(p) = \overset{+}{W}(p) + \overset{-}{W}(p)$ and choose a null basis l, n, L, N with associated bivectors $F, G, H \in \overset{+}{S}_p$ as given earlier and $\bar{G} = \frac{1}{\sqrt{2}}l \wedge L$, $\bar{F} = \frac{1}{2}(l \wedge n + L \wedge N)$ and $\bar{H} = \frac{1}{\sqrt{2}}n \wedge N$ members of $\overset{-}{S}_p$. Then $C_{abcd}G^{cd} = \overset{+}{W}_{abcd}G^{cd} = \sqrt{2}C_{abcd}l^c N^d$ and the first equation of (5.48) contracted with N^d gives $C_{abcd}l^b l^c N^d = (N \cdot q)l_a$ and so if $V_{ab} \equiv \overset{+}{W}_{abcd}G^{cd}$, $V_{ab}l^b = \sqrt{2}C_{abcd}l^b l^c N^d = \sqrt{2}(N \cdot q)l_a$. A similar argument using the tetrad member L and the bivector $\bar{G}$ gives, for $\bar{V}_{ab} = C_{abcd}\bar{G}^{cd} = \overset{-}{W}_{abcd}\bar{G}^{cd} = \sqrt{2}C_{abcd}l^c L^d$, $\bar{V}_{ab}l^b = \sqrt{2}(L \cdot q)l_a$. It follows that V is a linear combination of G and F, and $\bar{V}$ of $\bar{G}$ and $\bar{F}$. Now if l spans a repeated pnd of $C(p)$, q is a multiple of l and so $N \cdot q = L \cdot q = 0$. Thus V (respectively, $\bar{V}$) is a multiple of G (respectively, $\bar{G}$) and hence l spans a repeated pnd for $\overset{+}{W}(p)$ and $\overset{-}{W}(p)$, from lemma 5.6(*ii*). The converse is immediate from (5.27). If l spans a non-repeated pnd of $C(p)$ then (5.48) holds with $q \cdot l = 0$ but to avoid q being proportional to l (possibly zero), at least one of $N \cdot q$ and $L \cdot q$ is non-zero. Then the above, and lemma 5.7 show that l is a pnd for $\overset{+}{W}(p)$ and $\overset{-}{W}(p)$ but is non-repeated for at least one of them. Conversely, if l spans a pnd for $\overset{+}{W}(p)$ and $\overset{-}{W}(p)$ but is non-repeated for at least one of them, l satisfies (5.48) and is non-repeated from part (i). □

To take advantage of this result one recalls that it is (repeated and non-repeated) totally null principal 2−spaces that are important for $\overset{+}{W}(p)$ and $\overset{-}{W}(p)$. But it was shown earlier that if, with an abuse of language, $U \in \overset{+}{S_p}$ and $U' \in \overset{-}{S_p}$, with U and U' totally null, their blades must intersect in a single null direction spanned by, say, l at p. It follows that the map which associates the pair (U, U') with the null direction $U \cap U'$ is an *injective* map on an obvious subset of $\overset{+}{S_p} + \overset{-}{S_p}$, otherwise one would get the contradiction that two distinct members of either $\overset{+}{S_p}$ or $\overset{-}{S_p}$ have a non-trivial intersection. Then if U and U' are repeated principal 2−spaces for $\overset{+}{W}(p)$ and $\overset{-}{W}(p)$, respectively, the above lemma shows that l is a repeated pnd for $C(p)$, whereas if U and U' are principal 2−spaces for $\overset{+}{W}(p)$ and $\overset{-}{W}(p)$, respectively, with at least one of U, U' non-repeated, l is a non-repeated pnd for $C(p)$. Thus the repeated and non-repeated pnds for $C(p)$ are easily found from those of $\overset{+}{W}(p)$ and $\overset{-}{W}(p)$.

It is recalled from the Lorentz case and the Petrov classification that the number n of (repeated and non-repeated) pnds of the (assumed non-zero) Weyl conformal tensor satisfies $1 \leq n \leq 4$. In the case of neutral signature discussed here the situation is a little different. For example, if $C(p)$ is of type $(\mathbf{A}, \mathbf{B})$ with $\mathbf{A}$ and $\mathbf{B}$ each of either type $\mathbf{D_2}$ or that subclass of type $\mathbf{I_\mathbb{R}}$ for which no real principal 2−spaces arise, $C(p)$ admits no pnds. On the other hand if $C(p)$ is of type $(\mathbf{A}, \mathbf{O})$ for $\mathbf{A}$ any type except $\mathbf{D_2}$ and the above mentioned subclass of type $\mathbf{I_\mathbb{R}}$, infinitely many pnds arise since then $C(p) = \overset{+}{W}(p)$ (lemmas 5.6 and 5.7). In fact, only when exactly one of the type pairs for $C(p) \neq 0$ is $\mathbf{O}$ can this last situation occur. As other examples consider the case when $C(p)$ has type $(\mathbf{N}, \mathbf{N})$. In this case $\overset{+}{W}(p)$ and $\overset{-}{W}(p)$ each admit a single repeated principal 2−space and these lie in $\overset{+}{S_p}$ and $\overset{-}{S_p}$, respectively. Their intersection results in a (unique) repeated pnd for $C(p)$ and there are no non-repeated ones. This situation is written $(1, 0)$. If $C(p)$ has type $(\mathbf{N}, \mathbf{III})$ with $\overset{+}{W}(p)$ as above for the type $\mathbf{N}$ case and where $\overset{-}{W}(p)$ admits exactly one repeated and exactly one non-repeated principal 2−space, the appropriate intersections yield exactly one repeated and exactly one non-repeated pnd for $C(p)$, written $(1, 1)$. A fuller list can be found in [68].

Another analogy between the neutral signature classification of $C(p)$ and the Petrov classification in Lorentz signature arises from the following remarks. In the Petrov case attention centred on the tensor $\overset{+}{C}$ and its expression in terms of complex self-dual bivectors, that is, members of $\overset{+}{S}_p$ for that signature. The classification could equally well have been accomplished with respect to the tensor $\overset{-}{C}$ (the conjugate of $\overset{+}{C}$) and in terms of the anti-self dual bivectors in $\overset{-}{S}_p$ which are the conjugates of those in $\overset{+}{S}_p$. These approaches yield the eigenstructure of $\overset{+}{C}$ and $\overset{-}{C}$ in terms of bivectors in $\overset{+}{S}_p$ and $\overset{-}{S}_p$, respectively and, of course, lead to the same (Petrov) type. In this structure the complex null members of $\overset{+}{S}_p$ and $\overset{-}{S}_p$ play an important role and also figure prominently in the calculation of the Bel criteria for the Petrov types. Now reference to lemma 4.5 reveals that these complex null bivectors are totally null and contain (up to complex multiples) a *unique real* direction (which is necessarily null) called its principal null direction. Further, the intersection of the blades of a conjugate pair of complex null bivectors, necessarily one in each of $\overset{+}{S}_p$ and $\overset{-}{S}_p$, is, again up to complex multiples, their common real null direction (chapter 4). Thus the classification of $\overset{+}{C}$ and $\overset{-}{C}$ in Lorentz signature reveals "special" conjugate pairs of complex null bivectors whose (real null) intersections constitute the principal null directions of C. So the classifications of the Weyl conformal tensor in Lorentz and neutral signatures are mathematically similar but differ in the following sense. In the Lorentz case, each of $\overset{+}{C}$ and $\overset{-}{C}$ give the same algebraic type (unlike $\overset{+}{W}$ and $\overset{-}{W}$) with the real pnds for C being unique real null directions in the blades of complex null bivectors. In the case of neutral signature $\overset{+}{W}$ and $\overset{-}{W}$ have, in general, different algebraic types and the real pnds are fixed by blade intersections of principal 2−spaces for $\overset{+}{W}$ and $\overset{-}{W}$ [46].

Thus for neutral signature it is seen that the tensors $\overset{+}{W}$ and $\overset{-}{W}$ may each, at any $p \in M$, be classified into one of the 8 types $\mathbf{I}_{\mathbb{C}}$, $\mathbf{I}_{\mathbb{R}}$, $\mathbf{D_1}$, $\mathbf{D_2}$, $\mathbf{II}$, $\mathbf{III}$, $\mathbf{N}$ and $\mathbf{O}$. Then the Weyl conformal tensor C may be classified into the types $(\mathbf{A}, \mathbf{B})$ where each of $\mathbf{A}$ and $\mathbf{B}$ is one of the above types for $\overset{+}{W}$ and $\overset{-}{W}$, respectively. One can now establish a decomposition of M with respect to the Weyl conformal tensor by first considering the decompositions of M with respect to $\overset{+}{W}$ and $\overset{-}{W}$. First consider $\overset{+}{W}$ and let the symbol $\mathbf{I}_{\mathbb{C}}$, denote precisely that subset of points of M at which the algebraic type is $\mathbf{I}_{\mathbb{C}}$, and similarly for the other algebraic types of $\overset{+}{W}$. Then one has the disjoint decomposition $M = \mathbf{I}_{\mathbb{C}} \cup \mathbf{I}_{\mathbb{R}} \cup \mathbf{D_1} \cup \mathbf{D_2} \cup \mathbf{II} \cup \mathbf{III} \cup \mathbf{N} \cup \mathbf{O}$ from which, by taking interiors, one gets the disjoint decomposition

$$M = int\mathbf{I}_{\mathbb{C}} \cup int\mathbf{I}_{\mathbb{R}} \cup int\mathbf{D_1} \cup int\mathbf{D_2} \cup int\mathbf{II} \cup int\mathbf{III} \cup int\mathbf{N} \cup int\mathbf{O} \cup F \quad (5.50)$$

where, as before, int denotes the interior operator in the manifold topology of M and F is a closed subset of M defined by the disjointness of the decomposition. Now consider the matrix rank of the 3×3 matrix $\overset{+}{W}(p)$ (equal to the rank of the map $\overset{+}{f}_C$) easily obtainable from the original classification. If $\overset{+}{W}(p)$ is of type $\mathbf{I}_C$ or $\mathbf{I}_\mathbb{R}$ its rank is 2 or 3, if of type $\mathbf{D}_1$, $\mathbf{D}_2$ or $\mathbf{II}$ its rank is 3, if of type $\mathbf{III}$ its rank is 2, if of type $\mathbf{N}$ its rank is 1 and if of type $\mathbf{O}$ its rank is 0. It will now be shown that $\mathrm{int}F = \emptyset$. First it is recalled (chapter 4) that any simple root of the characteristic polynomial of $\overset{+}{W}(p)$ arises from a smooth function on some open neighbourhood U of p whose value at p is a simple root of $\overset{+}{W}(p)$ on U. It follows that the subsets $\mathbf{I}_C$ and $\mathbf{I} \equiv \mathbf{I}_C \cup \mathbf{I}_\mathbb{R}$ are open subsets of M and hence equal to their interiors. (Clearly this may not be true for $\mathbf{I}_\mathbb{R}$). Then using an obvious rank theorem (chapter 1) and the above rank results one sees that the subsets $A \equiv \mathbf{I}_C \cup \mathbf{I}_\mathbb{R} \cup \mathbf{D}_1 \cup \mathbf{D}_2 \cup \mathbf{II}$, $A \cup \mathbf{III}$ and $A \cup \mathbf{III} \cup \mathbf{N}$ are also open subsets of M. Now let $U \subset F$ be open and non-empty. It will be shown that this leads to a contradiction and hence that $\mathrm{int}F = \emptyset$. The disjointness of (5.50) and the known result $\mathbf{I}_C = \mathrm{int}\mathbf{I}_C$ shows that U is disjoint from $\mathbf{I}_C$ and hence that the (open) subset $U \cap \mathbf{I} = U \cap \mathbf{I}_\mathbb{R}$. If $U \cap \mathbf{I}_\mathbb{R}$ is not empty then neither is $U \cap \mathrm{int}\mathbf{I}_\mathbb{R}$ and this contradicts the disjointness of (5.50). Thus $U \cap \mathbf{I} = \emptyset$. Now define the open subset $U' \equiv U \cap A = U \cap (\mathbf{D}_1 \cup \mathbf{D}_2 \cup \mathbf{II})$. If $U' = \emptyset$, U is disjoint from A. Otherwise suppose that the open subset U' is not empty but that $U \cap \mathbf{II} = \emptyset$, which implies that $U' = U \cap (\mathbf{D}_1 \cup \mathbf{D}_2)$ and hence that $U' \subset \mathbf{D}_1 \cup \mathbf{D}_2$. But $U' \subset \mathbf{D}_1$ and $U' \subset \mathbf{D}_2$ are each impossible, otherwise U' and hence U would intersect non-trivially the sets $\mathrm{int}\mathbf{D}_1$ or $\mathrm{int}\mathbf{D}_2$, contradicting the disjointness of (5.50). Thus $U' \cap \mathbf{D}_1 \neq \emptyset$ and $U' \cap \mathbf{D}_2 \neq \emptyset$. Let $p' \in U' \cap \mathbf{D}_1$. Since the Segre type, including degeneracies, is fixed at $\{1(11)\}$ throughout U' there exists an open neighbourhood $V \subset U'$ of p' at each point of which the (locally smooth) simple eigenvalue of $\overset{+}{W}$ gives rise to its associated smooth eigenbivector Q [29] satisfying $|Q(p)| < 0$ (since $p \in \mathbf{D}_1$). Thus this algebraic type ($\mathbf{D}_1$) will be the same over some neighbourhood V' of p' contained in V and hence contained in $\mathbf{D}_1$ which leads to $U' \cap \mathrm{int}\mathbf{D}_1 \neq \emptyset$ and hence to $U \cap \mathrm{int}\mathbf{D}_1 \neq \emptyset$ which contradicts the disjointness of (5.50). It follows that $U \cap \mathbf{II} \neq \emptyset$. So choose $p'' \in U \cap \mathbf{II}$ so that the characteristic polynomial of $\overset{+}{W}$ has a simple root at p'' and gives rise to a smooth function γ on some open neighbourhood V'' of p'' with $V'' \subset U' \subset U$ and with the other root of this polynomial being, from the tracefree condition, $-\frac{1}{2}\gamma$. This gives rise to the smooth polynomial function $Z \equiv (\overset{+}{W} - \gamma I_3)(\overset{+}{W} + \frac{\gamma}{2}I_3)$ on V'' where I_3 is the unit 3×3 matrix. But then, the minimal polynomial structure of $\overset{+}{W}$ (see chapter 1) shows that Z vanishes at those points where the type is $\mathbf{D}_1$ or $\mathbf{D}_2$ (since these types have only simple elementary divisors) but not where the type is $\mathbf{II}$ (since then a non-elementary divisor arises). Thus Z does not vanish over some neighbourhood of p'' and hence $U \cap \mathbf{II}$ is non-empty and open and one achieves the contradiction that $U \cap \text{int} \mathbf{II} \neq \emptyset$. It follows that $U' = \emptyset$ and hence that U is disjoint from A. Next suppose that $U \cap \mathbf{III} \neq \emptyset$. Noting that $U \cap \mathbf{III} = U \cap (A \cup \mathbf{III})$ is open one then sees that $U \cap \text{int} \mathbf{III} \neq \emptyset$ and a contradiction follows. Thus $U \cap \mathbf{III} = \emptyset$ and one similarly shows that $U \cap \mathbf{N} = U \cap \mathbf{O} = \emptyset$ and finally gets the contradiction $U = \emptyset$. Thus $U = \emptyset$ and so $\text{int} F = \emptyset$. Thus the decomposition (5.50) is established and a similar one follows for $\overset{-}{W}$.

Now suppose one has the decomposition (5.50) for $\overset{+}{W}$ rewritten as $M = W_1 \cup ... \cup W_8 \cup F$ where $W_1 = \text{int} \mathbf{I}_\mathbf{C}, ..., W_8 = \text{int} \mathbf{O}$ and write a similar decomposition for $\overset{-}{W}$ as $M = W_1' \cup ... \cup W_8' \cup F'$ where $W_1' = \text{int} \mathbf{I}_\mathbf{C}', ..., W_8' = \text{int} \mathbf{O}'$, where primes denote the corresponding subsets of M with respect to $\overset{-}{W}$ and where F and F' are each closed subsets of M with $\text{int} F = \text{int} F' = \emptyset$. Let $E = F \cup F'$ so that E is a closed subset of M. It then follows (chapter 1) that $\text{int} E = \emptyset$. Now consider the open dense subsets $M \setminus F$ and $M \setminus F'$ of M so that $M \setminus F = \bigcup_{i=1}^{8} W_i$ and $M \setminus F' = \bigcup_{i=1}^{8} W_i'$. Let $X_{ij} = W_i \cap W_j'$ $(1 \leq i, j \leq 8)$ so that each X_{ij} is an open subset of M. Now $M \setminus (F \cup F') = (M \setminus F) \cap (M \setminus F')$ by the de Morgan laws (chapter 1) and so $M \setminus (F \cup F')$ is the union of the open sets X_{ij}. One thus has the disjoint decomposition

$$M = \bigcup_{i,j}^{8} X_{ij} \cup E \qquad (5.51)$$

where E is closed and $\text{int} E = \emptyset$. The following theorem is thus proved.

Theorem 5.3 *Let M be a 4-dimensional manifold admitting a smooth metric g of neutral signature and with associated Weyl conformal tensor C. The self dual and anti-self dual parts $\overset{+}{W}$ and $\overset{-}{W}$ of C admit disjoint algebraic decompositions of the form (5.50) whilst C admits a disjoint algebraic decomposition of the form (5.51).*

This gives a disjoint decomposition of open subsets of M one for each algebraic type of the Weyl conformal tensor C and which together comprise an open dense subset of M, together with the closed subset E with empty interior. This decomposition shows that each point in the open dense subset $M \setminus E$ of M lies in an open neighbourhood in which *the algebraic type of C is constant*. It follows [29] that the eigenvalues of C are locally smooth and that the eigenbivectors of C may be chosen to be locally smooth. Similar comments apply to $\overset{+}{W}$ and $\overset{-}{W}$, from (5.50), so that in the above canonical forms (5.32)–(5.37) the functions ρ_1 and ρ_2 and the bivectors G, H and F are locally smooth. It is remarked that the rank of $C(p) \neq 0$ is not necessarily an even integer (cf the Lorentz case) but can be any integer in the range $1 - 6$.

5.9   Curvature Structure

It is convenient to first consider the neutral signature equivalent of lemma 4.7.

Lemma 5.10 *Let h be a non-zero symmetric tensor, F a bivector and l, n, L, N and x, y, s, t null and orthonormal bases, respectively, for T_pM. Suppose h and F satisfy*

$$h_{ac}F^c{}_b + h_{bc}F^c{}_a = 0. \tag{5.52}$$

(*i*) *If U is a (real) eigenspace of F, then it is an invariant space for h. Thus if U is 1-dimensional, it gives an eigendirection for h. (This was given and proved in chapter 3 and is stated again here for convenience. It clearly applies also in the case when U is a complex subspace of the complexification of T_pM.)*

(*ii*) *If F is simple then the blade of F is an eigenspace of h.*

(*iii*) *If $F = \alpha(l \wedge n) + \beta(L \wedge N)$ with $\alpha, \beta \in \mathbb{R}$ and $0 \neq \alpha \neq \pm\beta \neq 0$, $l \wedge n$ and $L \wedge N$ are eigenspaces of h at p.*

(*iv*) *If $F = \alpha(l \wedge n) + \beta(L \wedge N)$ with $\alpha, \beta \in \mathbb{R}$ and $\alpha = \beta \neq 0$ (respectively, $\alpha = -\beta \neq 0$), $l \wedge L$ and $n \wedge N$ (respectively, $l \wedge N$ and $n \wedge L$) are invariant $2-$spaces of h at p.*

(*v*) *If $F = \alpha(x \wedge y) + \beta(s \wedge t)$ with $\alpha, \beta \in \mathbb{R}$ and $0 \neq \alpha \neq \pm\beta \neq 0$, $x \wedge y$ and $s \wedge t$ are eigenspaces of h at p.*

(*vi*) *If $F = \alpha(x \wedge y) + \beta(s \wedge t)$ with $\alpha, \beta \in \mathbb{R}$ and $\alpha = \beta \neq 0$ then the basis x, y, s, t may be chosen so that F retains exactly the given form in the new basis (as described in section 5.4) and $x \wedge t$ and $y \wedge s$ are invariant $2-$spaces for h (and similarly if $\alpha = -\beta \neq 0$).*

(*vii*) *If (5.52) holds for bivectors F and G, it holds for the bivector $[F, G]$.*

Proof The proof of part (*i*) is as given in lemma 3.6.

For part (*ii*) the proof can be done as in earlier chapters. Alternatively, it follows from part (*i*) by noting that the blade of $\overset{*}{F}$ is the $0-$eigenspace of F and hence invariant for h. Then the blade of F, which is orthogonal to the blade of $\overset{*}{F}$ is also invariant for h. Writing this last piece of information symbolically as $F = p \wedge q$ and $h(p) = ap + bq$ and $h(q) = cp + dq$ ($a, b, c, d \in \mathbb{R}$) and substituting into (5.52) gives $b = c = 0$ and $a = d$ and the result follows.

For (*iii*) it is noted that when regarded as a linear map F has eigenvector-eigenvalues pairs $l, (\alpha)$, $n, (-\alpha)$, $L, (\beta)$ and $N, (-\beta)$ and hence four distinct

1-dimensional eigenspaces. Part (i) then reveals that l, n, L and N are eigenvectors for h. Thus $h_{ab}l^b = al_a$ and $h_{ab}n^b = bn_a$ for $a, b \in \mathbb{R}$. Contractions of these equations, respectively, with n^a and l^a and use of the symmetry of h then show that $a = b$ and hence $l \wedge n$ is an eigenspace for h. One similarly handles $L \wedge N$ and the result follows.

For the proof of part (iv), say with $\alpha = \beta$, one notes that F has two 2-dimensional eigenspaces $l \wedge L$ and $n \wedge N$ with respective eigenvalues α and $-\alpha$ and which then become invariant for h, from part (i).

For the proof of part (v) note that $x \pm iy$ and $s \pm it$ are eigenvectors for F with (four) distinct eigenvalues $\pm i\alpha$ and $\mp i\beta$. Thus $x \pm iy$ and $s \pm it$ are eigenvectors for h and hence $x \wedge y$ and $s \wedge t$ are invariant 2$-$spaces for h. Since $x \wedge y$ has an induced positive definite metric from $g(p)$ one may select x and y to be orthogonal eigenvectors for h and a contraction of (5.52) with x^b reveals that $x \wedge y$ is an eigenspace for h. Similar comments apply to $s \wedge t$.

For part (vi) one sees that $x + iy$ and $s - it$ span the $i\alpha$ eigenspace of F and hence span an invariant 2$-$space for h, by part (i). Thus, in the usual symbolic notation, $h(x + iy)$ and $h(s - it)$ are complex linear combinations of $x + iy$ and $s - it$. This leads to (real) expressions for $h(x)$, $h(y)$, $h(s)$ and $h(t)$. One then applies the six conditions $h(x, y) = h(y, x)$, $h(x, s) = h(s, x), ..., h(s, t) = h(t, s)$ to get, with $a, b, c, d \in \mathbb{R}$,

$$h_{ab}x^b = ax_a + ct_a + ds_a, \qquad h_{ab}y^b = ay_a + cs_a - dt_a,$$
$$h_{ab}s^b = bs_a - cy_a - dx_a, \qquad h_{ab}t^b = bt_a - cx_a + dy_a. \tag{5.53}$$

From (5.53) it follows that $(cx - dy) \wedge t$ and $(cy + dx) \wedge s$ are real invariant 2$-$spaces for h. Defining $x' = K(cx - dy)$ and $y' = K(cy + dx)$, where $K = (c^2 + d^2)^{-\frac{1}{2}}$ (so that x', y', s, t is an orthonormal basis) one gets $F = \alpha(x' \wedge y' + s \wedge t)$ with $x' \wedge t$ and $y' \wedge s$ invariant 2$-$spaces for h. The case $\alpha = -\beta \neq 0$ is similar.

Part (vii) follows as in the previous cases. □

Now consider the curvature tensor $Riem(p)$ at $p \in M$ and the associated curvature map f at p. Then each member F of the Lie algebra $\overline{rgf(p)}$, the largest subalgebra of $o(2, 2)$ containing the range space $rgf(p)$ of f at p, satisfies (5.52) for $h = g$. For each curvature class of $Riem(p)$ except class A the possibilities for $\overline{rgf(p)}$ were given in section 5.6 above and the idea now is to use this and lemma 5.10 to gather information on $g(p)$. In particular suppose the curvature class of $Riem(p)$ is the "general" class A. Then it follows from section 5.6 that the possible subalgebra types for $\overline{rgf(p)}$ are $2a$, $2h$ $(\alpha\beta \neq 0)$, $2j$, $2l$, $3a$, $3b$, $3d$ $(\beta = 0 \neq \alpha)$, $3d$ $(\alpha\beta \neq 0)$, $3e$, $3f$, $4a$, $4b$, $4c$, $4d$, 5 and $o(2, 2)$.

So with $Riem(p)$ of class A, that is, $p \in A$, suppose that, in addition to the (neutral) metric g on M giving rise to $Riem$ on M, one also has a metric g' on M of arbitrary signature which gives rise to the same $Riem$ on M. Then, as explained earlier, (5.52) holds at p for $h = g$ and $h = g'$ and for any $F \in rgf(p)$. If $rgf(p)$ is of type $2a$ =Sp$(l \wedge N, l \wedge -L \wedge N)$ one sees from lemma 5.10 that $l \wedge N$ is an eigenspace of $g'(p)$ and, using the bivector $l \wedge n - L \wedge N$, that $n \wedge L$ is invariant for $g'(p)$. Thus, with all indices raised and lowered by g,

one has $g'_{ab}l^b = \alpha l_a$, $g'_{ab}N^b = \alpha N_a$, $g'_{ab}n^b = an_a + bL_a$ and $g'_{ab}L^b = cn_a + dL_a$
for $\alpha, a, b, c, d \in \mathbb{R}$. Using the symmetry relation $g'(l, n) = g'(n, l)$ one gets
$\alpha = a$, using $g'(L, N) = g'(N, L)$ one gets $\alpha = d$, using $g'(n, N) = g'(N, n)$
one gets $b = 0$ and using $g'(l, L) = g'(L, l)$, one gets $c = 0$. Thus l, n, L, N
span an eigenspace of g' with respect to g and hence $g'(p)$ is a multiple of
$g(p)$. The same result must also apply to the subalgebras $3a$, $3b$, $3e$, $3f$ and
all subalgebras of dimension ≥ 4 ($4a$, $4b$, $4c$, $4d$, 5 and $o(2, 2)$) since each of
these contains a subalgebra isomorphic to $2a$. Now suppose $\overline{rgf(p)}$ is of type
$2h = \text{Sp}(l \wedge N, \alpha(l \wedge n) + \beta(L \wedge N))$ for $\alpha, \beta \in \mathbb{R}$, $\alpha\beta \neq 0$, $\alpha \neq \pm\beta$. Then $l \wedge N$,
$l \wedge n$ and $L \wedge N$ are eigenspaces of g' (lemma 5.10) and again $g'(p)$ is a multiple
of $g(p)$. For type $2j = \text{Sp}(l \wedge N, \alpha(l \wedge n - L \wedge N) + \beta(l \wedge L))$ ($\alpha\beta \neq 0$), one
first sees that $l \wedge N$ is an eigenspace of g' so that $g'_{ab}l^b = \lambda l_a$, $g'_{ab}N^b = \lambda N_a$
($\lambda \in \mathbb{R}$). Then, noting that the bivector $\alpha(l \wedge n - L \wedge N) + \beta(l \wedge L))$ has
Segre type $\{22\}$ with real eigenvalues (section 5.4) whose only independent
eigenvectors are l and L with distinct eigenvalues $\pm\alpha$, use of lemma 5.10(i)
shows that L is also an eigenvector of g' which, since $N \cdot L = 1$, has the
same eigenvalue as N (and l). It follows that (5.52) holds for the bivector
$l \wedge L$ separately, and hence for the bivector $l \wedge n - L \wedge N$ alone. Thus $n \wedge L$
is invariant for g'. This information is then easily utilised to show that n
is also an eigenvector of g' with eigenvalue λ and so $g'(p)$ is a multiple of
$g(p)$. For type $2l = \text{Sp}(l \wedge N, \alpha(l \wedge n - L \wedge N) + \beta(l \wedge L + n \wedge N))$ (since
$x \wedge y - s \wedge t = l \wedge L + n \wedge N$) with $\alpha\beta \neq 0$, one sees that $l \wedge N$ is an eigenspace
of g' and that $l \pm iN$ and $n \pm iL$ are non-degenerate eigenvectors of the bivector
$\alpha(l \wedge n - L \wedge N) + \beta(l \wedge L + n \wedge N)$ (the latter admits four distinct complex
eigenvalues from section 5.4). Hence, by lemma 5.10(i) they are eigenvectors
of g'. Thus $n \wedge L$ is invariant for g' and use of the symmetry relations for g'
then easily shows that n and L are eigenvectors of g' with the same eigenvalue
as l (and N). It again follows that $g'(p)$ is a multiple of $g(p)$. The arguments
for $3d$ ($\beta = 0 \neq \alpha$) and $3d$ ($\alpha\beta \neq 0$, $\beta \neq \pm\alpha$) are similar and straightforward
and so g and g' are proportional on A. The argument given for theorem 3.3,
which is independent of the signature of g, then gives the following result.

Theorem 5.4 *Let M be a 4-dimensional manifold and let g be a smooth metric on M of neutral signature and with curvature tensor Riem. Suppose that
the subset A of points where the curvature class is A is an open dense subset
of M. Suppose also that g' is a smooth metric on M of arbitrary signature
which has the same curvature tensor Riem as g does on M. Then $g' = \alpha g$ for
$0 \neq \alpha \in \mathbb{R}$ and the Levi-Civita connections of g and g' are the same.*

For the other types for $\overline{rgf(p)}$ one can also obtain relations between $g(p)$
and $g'(p)$ just as described following theorem 3.3.

Now consider the tensor type $(1, 3)$ Weyl tensor C associated with g on
M. Again one may introduce the Weyl map f_C on bivectors, as before, to
achieve the Weyl class at each $p \in M$ just as was done for *Riem* and the
curvature class. If the Weyl class is D at p then the work in sections 5.7 and

5.8 shows that, at p, the Weyl tensor must be of type $(\mathbf{N}, \mathbf{O})$ (or $(\mathbf{O}, \mathbf{N})$), and conversely.

If the Weyl class at p is B, $rgf_C(p) = \mathrm{Sp}(P, Q)$ with $P \in \overset{+}{S}_p$, $Q \in \overset{-}{S}_p$, $[P, Q] = 0$ and P and Q having no common annihilator. The Abelian subalgebra $rgf_C(p)$ is then easily checked to be of type $2b$, $2c$, $2d$, $2e$ or $2f$. Then one writes at p

$$C_{abcd} = \alpha P_{ab} P_{cd} + \beta Q_{ab} Q_{cd} + \gamma(P_{ab} Q_{cd} + Q_{ab} P_{cd}) \qquad (5.54)$$

for $\alpha, \beta, \gamma \in \mathbb{R}$. Now, as shown in section 5.7, the subspaces $\overset{+}{S}_p$ and $\overset{-}{S}_p$ are invariant for f_C and on writing $C(p)$ as in (5.54) one sees that $\gamma = 0$. Thus, at p, one may write $C = \alpha PP + \beta QQ$ and so $^*C = \alpha PP - \beta QQ$ and $\overset{+}{W} = \alpha PP$. If $\alpha \neq 0$ this gives $P_{a[b} P_{cd]} = 0$ which forces P to be simple and similarly one shows that if $\beta \neq 0$, Q is simple. Thus for $\alpha\beta \neq 0$ it follows that P and Q are totally null members of $\overset{+}{S}_p$, and $\overset{-}{S}_p$, respectively, and hence their blades intersect non-trivially (in a null direction). This contradicts the class B assumption at p and so $\alpha\beta = 0$ which again contradicts the class B assumption. It follows that the Weyl tensor cannot be class B at any point.

If the class of $C(p)$ is C, $rgf_C(p)$ has dimension 2 or 3 and has a common annihilator. Thus $\overline{rgf(p)}$ is of type $2g$, $2h$ ($\alpha\beta = 0$), $2k$, $3c$ or $3d$ ($\alpha = 0$). Similar methods to those used in the last case reveal contradictions in each case except $2g$ and thus only for this type can the Weyl tensor be of class C. Its algebraic type is then $(\mathbf{N}, \mathbf{N})$, and conversely.

Thus the Weyl tensor at p, if it is not class A (and not zero), is either of type $(\mathbf{N}, \mathbf{O})$ (class D) or $(\mathbf{N}, \mathbf{N})$ (class C). It follows from lemma 5.8(i) that these two cases are collectively characterised by the condition that $C(p) \neq 0$ satisfies $C^a{}_{bcd} k^d = 0$ for $0 \neq k \in T_p M$ and then it follows that in each case k is necessarily null, being either a member of a certain fixed totally null $2-$space for type $(\mathbf{N}, \mathbf{O})$ or unique up to a scaling for type $(\mathbf{N}, \mathbf{N})$. It will be seen later that each of these special cases can occur. A similar argument to that given for the curvature classes in chapter 3 shows that the subset of all $p \in M$ where the Weyl class is A is open in M.

Theorem 5.5 *Suppose that g is a smooth metric on M of neutral signature and g' another smooth metric on M of arbitrary signature such that the respective type $(1, 3)$ Weyl tensors C and C' for g and g' are equal. Suppose also that the subset $U \subset M$ consisting of precisely those points of M at which the $(1, 3)$ Weyl tensor C of g is either zero or has one of the above algebraic types $(\mathbf{N}, \mathbf{O})$ (or $(\mathbf{O}, \mathbf{N})$) or $(\mathbf{N}, \mathbf{N})$ has empty interior in M. Then g and g' are conformally related.*

Proof It is first noted that on disjointly decomposing M into its Weyl classes as was done for the curvature classes and *Riem* and using the above discussion of Weyl classes, one finds $M = A \cup C \cup D \cup O = A \cup U$ where

$U = C \cup D \cup O$ and since A is open and $A \cap U = \emptyset$ and M is connected, U is closed (but not open unless it is empty). From the above the set C consists of precisely those points where the Weyl type is $(\mathbf{N}, \mathbf{N})$ and D of precisely those points where the Weyl types is $(\mathbf{N}, \mathbf{O})$ (or $(\mathbf{O}, \mathbf{N})$). If int$U = \emptyset$, A is open and dense in M. [Alternatively, one can note that $U = \{p \in M : |X| = 0$ for each $X \in rgf_C(p)\}$. To see this it is clear from the above discussion that all points in U satisfy this curly bracketed condition whilst for $p \in M \setminus U$ dimr$gf_C(p) \geq 2$ and so choosing a basis $X_i \in \Lambda_p M$, $(1 \leq i \leq 6)$ with $X_1, .., X_3 \in \overset{+}{S}_p$ and $X_4, .., X_6 \in \overset{-}{S}_p$, $f_C(X_i) \in \overset{+}{S}_p$ $(1 \leq i \leq 3)$ and $f_C(X_i) \in \overset{-}{S}_p$ $(4 \leq i \leq 6)$ and at least two of the $f_C(X_i)$ must be independent. If there are exactly two and they are such that one is in $\overset{+}{S}_p$ and one is in $\overset{-}{S}_p$ then they are each totally null and one immediately gets the contradiction that $p \in C$. It follows that whatever the dimension of $rgf_C(p)$ it contains (say) at least two members $F, G \in \overset{+}{S}_p$ and hence a member X with $|X| \neq 0$. Clearly then U is closed in M.] Continuing with the argument let $p \in M \setminus U$ so that dimr$gf(p) \geq 2$ and choose independent $F, G \in \overset{+}{S}_p$. Then since $\overset{+}{S}_p$ has Lorentz signature $(-, -, +)$ one may choose $|F| < 0$ and hence (lemma 5.5) a basis l, n, L, N with $F = l \wedge n - L \wedge N$. Now $l \wedge n - L \wedge N, l \wedge N, n \wedge L$ span $\overset{+}{S}_p$ and so one may take $G = \alpha X + \beta Y$, $(\alpha, \beta \in \mathbb{R}, \alpha^2 + \beta^2 \neq 0)$ with $X = l \wedge N$ and $Y = n \wedge L$. So (5.52) holds for F and G (with $h = g'$), the former revealing that $l \wedge N$ and $n \wedge L$ are invariant for g' at p. These immediately give $g'(l, l) = g'(N, N) = g'(l, N) = 0$ and $g'(n, n) = g'(L, L) = g'(n, L) = 0$. Thus one may write

$$g'_{ab} = \mu(l_a n_b + n_a l_b) + \nu(L_a N_b + N_a L_b) + \rho(l_a L_b + L_a l_b) + \sigma(n_a N_b + N_a n_b).$$
$$(5.55)$$

Also one has, using the bivector G

$$g'_{ac}(\alpha X^c{}_b + \beta Y^c{}_b) + g'_{bc}(\alpha X^c{}_a + \beta Y^c{}_a) = 0. \qquad (5.56)$$

If $\alpha \neq 0 \neq \beta$ contractions of (5.56) with $l^a l^b$, $n^a n^b$ and $l^a N^b$ give $g'(l, L) = g'(n, N) = 0 = g'(l, n) - g'(L, N)$ and when these are substituted into (5.55) one finds $\rho = \sigma = 0$ and $\mu = \nu$. If $\alpha\beta = 0$ an almost identical calculation again reveals the same results (in these cases $l \wedge N$ or $n \wedge L$ is an eigenspace for g' from lemma 5.10(*ii*)). Thus by the completeness relation g' is a multiple of g at p. Thus g' and g are conformally related on $M \setminus U$. If g and g' are not conformally related at some $p \in U$ then, by continuity, they are not conformally related on some open subset $W \neq \emptyset$ which then must be contained in U. This contradicts the fact that U has empty interior and so g and g' are conformally related on M [88]. □

It is remarked that in the event that int$C \neq \emptyset$ it can be checked (section 5.8) that if $p \in$int$C$ there exists an open coordinate neighbourhood $V \subset M$ with $p \in V \subset C$ and a smooth null vector field l on V which spans the repeated pnd for the Weyl tensor for g on V such that any other smooth metric g' on V with the same Weyl tensor as g on V satisfies $g'_{ab} = \phi g_{ab} + \lambda l_a l_b$ on V (with indices manipulated using the metric g) for smooth functions ϕ and λ on V. This common Weyl tensor is of type $(\mathbf{N}, \mathbf{N})$ for g and g'. Similar remarks apply if $\mathrm{int} D \neq \emptyset$ but in this case the potential expression for g' is more complicated.

To see that this result is "best possible" consider the following metric on some open coordinate domain of $\mathbb{R}^4$ with coordinates u, v, x, y

$$ds^2 = H(u, x, y)du^2 + 2dudv + dx^2 - dy^2 \qquad (5.57)$$

for some smooth function H. This metric clearly has neutral signature since it possesses a non-trivial null vector $\partial/\partial v$ and has positive determinant equal to 1. The only non-vanishing components of the type $(0, 4)$ Weyl conformal tensor are, up to index symmetries, $C_{1313} = C_{1414} = -\frac{1}{4}(H_{,xx} + H_{,yy})$ and $C_{1314} = -\frac{1}{2}H_{,xy}$, where a comma denotes a partial derivative. From these and (5.57) one can calculate the components g^{ab} and $C^a{}_{bcd} = g^{ae}C_{ebcd}$. It now follows that if the function $H(u, x, y)$ in (5.57) is replaced by a function of the form $H'(u, x, y) = H + \phi(u) + \phi'(u)x + \phi''(u)y$ for smooth functions ϕ, ϕ' and ϕ'' the metric g is, in general, changed to a metric g' not conformally related to g but the Weyl tensor $C^a{}_{bcd}$ is unchanged. The Weyl tensor of (5.57) is, in general, of algebraic type $(\mathbf{N}, \mathbf{N})$ (class C) but for certain special choices of H it may have algebraic type $(\mathbf{N}, \mathbf{O})$ (class D), for example, $H(u, x, y) = f(u)e^{x+y}$ for some smooth function f. This justifies the "best possible" remark above [88].

It is remarked here that from the discussion of the Weyl conformal tensor in this and the previous two chapters one can state, independently of signature, that if g and g' are smooth metrics on M which have the same type $(1, 3)$ Weyl conformal tensor C and if the (closed) subset $U \subset M$ on which the equation $C^a{}_{bcd}k^d = 0$ has a non-trivial solution for $k \in T_pM$ satisfies $\mathrm{int} U = \emptyset$, then g and g' are conformally related on M. [This follows in the positive definite case since then the Weyl tensor vanishes at p if and only if $p \in U$ and is thus non-vanishing on the open dense subset $M \setminus U$ (chapter 3).]

5.10   Sectional Curvature

Sectional curvature has already been discussed for positive definite and Lorentz metrics and now this topic can be dealt with for the case of neutral signature. One now considers (M, g) as usual with g smooth and of neutral signature and lets G_p (equivalently, the manifold of projective simple bivectors at p) denote the 4-dimensional Grassmann manifold of all $2-$spaces at $p \in M$ consisting of its subsets of spacelike, timelike, null and totally null $2-$spaces at p, denoted by S_p^2, T_p^2, N_p^2 and TN_p^2, respectively, and with S_p^2, T_p^2 and

$S_p^2 \cup T_p^2$ each 4-dimensional, open submanifolds of G_p and $\widetilde{N}_p^2 \equiv N_p^2 \cup TN_p^2$ a closed subset of G_p.

The sectional curvature function σ_p at $p \in M$ is defined as in chapters 3 and 4 but this time on the subset $\bar{G}_p \equiv S_p^2 \cup T_p^2 = G_p \setminus \widetilde{N}_p^2$ (since a representative simple bivector F for any of the 2−spaces in G_p satisfies $|F| = 0$ if and only if the 2−space is null or totally null) and which is neither connected nor closed. As in the Lorentz case if σ_p is constant on $\bar{G}_p$ it is trivially (continuously) extendible to a constant function on G_p and the constant curvature condition holds at p. If, however, σ_p not constant on $\bar{G}_p$ it turns out that *it cannot be continuously extended to any point of $\widetilde{N}_p^2$*. The proof of this fact is given in [59] and is similar to the proof in the Lorentz case given in [61]. Thus the existence of a continuous extension of σ_p to any member of $\widetilde{N}_p^2$ forces the constant curvature condition at p and if σ_p is not constant on $\bar{G}_p$ the subset $\widetilde{N}_p^2$ of G_p is precisely the *complement of its domain of definition*.

So suppose that σ_p is not constant on $\bar{G}_p$ and suppose further that F and F' are representative bivectors of two distinct members of $\widetilde{N}_p^2$. Then the bivector $F + \lambda F'$ for *some* $0 \neq \lambda \in \mathbb{R}$ also represents a member of $\widetilde{N}_p^2$ if and only if the blades of F and F' intersect in a null direction and then $F + \lambda F'$ represents a member of $\widetilde{N}_p^2$ for *all* such λ. To see this note that if F and F' represent independent members of $\widetilde{N}_p^2$ with blades intersecting in the null direction spanned by the null vector l then one may write $F = l \wedge u$ and $F' = l \wedge v$ for $u, v \in T_pM$ with $l \cdot u = l \cdot v = 0$ and l, u, v independent and clearly $F + \lambda F'$ represents a member of $\widetilde{N}_p^2$ for all $\lambda \in \mathbb{R}$. Conversely, suppose F, F' *and* $F + \lambda F'$ represent members of $\widetilde{N}_p^2$ with F and F' independent, $F = p \wedge q$, $F' = r \wedge s$ and $F + \lambda F' = e \wedge f$ for non-zero $p, q, r, s, e, f \in T_pM$ and $0 \neq \lambda \in \mathbb{R}$. From the general theory of bivectors for this signature one may always choose $p \cdot p = p \cdot q = r \cdot r = r \cdot s = e \cdot e = e \cdot f = 0$ and a contraction of $F + \lambda F' = e \wedge f$ with e gives $(e \cdot q)p - (e \cdot p)q + \lambda(e \cdot s)r - \lambda(e \cdot r)s = 0$. This shows that p, q, r and s are dependent members of T_pM otherwise one would get the contradiction that e is orthogonal to each member of the basis p, q, r, s for T_pM. It follows that the blades of F and F' intersect in a direction spanned by, say, $k \in T_pM$ and so one may write $F = k \wedge r'$ and $F' = k \wedge s'$ for k, r', s' independent members of T_pM. If k is not null one may choose r' and s' null with each orthogonal to k. Then the conditions $|F| = |F'| = |F + \lambda F'| = 0$ give $F \cdot F' = 0$, then $|k|r' \cdot s' = 0$ and so $r' \cdot s' = 0$ and then $r' \wedge s'$ is totally null. But then the conditions $k \cdot r' = k \cdot s' = 0$ imply that $k \in r' \wedge s'$ which contradicts the fact that k is not null (or the fact that k, r', s' are independent). It follows that k is null and hence $F + \lambda F'$ is null or totally null for each such λ. The conclusion is that if F, F' and $F + \lambda F'$, for some $0 \neq \lambda \in \mathbb{R}$, each represent members of $\widetilde{N}_p^2$ their blades intersect in a null direction at p and then *all* subsets of the form $F + \lambda F'$ ($\lambda \in \mathbb{R}$) lie in $\widetilde{N}_p^2$ and together determine the collection of all null directions (and hence the collection of all null vectors) at p. [Clearly not all such subsets lie in $\widetilde{N}_p^2$ as the choice (in a hybrid basis at p)

of $F = l \wedge y$ and $F' = n \wedge y$, each being in $\widetilde{N}^2_p$ and with intersecting blades, shows.]

Now suppose that g' is another smooth metric on M of neutral signature with sectional curvature function σ'_p. Suppose also that the sectional curvature function σ_p is identical to σ'_p at each $p \in M$ but is such that this common function is not a constant function on $\bar{G}_{p'}$ on some open dense subset of M. Then the argument of the previous paragraph shows that the sets $\widetilde{N}^2_p$ for g and g' of all null and totally null 2–spaces at p which are the complements of the domains of σ_p and σ'_p $(= \sigma_p)$ are the same and hence that g and g' share the same set of null vectors at each $p \in U$. Thus g and g' are *conformally related* on this open dense subset and hence on M (and hence g' is also of neutral signature). Then one has, from a similar argument to that given in chapter 3, at each $p \in M$ and in an obvious notation using primes to denote geometrical quantities arising from g',

$$(i)\, g' = \phi g, \quad (ii)\, R'_{abcd} = \phi^2 R_{abcd}, \quad (iii)\, R'^a{}_{bcd} = \phi R^a{}_{bcd} \quad (5.58)$$
$$(iv)\, R'_{ab} = \phi R_{ab}, \quad (v)\, R' = R, \quad (vi)\, C' = \phi C$$

for some smooth nowhere-zero function $\phi : M \to \mathbb{R}$. Again it is noted that since g and g' are conformally related, the type $(1,3)$ Weyl tensors are equal, $C' = C$; the last equation in (5.58) is, however, deduced from the others in (5.58) and it is noted that one may have $C' = C = 0$ at some points.

One can now continue with the set-theoretic discussion given in chapters 3 and 4 and with the same notation [59]. Thus $X \subset M$ is the closed subset on which σ_p is constant and so $M \setminus X \subset M$ is the open and dense subset of M on which σ_p is not constant, $V \subset M$ the open subset on which $d\phi \neq 0$, $Y \subset M$ the subset on which $\phi = 1$ and $U \subset M$ the open subset on which the common value of the Weyl tensors C and C' is non-zero. All the points at which *Riem* and *Riem'* vanish are contained in X. Thus $\phi = 1$ on U (and hence $U \subset Y$)) and $U \cap V = \emptyset$. The rest of the argument is as in chapter 3 and 4 and one achieves the disjoint decomposition $M = V \cup \text{int} Y \cup K$ where $K \subset M$ is the closed subset defined by the decomposition and with empty interior. Clearly ϕ is a non-zero constant on each component of $\text{int}(M \setminus V)$ and $g' = g$ on $\text{int} Y$. Since $U \cap V = \emptyset$ the open subset V is conformally flat for g' and g.

As in the previous cases the most interesting part of this decomposition is the open subset V, provided it is not empty. The analysis of this subset is almost the same as that given in the Lorentz case. In fact one can show that $d\phi$ is a nowhere-zero null (co)vector field on V with respect to both metrics [59], that the Ricci tensors of each metric is either zero or of Segre type $\{(211)\}$ with null eigen(co)vector proportional to $d\phi$ and eigenvalue zero on V (and hence the Ricci scalars of each metric vanish on V). The Riemann tensor satisfies $R^a{}_{bcd} \phi_a = 0$ on V. Further each metric admits a local parallel null (co)vector field, proportional to $d\phi$. On the open dense subset of V on which *Riem* does not vanish the conditions of Walker's non-simple K^*_n spaces are satisfied [81] and so about any point p of this subset of V there exists an open neighbourhood W contained in V in which one may choose coordinates x, y, u, v so that g takes the form (5.57) with $H(u, x, y) = \delta(u)(x^2 - y^2)$ (from the conformally flat condition) for some smooth function δ on W. Here du is the parallel (co)vector field. This is the neutral signature analogue of a conformally flat plane wave in general relativity.

One thus obtains the following theorem [59].

Theorem 5.6 *Let M be a 4-dimensional manifold admitting smooth metrics g and g' of neutral signature. Suppose that g and g' have the same sectional curvature function at each $p \in M$ and which is not a constant function at any point over some open dense subset of M. Then one may decompose M as above according to $M = V \cup int Y \cup K$ where $int K = \emptyset$, $g = g'$ on $int Y$ and where V is an open subset of M on which g and g' are (locally) conformally related, conformally flat "plane waves".*

The finding of non-trivially conformally related metric pairs g and g', as described above, is similar to the Lorentz case and can be found by modifying the work in [63].

5.11 The Ricci-Flat Case

Now impose the condition that $Ricc \equiv 0$ on (M, g) so that $Riem \equiv C$ on M. If (M, g) is non-flat, $Riem$ (and hence C) does not vanish over any non-empty open subset of M. Thus the sectional curvature function is not a constant function on any non-empty open subset of M, otherwise C and hence $Riem$ would vanish on this subset. Then in the above notation $U \subset Y$ with U open and dense and $\phi = 1$ on Y and so $\phi = 1$ on M. Thus one has the following theorem.

Theorem 5.7 *Let g be a smooth metric of neutral signature on M. Suppose (M, g) is Ricci flat and non-flat. If g' is any other smooth metric on M of neutral signature and with the same sectional curvature function on M as g, then $g' = g$. Thus the sectional curvature uniquely determines the metric and its Levi-Civita connection.*

One may continue in this way to obtain analogues of theorems 10 and 11 in chapter 4 and where in the former the reference to "pp-waves" is replaced by the "neutral signature equivalent of pp-waves" By this is meant that each point of W admits a coordinate neighbourhood in which g takes the general form (5.57) and where the Ricci-flat condition imposes the extra constraint $\partial^2 H / \partial x^2 = \partial^2 H / \partial y^2$. The Weyl conformal tensor satisfies $C^a{}_{bcd} \phi^d = 0$ with $\phi^a \equiv g^{ab} \phi_b$ non-zero. Thus (section 5.8) ϕ^a is null and the algebraic type of C is $(\mathbf{N}, \mathbf{N})$ or $(\mathbf{N}, \mathbf{O})$ (or $(\mathbf{O}, \mathbf{N})$).

5.12   Algebraic Classification Revisited

By way of a brief appendix this final section shows a link between the classification of second order symmetric tensors described in sections 4.2 and 5.3 and the classification of the Weyl tensor in sections 4.5, 5.7 and 5.8. No proofs will be given——they can be found in [82]. Thus for the usual pair (M, g) with $\dim M = 4$ and g of neutral signature (the easier Lorentz case will be described at the end of this section) let T be a second order, symmetric, *tracefree* tensor with components T_{ab} (and $T^c{}_c = 0$) and with associated linear map f on $T_p M$ and construct the type $(0, 4)$ tensor E' given by

$$E'_{abcd} = \frac{1}{2}\{T_{ac}g_{bd} - T_{ad}g_{bc} + T_{bd}g_{ac} - T_{bc}g_{ad}\}, \qquad (\Rightarrow E'^c{}_{acb} = T_{ab}). \quad (5.59)$$

The tensor has all the index symmetries of the tensor E in (3.8), (3.9) (3.10) and (3.17) and in the case when $E' = E$, T becomes the tracefree Ricci tensor. Given g the correspondence between E' and T is bijective and it is noted that T has the same Segre type, including degeneracies, of any second-order symmetric tensor whose tracefree part is T. One is thus led to the linear map $f_{E'}$ on $\Lambda_p M$, for $p \in M$, just as was described for the curvature and Weyl maps in sections 3 and 4. Thus if F is a real or complex bivector in (the complexification of) $\Lambda_p M$ it is an *eigenbivector* of E' with eigenvalue $\lambda \in \mathbb{C}$ if $E'_{abcd}F^{cd} = \lambda F_{ab}$. The algebraic situation for E' is expressed in the following collection of results and T may then be found from (5.59) [82]:

(a) $B \in \Lambda_p M$ is (real and) simple the blade of B is an invariant $2-$space of f if and only if B is a (real) simple eigenbivector of $f_{E'}$. Since (chapter 3) f always admits an invariant $2-$space, it follows that $f_{E'}$ always admits a real simple eigenbivector.

(b) If E' admits a real eigenbivector B with eigenvalue $\alpha(\neq 0) \in \mathbb{R}$ then B is simple and $\overset{*}{B}$ is (independent of B and) also a simple eigenbivector of E' with eigenvalue $-\alpha$ and the blades of B and $\overset{*}{B}$ are invariant for f.

(c) The dual properties of E' reveal that, at $p \in M$, $f_{E'}(\overset{+}{S}_p) \equiv \overset{-}{A} \subset \overset{-}{S}_p$ and $f_{E'}(\overset{-}{S}_p) \equiv \overset{+}{A} \subset \overset{+}{S}_p$. Further, $\dim\overset{+}{A} = \dim\overset{-}{A}$ and the rank of $f_{E'}$ equals $\dim\overset{+}{A} + \dim\overset{-}{A}$. Hence the rank of $f_{E'}$ is even.

(d) If B is a real or complex eigenbivector of E' with $|B| = 0$ then either B arises from a non-simple elementary divisor or its eigenvalue is degenerate. If B is a real or complex multiple of a member of $\widetilde{S}_p$ its eigenvalue is zero. Otherwise if an eigenbivector B is such that B and $\overset{*}{B}$ are independent and arises from a non-simple elementary divisor of order n then so also does $\overset{*}{B}$.

(e) Let B be a complex eigenbivector of E', $E'B = zB$, with $z = a + ib$ for $a, b \in \mathbb{R}$ and $b \neq 0$. If $a \neq 0$, B, $\bar{B}$, $\overset{*}{B}$ and $\overset{*}{\bar{B}}$ are independent eigenbivectors. If, however, the eigenvalues arising from f are $\pm ib$ and $\pm id$ with $0 \neq b \neq \pm d \neq 0$, again four independent (complex) eigenbivectors occur (but recall (a) above).

(f) From the above it can be deduced that many of the potential Segre types for $f_{E'}$, whatever the eigenvalue degeneracies, are impossible. In fact only nine possible types remain and they are (where a positive integer $n' > 1$ carrying a prime in the Segre type refers to a complex eigenvalue arising from a non-simple elementary divisor of order n): $\{111111\}$, $\{1111z\bar{z}\}$, $\{11z\bar{z}w\bar{w}\}$, $\{2211\}$, $\{2'2'11\}$, $\{3111\}$, $\{31z\bar{z}\}$, $\{51\}$ and $\{33\}$ and correspond to the nine types for T in theorem 5.1. The first six of these each admit an orthogonal pair of simple timelike eigenbivectors. This enables them to be handled together conveniently [82].

(g) The diagonalisable over $\mathbb{R}$ types $\{111111\}$ for $f_{E'}$ correspond to the diagonalisable over $\mathbb{R}$ types $\{1111\}$ for f whilst the diagonalisable over $\mathbb{C}$ types $\{1111z\bar{z}\}$ and $\{11z\bar{z}w\bar{w}\}$ for $f_{E'}$ correspond to the diagonalisable over $\mathbb{C}$ types $\{z\bar{z}11\}$ and $\{z\bar{z}w\bar{w}\}$ for f (from (5.59)). The other corresponding types are: $\{2211\}$—$\{211\}$, $\{2'2'11\}$—$\{2z\bar{z}\}$, $\{3111\}$—$\{22\}$, $\{31z\bar{z}\}$—$\{2'2'\}$, $\{51\}$—$\{4\}$ and $\{33\}$—$\{31\}$.

The Lorentz case is much easier and the above remarks still hold where applicable (recalling that $^*E' = -E'^*$ is true for all signatures) [44, 45]. One finds that the only possible types for E' (excluding degeneracies) are $\{111111\}$, $\{11z\bar{z}w\bar{w}\}$, $\{2211\}$ and $\{33\}$ and that these correspond to the cases for T given by $\{1111\}$, $\{z\bar{z}11\}$, $\{211\}$ and $\{31\}$, respectively (chapter 4). The situation for positive definite signature is straightforward.

It is remarked that one may also define repeated and non-repeated principal null directions for E' for either Lorentz or neutral signature, by analogy with those arising from the Bel criteria and the tensor C and described earlier. In particular, call $k \in T_pM$ a *repeated principal null direction for E'* if $E'_{abcd}k^a k^c = \alpha k_b k_d$ for $\alpha \in \mathbb{R}$. (If $\alpha \neq 0$ k is clearly null and if $\alpha = 0$ one must assume k is null since it may not be.) Then this statement is equivalent to k being a null eigenvector of T. Other similar criteria are also available [82].

Chapter 6

A Brief Discussion of Geometrical Symmetry

6.1 Introduction

This chapter is devoted to a brief discussion of geometrical symmetry on an $n-$dimensional, smooth, connected, Hausdorff, second countable manifold M of dimension $n > 2$ and which admits a smooth metric g of arbitrary signature with Levi-Civita connection ∇. It will be kept brief, with few proofs, but with references to where the proofs can be found. It closely follows [13]. Such symmetries are usually, and can naturally be, described in terms of local smooth diffeomorphisms on M, that is, smooth bijective maps $f : U \to V$ for U and V open subsets of M (and with $f^{-1} : V \to U$ also smooth) where the map f simulates some particular symmetry property between U and V. For symmetries which apply to the whole of M one would require many such local maps which would be complicated to deal with. Thus the symmetries described here will be handled in a more convenient and conventional way in terms of global, smooth vector fields on M. For such a vector field X it is recalled from section 2.6 that given $p' \in M$ there exists an open neighbourhood U of p' and an open interval $I \subset \mathbb{R}$ containing 0 such that there exists an integral curve c_p of X with domain I starting from any $p \in U$. Thus for each $t \in I$ there is a map $\phi_t : U \to \phi_t(U)$ defined by $\phi_t(p) = c_p(t)$ for each $p \in U$ and so each point of U is "moved" a parameter distance t along the integral curve of X through that point. Each map $\phi_t : U \to \phi_t(U)$ is then a smooth local diffeomorphism between the open submanifolds U and $\phi_t(U)$ of M [22, 89]. The maps ϕ_t above are called the *local flows* of X and the idea is to describe a symmetry on M in terms of certain vector fields on M through the action of their local flows. Such an approach supplies local diffeomorphisms everywhere on M and the symmetry then often gives rise to convenient differential conditions on X which are easier to handle mathematically than the actual local flows themselves (but the local flows are indispensable for the geometrical interpretation of the symmetry).

To see this idea work in a particular situation let X be a global, smooth vector field on M and T a global, smooth tensor field on M. Let $f : U \to V$ be a local flow of X with U chosen as a coordinate domain of M with coordinates

x^a $(1 \leq a \leq n)$ and $V = f(U)$. Then V is also a coordinate domain of M with coordinate functions $y^a = x^a \circ f^{-1}$ and one may compute (chapter 2) for $p \in U$ and $q = f(p) \in V$

$$f_*(\partial/\partial x^a)_p = (\partial/\partial y^a)_q, \qquad f^{-1*}(dx^a)_p = (dy^a)_q. \tag{6.1}$$

Then, with T_q denoting $T(q)$, etc,

$$(f^*T_q)_p(\partial/\partial x^a, ..., dx^b, ...) = T_q(f_*(\partial/\partial x^a)_p, ..., f^{-1*}(dx^b)_p, ...)$$
$$= T_q((\partial/\partial y^a)_q, ..., (dy^b)_q, ...). \tag{6.2}$$

Thus the components of the tensor T at q in the coordinates y^a equal the components of the pullback tensor f^*T at p in the coordinates x^a. Suppose one defines f to be a "symmetry" of T if the components of T at $p \in U$ in the coordinates x^a equal those of T at q in the coordinates y^a for each $p \in U$, that is, if $f^*T = T$ for each $p \in U$. Thus f is a "symmetry" of T if, given coordinates x^a and then defining coordinates $y^a = x^a \circ f^{-1}$ in $V = f(U)$ using f, the components of T are equal at points linked by f in these coordinates. Finally suppose this is true for *every* local flow of X so that it is X that is giving rise to the symmetry in some open neighbourhood of any point of M. It will be shown in the next section that this situation gives rise to differential relations between the components of T and X and which are more convenient to handle mathematically.

Of course not all symmetries may be so formulated but many can be handled in this (or a similar) fashion and this chapter will describe some of the main ones. Sometimes the above condition $f^*T = T$ for each local flow of X will be weakened but in each case the geometrical spirit of the role of the local flows of X is maintained.

6.2　The Lie Derivative

Let X be a global, smooth vector field on M with local flows ϕ_t. There is a type of derivative arising from X and its local flows and which is reminiscent of the Newton quotient. It applies to any smooth tensor field T on M and is written $\mathcal{L}_X T$ and called the *Lie derivative* (of T along X). It is defined at $p \in M$ by

$$(\mathcal{L}_X T)(p) = \lim_{t \to 0} \{\frac{1}{t}[(\phi_t^* T)(p) - T(p)]\}. \tag{6.3}$$

This limit always exists (but not obviously so) [22] and gives rise to a smooth tensor field on M of the same type as T. It has the following properties for smooth tensor fields S and T on M, smooth vector fields X and Y on M, $a, b \in \mathbb{R}$ and a smooth function f on M (and with the obvious modifications if $S, ..., f$ are only defined on open subsets of M) and where the first is to be regarded as the definition of $\mathcal{L}_X f$ [10, 22, 89]. The operator $\mathcal{L}_X$ is clearly $\mathbb{R}$-linear.

$$\mathcal{L}_X f = X(f), \tag{6.4}$$

$$\mathcal{L}_X Y = [X, Y], \tag{6.5}$$

$$\mathcal{L}_X(S \otimes T) = \mathcal{L}_X(S) \otimes T + S \otimes \mathcal{L}_X(T), \tag{6.6}$$

$$\mathcal{L}_{aX+bY} T = a\mathcal{L}_X T + b\mathcal{L}_Y T, \tag{6.7}$$

$$\mathcal{L}_{[X,Y]} T = \mathcal{L}_X(\mathcal{L}_Y T) - \mathcal{L}_Y(\mathcal{L}_X T), \tag{6.8}$$

$$\mathcal{L}_X(fT) = X(f)T + f(\mathcal{L}_X T). \tag{6.9}$$

The operator $\mathcal{L}_x$ commutes with the contraction operator and, if a real-valued function is regarded as a type $(0,0)$ tensor, the last of these follows from the first. In any coordinate domain of M one can then find an expression for the components of the tensor $\mathcal{L}_X T$ (using a comma to denote a partial derivative) as

$$(\mathcal{L}_X T)^{a...b}_{c...d} = T^{a...b}_{c...d,e} X^e - T^{e...b}_{c...d} X^a_{,e} \cdots - T^{a...e}_{c...d} X^b_{,e}$$
$$+ T^{a...b}_{e...d} X^e_{,c} \cdots + T^{a...b}_{c...e} X^e_{,d}. \tag{6.10}$$

A traditional abuse of notation is often used to denote the left hand side of (6.10) as $\mathcal{L}_X T^{a...b}_{c...d}$.

Now in the previous section attention was drawn to the situation when for a smooth tensor T and smooth vector field X the condition $\phi^* T = T$ holds for each local flow ϕ_t of X. This can be shown [10] to be equivalent to the condition $\mathcal{L}_X T = 0$ on M and which, from (6.10), gives the differential relations between T and X as promised.

6.3   Symmetries of the Metric Tensor

A local diffeomorphism f on M is called a *(local) isometry* of the metric g on M if $f^* g = g$ on the domain of f. Suppose that X is a global, smooth vector field on M such that *each* local flow ϕ_t of X is a local isometry of g on M, that is, $\phi_t^* g = g$ for each such ϕ_t. Then X gives rise to a special symmetry of g and satisfies, from the above remarks, $\mathcal{L}_X g = 0$ on M. Such a vector field X is called a *Killing vector field* and this latter equation is referred to as *Killing's equation*. This equation may be rewritten in either of the following two ways, from (6.10) (and recalling the use of a semi-colon to denote a covariant derivative with respect to ∇),

$$g_{ab,c} X^c + g_{cb} X^c_{,a} + g_{ac} X^c_{,b} = 0, \qquad X_{a;b} + X_{b;a} (\equiv \mathcal{L}_X g_{ab}) = 0. \tag{6.11}$$

The second of these equations involves the covariant derivative with respect to ∇ but it is stressed that the connection on M is not required for the definition of the Lie derivative. However, if one exists, the second equation in (6.11) is a convenient alternative to the first one (and this alternative is easily generalised to the equation $\mathcal{L}_X T = 0$ for any tensor T).

From (6.11) it is convenient to define the *Killing bivector* F associated with X, in components in any coordinate domain, by $F_{ab} = X_{a;b} = -F_{ba}$. The Ricci identity on X then gives $X_{a;bc} - X_{a;cb} = X^d R_{dabc}$ and so $F_{ab;c} - F_{ac;b} = X^d R_{dabc}$. Permuting the indices in this last equation according to $a \to b \to c \to a$ gives, after a change in sign, $F_{ba;c} - F_{bc;a} = -X^d R_{dbca}$ and another such permutation (but with no change of sign) then gives $F_{cb;a} - F_{ca;b} = X^d R_{dcba}$. Adding these last three equations and using $R_{a[bcd]} = 0$ finally gives

$$F_{ab;c} = R_{abcd} X^d \qquad (\Rightarrow X_{a;bc} = R_{abcd} X^d). \qquad (6.12)$$

The equation (6.11), rewritten as $X_{a;b} = F_{ab}$, and (6.12) give rise to a Cauchy system of first order differential equations in the following way. Let $p \in M$ and $c : I \to M$ be a smooth path with $c(I)$ lying in some coordinate domain of M with coordinates x^a, nowhere-zero tangent $k^a = d/dt(x^a \circ c)$, where t is the parameter of c, I an open interval of $\mathbb{R}$ containing 0 and $c(0) = p$. A contraction of (6.11) with k^b and (6.12) with k^c yield first order differential equations of the form $dX_a/dt = G_a$ and $dF_{ab}/dt = G_{ab}$ for smooth functions G_a and G_{ab} on $c(I)$ determined entirely by $X_a(t)(= (X_a \circ c)(t))$ and $F_{ab}(t)(= (F_{ab} \circ c)(t))$ (apart from geometrical quantities derived from the given metric g and ∇). Thus if X and F are globally defined on M and satisfy (6.11) and (6.12) on M and if X' and F' are similarly globally defined on M also satisfying (6.11) and (6.12) on M and if X and X', and F and F' agree at $t_0 \in I$, that is, they agree at the point $c(t_0)$ on the path c, they will, from the theory of first order differential equations, agree on some open subinterval of I containing t_0. On the other hand, if this agreement fails at some t'_0 then, by continuity, it will fail on some open subinterval of I containing t'_0. Since I is connected and if agreement occurs at $p = c(0)$, X and X', and F and F' agree on I, that is, on $c(I)$. The freedom of choice of smooth paths through any point of M shows that if X and X', and if F and F' agree at some point of M they will agree on some open neighbourhood of that point (and, by continuity a similar statement holds if this agreement is replaced by a failure to agree). It easily follows that M is a disjoint union of two open subsets of M on one of which agreement (in the above sense) occurs and on the other disagreement occurs. Since M is connected and given that agreement occurs at some point of M, $X = X'$ and $F = F'$ on M. It follows that any smooth, global Killing vector field on M is uniquely determined by the n values $X_a(p)$ and the $\frac{1}{2}n(n-1)$ values $F_{ab}(p)$, that is, by the values of $X(p)$ and $\nabla X(p)$ (from which it follows that if X vanishes on some non-empty open subset of M it vanishes on M). Since it is clear, by definition, that the collection of global, smooth Killing vector fields on M, denoted by $K(M)$, is a vector space it follows that $K(M)$ is finite-dimensional and $\dim K(M) \leq \frac{1}{2}n(n+1)$. Finally, if $X, Y \in K(M)$, it follows from the above that $\mathcal{L}_{[X,Y]}g = 0$ and so $[X, Y] \in K(M)$. Thus $K(M)$ is a Lie algebra under the bracket operation and is called the *Killing algebra* (of (M, g)). [It is remarked that if $\dim K(M) = \frac{1}{2}n(n+1)$, (M, g) is necessarily of constant curvature, that is *Riem* satisfies

$$R_{abcd} = \frac{R}{n(n-1)}(g_{ac}g_{bd} - g_{ad}g_{bc}) \tag{6.13}$$

at each point of M and so the Einstein space condition $R_{ab} = \frac{R}{n}g_{ab}$ also holds on M and hence, since $n > 2$, R is constant on M (chapter 3). Conversely if (M, g) is of constant curvature and $p \in M$ there exists an open neighbourhood U of p such that the smooth Killing vector fields on U arising with respect to the restricted metric on U from g give rise to a (local) Killing algebra $K(U)$ of dimension $\frac{1}{2}n(n+1)$. Proofs of these facts may be gleaned from [37].]

If $X \in K(M)$ and $p \in M$ with $X(p) \neq 0$ a standard result [9] shows that one may always choose an open coordinate domain U of p with coordinates x^a on which $X = \partial/\partial x^1$, that is, $X^1 = 1, X^2 = ... = X^n = 0$ and then (6.11) shows that the functions g_{ab} are independent of x^1 on U. In this case x^1 is referred to as an *ignorable coordinate* for g (on U). Conversely it is easily checked that if, on a coordinate domain U, x^1 is an ignorable coordinate for g, $\partial/\partial x^1$ is a local Killing vector field for (the restriction of) g to U.

Another symmetry arising from g is when each local flow of the global, smooth vector field X on M is a *(local) conformal diffeomorphism* of g on M, that is, it satisfies $\phi_t^* g = fg$ for some smooth function f on the domain of ϕ_t. This can be shown to be equivalent to the differential condition $\mathcal{L}_X g = 2\kappa g$ for some smooth function κ on M (called the *conformal function*) and then to the condition

$$X_{a;b} = \kappa g_{ab} + F_{ab} \tag{6.14}$$

where F is a smooth bivector field on M called the *conformal bivector* of X and X is called a *conformal vector field*, on M. The collection of all global smooth vector fields X on M satisfying (6.14) is, as in the Killing case, a Lie algebra under the bracket operation called the *conformal algebra* (of (M, g)) and denoted by $C(M)$. It can then be checked that one is thus led to a set of first order differential equations for the quantities X, κ, $d\kappa$ and F and so, by a similar argument to the Killing case, a global conformal vector field X on M is uniquely determined by $X(p)$, $\kappa(p)$, $d\kappa(p)$ and $F(p)$ at any $p \in M$ and so $\dim C(M) \leq (n) + (1) + (n) + (\frac{1}{2}n(n-1)) = \frac{1}{2}(n+1)(n+2)$. The vector field X can also be shown to be uniquely determined by the values $X(p)$, $\nabla X(p)$ and $\nabla(\nabla X)(p)$ at any $p \in M$ and so if X vanishes over a non-empty open subset of M, it vanishes on M. If $\dim C(M) = \frac{1}{2}(n+1)(n+2)$, (M, g) is conformally flat whilst if (M, g) is conformally flat and if $p \in M$ some open neighbourhood of p admits a (local) conformal algebra of this maximum dimension. If g and g' are smooth, conformally related metrics on M it is easily checked that their conformal algebras are equal.

A special case of the work of the previous paragraph occurs when each local flow of X is a (local) *homothety*, that is $\phi_t^* g = ag$ for some constant $a \in \mathbb{R}$. In this case (6.14) becomes

$$X_{a;b} = cg_{ab} + F_{ab} \qquad (\Leftrightarrow \mathcal{L}_X g = 2cg) \qquad (6.15)$$

where c is a real constant (the *homothetic constant*). Such a global smooth vector field X is called a *homothetic vector field* and F is the (smooth) *homothetic bivector* (field) of X. An identical argument to that in the Killing case reveals that if X is homothetic, (6.12) holds on M. A conformal vector field which is not homothetic is called *proper conformal* and a homothetic vector field with $c \neq 0$ is called *proper homothetic* (and clearly if $c = 0$, X is Killing.) Again the set of homothetic vector fields is Lie algebra under the bracket operation called the *homothetic algebra* and denoted by $H(M)$ (and so $K(M) \subset H(M) \subset C(M)$). The homothetic algebra gives rise to a set of first order differential equations in the quantities X, F and c and any such vector field is uniquely determined by $X(p)$, $F(p)$ and c (or by $X(p)$ and $\nabla X(p)$) at any $p \in M$. Hence $\dim H(M) \leq n + \frac{1}{2}n(n-1) + 1 = \frac{n}{2}(n+1) + 1$ and if this maximum is achieved (M, g) is flat. If (M, g) is flat this maximum dimension is achieved locally. A member of $H(M)$ which vanishes over some non-empty open subset of M vanishes on M. It is easily checked that $\dim H(M) \leq \dim K(M) + 1$ because if $X, Y \in H(M)$ with $\mathcal{L}_X g = 2c_1 g$ and $\mathcal{L}_Y g = 2c_2 g$ ($c_1, c_2 \in \mathbb{R}$) then, with $Z = c_2 X - c_1 Y$, one easily finds that $\mathcal{L}_Z g = 0$ and so $Z \in K(M)$. Further, if $X, Y \in H(M)$, $\mathcal{L}_{[X,Y]} g = 0$ and so $[X, Y] \in K(M)$.

6.4   Affine and Projective Symmetry

Another symmetry of interest arises when a global, smooth vector field X on M satisfies the condition that each of its local flows ϕ_t preserves geodesics, that is, if for any geodesic c of ∇, $\phi_t \circ c$ is also a geodesic for ∇. To examine this possibility further one requires the definition of the Lie derivative of the connection ∇ associated with g [90]. This is denoted by $\mathcal{L}_X \nabla$ and acts on two smooth vector fields Y and Z defined on some (any) open subset $U \subset M$ and gives rise to a smooth vector field on U as follows

$$\mathcal{L}_X \nabla(Y, Z) = [X, \nabla_Y Z] - \nabla_{[X,Y]} Z - \nabla_Y [X, Z] \qquad (6.16)$$

and which, following techniques used in section 2.9 to establish the tensor nature of the curvature tensor, gives rise to a type $(1, 2)$ tensor field on M with components D^a_{bc} given by

$$\mathcal{L}_X \nabla(\partial/\partial x^b, \partial/\partial x^c) = D^a_{bc} \partial/\partial x^a \qquad D^a_{bc} = X^a_{;bc} - R^a{}_{bcd} X^d. \qquad (6.17)$$

The condition that each local flow of X preserves geodesics is, in the above notation [90, 91], equivalent to the existence of a global smooth $1-$form field ψ on M such that for arbitrary vector fields Y and Z defined on some open subset of M

$$\mathcal{L}_X \nabla(Y, Z) = \psi(Y)Z + \psi(Z)Y \tag{6.18}$$

and which, in any coordinate domain in M, is using (6.17)

$$X^a_{;bc} = R^a{}_{bcd}X^d + \delta^a_c \psi_b + \delta^a_b \psi_c. \tag{6.19}$$

Now if one decomposes ∇X in any coordinate domain in M as

$$X_{a;b} = \frac{1}{2}h_{ab} + F_{ab}, \qquad (h_{ab} = \mathcal{L}_X g_{ab} = h_{ba}, \qquad F_{ab} = -F_{ba}) \tag{6.20}$$

and then substitutes into (6.19), taking the symmetric and skew-symmetric parts with respect to the indices a, b, one gets in any coordinate domain [37]

$$h_{ab;c} = 2g_{ab}\psi_c + g_{ac}\psi_b + g_{bc}\psi_a \tag{6.21}$$

and

$$F_{ab;c} = R_{abcd}X^d + \frac{1}{2}(g_{ac}\psi_b - g_{bc}\psi_a). \tag{6.22}$$

A contraction of (6.21) with g^{ab} shows that $\psi_a = \frac{1}{2(n+1)}(g^{bc}h_{bc})_{,a}$ and hence that the $1-$form ψ is the global gradient of the smooth function $\frac{1}{2(n+1)}g^{ab}h_{ab}$ on M and is a consequence of ∇ being a metric connection (see, e. g. [98]). This $1-$form is called the *projective* 1-form and F the *projective bivector* associated with X, and X is called a *projective vector field* on M. In fact it may be checked that given the decomposition (6.20) and (6.21), the condition (6.19) and hence (6.22) then follow (see, e.g.[13]) and so X is projective if and only if either of the equivalent conditions (6.19) or (6.21) holds. It can also be shown from all this that a system of first order differential equations in the quantities X, h, F and ψ arises on M and so, as earlier, a global, projective vector field on M is uniquely determined by the values $X(p)$, $h(p)$, $F(p)$ and $\psi(p)$ at any $p \in M$. The collection of all global, projective vector fields on M then becomes a vector space (in fact a Lie algebra under the bracket operation) called the *projective algebra* of M and is denoted by $P(M)$ and $\dim P(M) \leq n + \frac{n}{2}(n+1) + \frac{n}{2}(n-1) + n = n(n+2)$. In addition, the quantities X, h, F and ψ are uniquely determined on M if X, ∇X and $\nabla(\nabla X)$ are given at some (any) $p \in M$ and thus if X vanishes on some non-empty open subset of M it vanishes on M (see e.g.[13]).

A special case of this arises when each local flow ϕ_t of X preserves not only the geodesics of ∇ but also their affine parameters, that is, whenever c is an affinely parametrised geodesic of ∇ so also is $\phi_t \circ c$. This can be shown to be equivalent to the condition $\mathcal{L}_X \nabla = 0$ [90, 91, 10] and hence, from (6.17) to the bracketed condition in (6.12) (and then to the rest of (6.12)) and from which, after symmetrising this equation in the indices a, b, leads to (6.20) with

$\nabla h = 0$ and so, from (6.21), to $\psi = 0$ on M. Such a global vector field X is called an *affine vector field* and F the *affine bivector* and this is equivalent to (6.20) together with $\nabla h = 0$ on M, or to the projective condition (6.19) together with the condition $\psi = 0$ on M. The collection of all such vector fields is denoted by $A(M)$ and is a Lie algebra under the bracket operation called the *affine algebra* of M. Again the quantities X, h and F give rise to a system of first order differential equations on M and thus X, h and F are uniquely determined on M by the values of $X(p)$, $h(p)$ and $F(p)$ at any $p \in M$. It follows that $\dim A(M) \leq n + \frac{n}{2}(n+1) + \frac{n}{2}(n-1) = n(n+1)$ with the maximum arising if (M, g) is flat. If (M, g) is flat this maximum is achieved only locally. Also X, h and F are uniquely determined on M by the values of $X(p)$ and $\nabla X(p)$ at any $p \in M$ and so if X vanishes on some non-empty open subset of M, then it vanishes on M. By definition $A(M) \subset P(M)$ and since (6.12) holds for all members of $K(M)$ and $H(M)$ one has $K(M) \subset H(M) \subset A(M) \subset P(M)$. The members of $A(M) \setminus H(M)$ are called *proper affine* and the members of $P(M) \setminus A(M)$ are called *proper projective*.

If $X \in H(M)$ it is easily checked that $\mathcal{L}_X Riem = 0$ and $\mathcal{L}_X Ricc = 0$ whilst if C is the type $(1, 3)$ Weyl tensor and $X \in C(M)$, $\mathcal{L}_X C = 0$. If $X \in P(M)$ then, with the usual abuse of notation,

$$\mathcal{L}_X R^a{}_{bcd} = \delta^a_d \psi_{b;c} - \delta^a_c \psi_{b;d}, \qquad \mathcal{L}_X R_{ab} = (1 - n)\psi_{a;b} \qquad (6.23)$$

and so for $X \in A(M)$, $\mathcal{L}_X Riem = 0$ and $\mathcal{L}_X Ricc = 0$ (see e.g. [13]). It is noted that a projective vector field X satisfying (6.20) and (6.21) is affine if and only if the projective $1-$form ψ vanishes on M and thus ψ controls the preservation of affine parameters on the geodesics of ∇. Also, if M admits a *proper* affine vector field it necessarily admits the global, nowhere-zero, second order, symmetric, covariantly constant tensor field h and this seriously limits the existence of such vector fields on M.

6.5   Orbits and Isotropy Algebras for $(K(M)$

Although these few remarks are here given for $K(M)$ they may be applied equally well for $H(M)$, $C(M)$, $A(M)$ and $P(M)$ where it is important to note that each of these Lie algebras of smooth, global vector fields on M is finite-dimensional [13].

The discussion of section 2.8 shows that $K(M)$, if non-trivial, gives rise to a generalised distribution on M and, from section 6.3, the subset W of points $p \in M$ at which the subspace $\{X(p) : X \in K(M)\}$ of $T_p M$ is trivial has empty interior. Thus through each point of the open dense subset $M \setminus W$ of M a unique maximal integral manifold of $K(M)$ passes. Now let k be a positive integer, let $X_1, ..., X_k \in K(M)$ and let $\phi^1_t, ..., \phi^k_t$ be local flows of $X_1, ..., X_k$, respectively, and consider the local diffeomorphism between open subsets of M (where defined) given for $p \in M$ by

$$p \to \phi_{t_1}^1(\phi_{t_2}^2(\cdots\phi_{t_k}^k(p)\cdots)) \tag{6.24}$$

for all choices of k, $X_1, ..., X_k$ and $(t_1, ..., t_k) \in \mathbb{R}^k$ under the usual rules of compositions and inverses. Define an equivalence relation on M given for $p_1, p_2 \in M$ by $p_1 \sim p_2$ if and only if some map of the form (6.24) maps p_1 to p_2. The resulting equivalence classes are called *Killing orbits* of M (associated with $K(M)$) and each such orbit is a maximal integral manifold (of $K(M)$) [23, 24, 25]. The orbit through $p \in M$ is labelled O_p.

Now define the subset $K_p^* \equiv \{X \in K(M) : X(p) = 0\}$. Clearly K_p^* is a subspace of $K(M)$ and also a Lie subalgebra of $K(M)$ since if $X, Y \in K_p^*$, $[X, Y] \in K_p^*$. The subalgebra K_p^* is called the *(Killing) isotropy* algebra of $K(M)$ at p and it is noted that if $X \in K_p^*$ and ϕ_t a local flow of X, $\phi_t(p) = p$ and $g_p(u, v) = (\phi_t^* g)_p(u, v) = g_p(\phi_{t*}u, \phi_{t*}v)$ so that ϕ_{t*} is a member of the identity component of the orthogonal group at p. Further, a consideration of the linear map $K(M) \to T_p M$ given for $X \in K(M)$ by $X \to X(p)$ with kernel K_p^* and range equal to the subspace of $T_p M$ tangent to the Killing orbit O_p through p (that is to the maximal integral manifold of $K(M)$ through p) shows that $\dim K(M) = \dim K_p^* + \dim O_p$, for each $p \in O_p$ (chapter 1).

Now consider the linear map $f : K_p^* \to \Omega_p M$ with $\Omega_p M$ the usual Lie algebra of bivectors at p under commutation and which maps $X \in K_p^*$ into the negative of its Killing bivector at p. Clearly f is injective since the member of K_p^* with zero Killing bivector is the zero member of $K(M)$. Thus f is a vector space isomorphism from K_p^* onto its range in $\Omega_p M$. Then if $X, Y \in K_p^*$ with respective Killing bivectors F and G at p and if $Z = [X, Y]$, its Killing bivector at p is easily calculated to have components $G_{ac}F_b^c - F_{ac}G_b^c$ (chapter 2). But the commutator of $f(X)(= -F)$ and $f(Y)(= -G)$ is $F_{ac}G_b^c - G_{ac}F_b^c$ and is the negative of the Killing bivector of Z at p, that is $f(Z)$. Thus $f([X, Y]) = [f(X), f(Y)]$ and so f is a Lie algebra isomorphism and K_p^* is Lie isomorphic to a subalgebra of $\Omega_p M$, that is, to a subalgebra of the appropriate orthogonal algebra of $g(p)$.

Chapter 7

Projective Relatedness

7.1 Recurrence and Holonomy

In this section M is the usual $n-$dimensional, smooth, connected, Hausdorff, second countable manifold ($n > 2$) admitting a smooth metric g of arbitrary signature and with Levi Civita connection ∇. As before, when component notation is used, a semi-colon denotes a ∇-covariant derivative and a comma a partial derivative. A smooth, real tensor field T defined and nowhere-zero on some non-empty, open, connected subset U of M is called *recurrent* (on U) if $\nabla T = T \otimes P$ on U for some smooth, real 1-form P on U called the *recurrence 1-form* of, or associated with, T (see, for example, [21]). This is equivalent to the statement that in any coordinate domain in U, $T^{a...b}_{c...d;e} = T^{a...b}_{c...d} P_e$. If T is recurrent on U and $V \subset U$ is open and connected, T is recurrent on V and if $\rho : U \to \mathbb{R}$ is smooth and nowhere-zero, ρT is recurrent on U with recurrence 1-form $P + d(\log|\rho|)$. The condition that T is recurrent on U as described above is equivalent to the statement that if $p, p' \in U$ and c any smooth path in U from p to p', then $T(p')$ is proportional to the parallel transport of $T(p)$ along c at p' with proportionality ratio depending on p, p' and c. To see this, suppose that T is recurrent on U and let c be a smooth path $p \to p'$ with tangent vector $\tau(t)$ at $c(t)$ where t is a parameter along c. Let $B(p) \equiv x(p), ..., z(p)$ be a basis for T_pM and $B'(p) \equiv x'(p), ..., z'(p)$ be a basis for the cotangent space $\overset{*}{T}_pM$. The values of T along c are denoted by $T(t) = T(c(t))$ and $T'(t)$ similarly denotes the values of the parallel transport of $T(p)$ from p along c to p'. Consider the components $\alpha(t)$ of $T(t)$ in the bases $B(t)$ and $B'(t)$ at $c(t)$ obtained from $B(p)$ and $B'(p)$ by parallel transport of each of these basis members from p along c to p' (so that each of these basis members is smooth on c as in chapter 2). One gets, in any coordinate domain in M containing p, $\dot{\alpha} = \alpha_{,c}\tau^c = \alpha_{;c}\tau^c = \alpha(P_c\tau^c)$, where a dot means d/dt and so $\alpha(t)(= \alpha(c(t))) = \alpha(p)e^{\int P_c\tau^c}$ with the integral taken along the appropriate segment of c. However, the components $\alpha'(c(t))$ of the tensor T' in these parallel transported bases along c satisfy $\alpha'(c(t)) = \alpha(p)$. Now since the range of c between p and p' is the image under the continuous map c of a closed bounded interval of real numbers, which is compact (chapter 1), this range space is a compact subspace of M. Thus, covering this range with finitely many such

DOI: 10.1201/ 9781003023166-7

coordinate domains, as one can, one finds $T(t) = e^{\int P_c \tau^c} T'(t)$ along c from p to p'. Conversely, if T is nowhere zero on U and has the above property along any smooth path in U then, on any such path $T(c(t)) = \gamma(t)T'(c(t))$ for some smooth, nowhere-zero function γ on (the appropriate range of) c. Then, in the above parallel transported bases, $\nabla_\tau T' = 0$ and $\nabla_\tau T = \dot\gamma(t)T'$ where, as described before, ∇_τ is the covariant derivative along c. Thus in a coordinate domain of the type used above, $T^{a...b}_{c...d;e}\tau^e = \frac{\dot\gamma(t)}{\gamma(t)}T^{a...b}_{c...d}$. Since p, p' and c are arbitrary, T is recurrent on U (cf [93]).

Suppose that $p \in M$ and that $V \subset T_pM$ is a 1−dimensional subspace of T_pM which is holonomy invariant (chapter 2). (Here the holonomy group refers to that arising from the Levi-Civita connection ∇ of g.) Then V gives rise to a 1−dimensional, smooth, integrable distribution on M. Thus there exists a connected, open neighbourhood W of p and a smooth vector field X on W which spans this distribution on W and whose "direction" is unchanged by parallel translation in W. It follows from the above argument that X is recurrent on W and thus if a 1−dimensional holonomy invariant distribution arises on M each $p \in M$ admits an open neighbourhood W on which a (local) recurrent vector field is admitted.

A special case of recurrence arises when the tensor T is recurrent on an open, connected subset W but with recurrence 1-form P identically zero on W, that is, $\nabla T = 0$. In this case T is called *parallel* or *covariantly constant* on W. [Thus the metric tensor is parallel on M.] This leads to a special type of recurrence for T. First recall that if T is recurrent on W and if $\psi : U \to \mathbb{R}$ is a nowhere zero function on W, then ψT is also recurrent on W with recurrence 1-form $P + d(\log|\psi|)$. Now suppose T is recurrent on W with recurrence 1-form P but that there exists a nowhere zero function $\psi : W \to \mathbb{R}$ such that the scaled (recurrent) tensor ψT is parallel. Then $P = -d(\log|\psi|)$, that is, P is a gradient on W. Conversely, if P is a gradient on W, $p = d\beta$, for some smooth $\beta : W \to \mathbb{R}$, then it is easily checked that $e^{-\beta}T$ is parallel on W. [The author has learned that some results in this direction were given in [94, 95]]. It is convenient to have the concept of a tensor being "properly recurrent" on W in the sense that it does not become parallel (or could be scaled, as above, to become parallel) on some non-empty open subset of W. To do this one invokes the Ricci identities (chapter 2). Let X be a recurrent vector field on W, $\nabla X = X \otimes P$. Then clearly the covector field associated with X is recurrent and use of the recurrence of X and the appropriate Ricci identity gives in any coordinate domain in W

$$X_{a;bc} - X_{a;cb} = X^d R_{dabc} = X_a(P_{b;c} - P_{c;b}). \qquad (7.1)$$

If P vanishes on W, X is parallel on W. Otherwise there exists $p \in W$ and an open coordinate domain $V \subset W$ with P nowhere zero on V. In this latter case, if the second of the expressions in (7.1) vanishes on V, $P_{a;b} = P_{b;a}$ on V (and conversely) and so one may choose V such that $P = d\psi$ for a nowhere zero function ψ on V and then $e^{-\psi}X$ is parallel on V. This leads to the definition that X is *properly recurrent on* W if X is recurrent on W and if the open subset $\{p \in W : X^d R_{dabc}(p) \neq 0\}$ is dense in W [21]. This definition is easily extended to the concept of proper recurrence for any smooth tensor, using the appropriate Ricci identity. Here it will just be stated for a second order symmetric tensor h on W with components h_{ab}. If h is recurrent on W, $\nabla h = h \otimes P$, the Ricci identity gives

$$h_{ab;cd} - h_{ab;dc} = h_{ae}R^e{}_{bcd} + h_{be}R^e{}_{acd} = h_{ab}(P_{c;d} - P_{d;c}) \qquad (7.2)$$

and the condition for proper recurrence is that the subset $\{p \in W : (h_{ae}R^e{}_{bcd} + h_{be}R^e{}_{acd})(p) \neq 0\}$ is (open and) dense in W. It is remarked that proper recurrence is not possible for any tensor if (M, g) is flat.

Some special cases may now be discussed. Consider the recurrent tensor field T on W satisfying $\nabla T = T \otimes P$. Define the function $\delta \equiv T^{a...b}_{c...d}T^{c...d}_{a...b}$ and suppose it is nowhere zero on a coordinate domain $V \subset W$. Then an obvious contraction of the recurrence condition $T^{a...b}_{c...d;e} = T^{a...b}_{c...d}P_e$ with $T^{c...d}_{a...b}$ reveals that $P = d\nu$ on V where $\nu = \frac{1}{2}\log|\delta|$ and so $e^{-\nu}T$ is parallel on V. Further, if X is a recurrent vector field on W which is spacelike (respectively timelike or null) at some point of W it is spacelike (respectively timelike or null) at *all* points of W. Putting these last two results together one sees that if g is positive definite and T is a recurrent tensor field on U (hence nowhere zero on U), then δ is nowhere zero on W and any $p \in W$ admits an open neighbourhood V on which T may be scaled so as to be parallel on V, that is, it cannot be properly recurrent. If g is not positive definite and if X is a recurrent vector field on W it can only be properly recurrent (but need not be) if X is null on W. It is also remarked that if X and Y are properly recurrent and null on W and if $X \cdot Y$ is a non-zero constant on W the recurrence 1-forms of X and Y differ only in sign. [One may extend the concept of recurrence to a nowhere-zero complex tensor field on M, the recurrence 1-form then becoming complex. However, in what is to follow, all tensors will be assumed real and if recurrence is required for a complex tensor, it will be clearly pointed out.]

Consider a (real), smooth type $(0, 2)$ symmetric tensor field T which is recurrent on some open, connected subset $W \subset M$ with recurrence 1-form P, so that $\nabla T = T \otimes P$ on W. Then in any (connected) coordinate domain $V \subset W$ $T_{ab;c} = T_{ab}P_c$. Let $p, p' \in V$ and c a smooth path from p to p' and let $0 \neq k \in T_pM$ be a real or complex eigenvector of T at p so that $T^a{}_b k^b = \alpha k^a$ $(T^a{}_b k^b - \alpha k^a = 0)$ at p for $\alpha \in \mathbb{C}$. Let $T'(t)$ and $k'(t)$ be obtained from $T(p)$ and $k(p)$ by parallel transport along c to some point $c(t)$ on c and define a constant function, also called α, on V which maps each $p \in V$ to α. Then $T'^a{}_b k'^b - \alpha k'^a$ is parallel transported along c and is hence zero at each point $c(t)$ showing that $k'(t)$ is an eigenvector of $T'(t)$ with eigenvalue α at each of these points. From earlier remarks it follows that $k'(t)$ is an eigenvector of $T(t) \equiv T(c(t))$ along c with eigenvalue $\alpha e^{\int_c P_a \tau^a}$. It is then clear that since V is connected, hence path-connected, the Segre type of T, including degeneracies, is the same at each point of W (cf [93]).

Now consider the eigenspaces of T at p with associated (finitely many) distinct eigenvalues $\alpha, \beta, ..., \gamma \in \mathbb{C}$ and suppose at least one is non-zero so that one may order them as $|\alpha| \leq |\beta| \leq ... \leq |\gamma|$ with $|\gamma| > 0$ and for distinct paths c and c' from p to p' let $\mu = e^{\int_c P_a \tau^a}$ and $\nu = e^{\int_{c'} P_a \tau^a}$ with $\mu, \nu \geq 0$. Suppose $\mu \neq \nu$ so that one may choose $\nu < \mu$. Then the above shows that the resulting eigenvalues at p' are $\alpha\mu, \beta\mu, ..., \gamma\mu$ (along c) and $\alpha\nu, \beta\nu, ..., \gamma\nu$ (along c'). But these finite collections, which must be the same since T is defined on V, cannot be since $|\mu\gamma|$ is greater than each of $|\alpha\nu|, |\beta\nu|, ..., |\gamma\nu|$ and this contradiction shows that $\mu = \nu$, that is, $e^{\int_c P_a \tau^a}$ is independent of the path from p to p', that is, $\oint_c P_a \tau^a = 0$ for any smooth, closed path c at any point of V. From this it follows from a version of the Poincaré lemma (see, e.g. [22]) that P is a gradient on some connected, open neighbourhood U of p and hence that T is not properly recurrent on V. Thus in order that T be properly recurrent, all its eigenvalues vanish at each $p \in U$. Thus again, if g is positive definite, T cannot be properly recurrent since this would force T to be identically zero on V. In the the most relevant case here, where $\dim M = 4$, if g has Lorentz signature T must have Segre type $\{(211)\}$ or $\{(31)\}$ at each point of V and if g has neutral signature, T must have Segre type $\{(211)\}$, $\{(31)\}$, $\{(22)\}$ or $\{4\}$ at each point of U and where for each of these the eigenvalue is zero. It is remarked that not all of these candidates can actually be properly recurrent. In fact if T is recurrent with recurrence 1-form P then so also is the symmetric tensor $T_{ac}T^c{}_d...T^d{}_b$ (m products) with recurrence 1-form mP, for each $m \geq 1$. This may be used to show that the Segre type $\{4\}$ in neutral signature cannot be recurrent (or parallel) [21]. However, all the other Segre types could be. If T is parallel on W its Segre type, including degeneracies, is the same at each point of W and its eigenvalues are constant.

Returning to the general case of $\dim M = n$ with metric g of arbitrary signature and associated Levi-Civita connection ∇, one can relate the concept of recurrent (including parallel) vector fields to the holonomy structure of M. Let Φ be the holonomy group of (M, g) and Φ^0 its associated, restricted holonomy group, each with Lie algebra ϕ. Then Φ^0 is a connected Lie group. Now let $\widetilde{M}$ denote the $n-$dimensional universal covering manifold of M with smooth natural projection $\pi : \widetilde{M} \to M$. Then the pullback π^*g of g is a smooth metric on $\widetilde{M}$ of the same signature as g and the holonomy group $\widetilde{\Phi}$ of $(\widetilde{M}, \pi^*g)$ arising from the Levi-Civita $\widetilde{\nabla}$ on $\widetilde{M}$, from π^*g, equals its restricted holonomy group $\widetilde{\Phi^0}$ since $\widetilde{M}$ is simply connected. It is also true that $\widetilde{\Phi^0} = \Phi^0$ [96]. Thus $\widetilde{\Phi} = \widetilde{\Phi^0}$ is a connected Lie group with holonomy algebra ϕ and hence (chapter 2) each member of $\widetilde{\Phi}$ is a finite product of exponentials of members of ϕ, the members of the latter regarded as bivectors at, say, $\tilde{p} \in \widetilde{M}$. Now if $p = \pi(\tilde{p})$ there exist (chapter 2) connected, open neighbourhoods U of p and $\widetilde{U}$ of $\tilde{p}$ such that $\pi : \widetilde{U} \to U$ is a smooth diffeomorphism and where U and $\widetilde{U}$ have the metrics restricted from g and π^*g, respectively. Thus π is an *isometry* between $(\widetilde{U}, \pi^*g)$ and (U, g) and one may regard ϕ as consisting of bivectors at p. Now suppose that $0 \neq k \in T_{\tilde{p}}\widetilde{M}$ is an eigenvector of each member of ϕ and $\tilde{c}$ is a closed path at $\tilde{p}$. Then the action of any member of $\widetilde{\Phi}$, that is, parallel transport of k around $\tilde{c}$, is a finite product of exponentials of members of ϕ and is hence a non-zero multiple of k (section 2.11). It follows that k gives rise to a $1-$dimensional, smooth, holonomy invariant distribution on $\widetilde{M}$ and hence to a recurrent vector field with respect to $\widetilde{\nabla}$ on some connected, open neighbourhood $\widetilde{V}$ of $\tilde{p}$, chosen so that $\widetilde{V} \subset \widetilde{U}$. Use of the smooth diffeomorphism π then reveals a smooth, recurrent vector field with respect to ∇ on the connected, open neighbourhood $\pi(\widetilde{V})$ of p in M which equals $\pi_* k$ at p. Further, if the original $k \in T_{\tilde{p}} \widetilde{M}$, which is an eigenvector of each member of ϕ, is such that each of the resulting eigenvalues zero the action of any member of $\widetilde{\Phi}$ on k yields the same vector $k \in T_{\tilde{p}} \widetilde{M}$ and, as a consequence, leads to a *global*, parallel, vector field on $\widetilde{M}$. The diffeomorphism π then leads to a parallel vector field on some connected, open neighbourhood of p whose value at p is $\pi_* k$. These remarks lead to the following lemma (which is only needed in the special case dim$M = 4$).

Lemma 7.1 *Let M be a $4-$dimensional manifold with smooth metric g of arbitrary signature, associated Levi-Civita connection ∇, holonomy group Φ and holonomy algebra ϕ. Then if $p \in M$ and $k \in T_p M$ is an eigenvector of each member of ϕ there exists a connected open neighbourhood U of p and a smooth recurrent vector field X on U such that $X(p) = k$. If each of the eigenvalues for k above are zero, X may be chosen parallel on U.*

It is remarked that this result also holds for a complex k (in the complexification of $T_p M$) and leads, in an obvious sense, to a recurrent complex vector field with complex recurrence 1-form.

Thus for $p \in M$, the holonomy group at p, Φ_p, is a Lie subgroup of $o(4)$, $o(1,3)$ or $o(2,2)$ depending on the signature of g and the associated Lie algebra ϕ is represented at any $p \in M$ as a Lie algebra of bivectors $F^a{}_b$ under the bracket operation in some basis of $T_p M$ each of which satisfies

$$g_{ac} F^c{}_b + g_{bc} F^c{}_a = 0 \tag{7.3}$$

at p and which includes all bivectors in the range space of the curvature map f at p. This follows since the infinitesimal holonomy algebra at p contains $rgf(p)$ and is a subalgebra of ϕ (section 2.13). Thus, for example, if dim$\phi = 1$ there exists $p \in M$ and a coordinate domain U containing p such that *Riem* is nowhere zero on U and hence, on U,

$$R_{abcd} = \alpha G_{ab} G_{cd} \qquad (G_{ab} = g_{ac} G^c{}_b) \tag{7.4}$$

where $\alpha : U \to \mathbb{R}$ and G is a bivector field on U spanning ϕ at each $p \in U$. It follows that α and G may be chosen smooth on some open neighbourhood of p. To see this choose a bivector H at p such that $R_{abcd} H^{cd} \neq 0$ and extend H to a smooth bivector (also labelled H) on some neighbourhood $V \subset U$ of p so that $R_{abcd} H^{cd}$ is nowhere zero, smooth and proportional to G on V. It follows that G may be chosen smoothly on V and, as a consequence of the smoothness of $Riem$, α is then also smooth on V. The curvature class of $Riem$ is then either O or D on M and where it is D, the bivector G, as in (7.4), is simple (chapter 3 or 4).

7.2   Projective Relatedness

In this section the general case when $\dim M = n > 2$ and g is an arbitrary metric on M is studied and the geodesics arising from the Levi-Civita connection ∇ which, in turn, arise from g are considered. Such a geodesic (chapter 2) is a smooth path $c : I \to M$ with I an open interval of $\mathbb{R}$ and if $p \in c(I)$ and I is adjusted so that $0 \in I$ and $c(0) = p$, this geodesic is said to *start from* p, or to have *initial point* p. Choosing a coordinate domain containing p with coordinate functions x^a then on the path c (that is, on $c(I)$) the coordinates $x^a(t) \equiv x^a \circ c(t)$ satisfy the geodesic equations

$$\frac{d^2 x^a(t)}{dt^2} + \Gamma^a_{bc}(t)\frac{dx^b(t)}{dt}\frac{dx^c(t)}{dt} = \lambda(t)\frac{dx^a(t)}{dt} \tag{7.5}$$

where the Γ^a_{bc} are the Christoffel symbols arising from ∇ (and $\Gamma^a_{bc}(t) = \Gamma^a_{bc}(x^a \circ c(t))$) and where λ is a smooth function on the appropriate domain. The function λ reflects the parametrisation chosen for the geodesic and may be set to zero if affine parametrisation is required. However, the more general (parameter-independent) parametrisation (7.5) will be retained here and which, in the above chart domain, may be rewritten

$$k^a{}_{;b}k^b = \lambda k^a \tag{7.6}$$

where $k^a(t) = \frac{dx^a(t)}{dt}$ is the tangent vector to c. The work in section 2.9 shows, with an abuse of notation, that the range $c(I)$ of the geodesic c through p is determined by p and the *direction* of the (non-zero) tangent to c at p.

In this chapter a situation is considered where a manifold M admits two metrics and associated Levi-Civita connections and which are such that their geodesics coincide. This subject has received significant coverage over the past few years and much information, history and bibliographical detail of the subject can be found in [97]. In particular, the early work of Eisenhart [37] and Petrov [49] are very important. There is some overlap between [97] and parts of this chapter but the methods and general approach used here are independent and quite different and which, after some generalities, will concentrate on the 4-dimensional case and for each of the three possible signatures.

So suppose that g and g' are smooth metrics on M with Levi-Civita connections ∇ and ∇' and Christoffel symbols Γ^a_{bc} and Γ'^a_{bc}, respectively. Suppose also that *the geodesics arising from ∇ and ∇' are the same*, that is, if for some open interval $I \subset \mathbb{R}$, $c : I \to M$ is a geodesic as defined in chapter 2 satisfying (2.23) (or (7.5) or (7.6) above) with respect to ∇ for some function λ it satisfies (2.23) (or (7.5) or (7.6)) with respect to ∇' for some function λ'. Thus for each $p \in M$ and each $0 \neq v \in T_pM$, there exists an open interval $I \subset \mathbb{R}$ containing 0 and a smooth path $c : I \to M$ starting from p and with $\dot{c}(0) = v$ which is geodesic *for both* ∇ *and* ∇'. In this case (M, g) and (M, g'), or ∇ and ∇', or their respective metrics g and g', are said to be *projectively (or geodesically) related* or *projectively (or geodesically) equivalent*. It is noticed that no restrictions are placed on the parameters of c and thus, for example, a particular parameter on c could be affine for each of, exactly one of, or neither of ∇ and ∇'. Sometimes the range space $c(I) \subset M$ is referred to as an *unparametrised* geodesic (for ∇ and/or ∇'). Then in coordinate language the tangent vector $k^a(t)$ to c at $c(t)$ satisfies $k^a_{;b}k^b = \lambda k^a$ and $k^a_{|b}k^b = \lambda' k^a$ for smooth functions λ and λ' and where a semi-colon and a vertical stroke denote, respectively, a covariant derivative with respect to ∇ and ∇'. It follows that $k^a_{;b}k^b k^e = k^e_{;b}k^b k^a$ and similarly $k^a_{|b}k^b k^e = k^e_{|b}k^b k^a$ hold on $c(I)$. A subtraction of these last two equations removes the partial derivatives and gives $P^a_{bc}k^b k^c k^e = P^e_{bc}k^b k^c k^a$, where $P^a_{bc} \equiv \Gamma'^a_{bc} - \Gamma^a_{bc} = P^a_{cb}$. This last equation may be rewritten to expose the tangent vector components as

$$(\delta^e_d P^a_{bc} - \delta^a_d P^e_{bc})k^b k^c k^d = 0 \tag{7.7}$$

and is true for each initial choice of $k = v \in T_pM$ and for each $p \in M$. Thus the bracketed term, when symmetrised over the indices b, c and d, is zero and so, since $P^a_{bc} = P^a_{cb}$,

$$\delta^e_d P^a_{bc} - \delta^a_d P^e_{bc} + \delta^e_b P^a_{cd} - \delta^a_b P^e_{cd} + \delta^e_c P^a_{db} - \delta^a_c P^e_{db} = 0 \tag{7.8}$$

and a contraction over the indices d and e finally gives in any coordinate domain in M

$$P^a_{bc} = \Gamma'^a_{bc} - \Gamma^a_{bc} = \delta^a_b \psi_c + \delta^a_c \psi_b \tag{7.9}$$

where $\psi_a = \frac{1}{n+1}P^b_{ab}$. Now it is easily seen that the P^a_{bc} are the components of a global, smooth, type $(1,2)$ tensor field on M and hence that the ψ_a are the components of a global, smooth 1-form on M. Conversely, if (7.9) holds in any coordinate domain of M and for some global, smooth 1-form field ψ on M then, in that domain, $P^a_{bc}k^b k^c k^e = P^e_{bc}k^b k^c k^a$ and recalling the earlier derivation of (7.7) this shows that if the geodesic equation for k and ∇ is satisfied, that is, $k^a_{;b}k^b$ is a multiple of k^a along c, $k^a_{|b}k^b k^e = k^e_{|b}k^b k^a$ holds along c and hence the geodesic equation for ∇' is satisfied, and vice versa. Thus ∇ *and* ∇' *are projectively related if and only if (7.9) holds for each coordinate domain in M for some global, smooth 1-form ψ on M*. The 1-form ψ, which is determined by ∇ and ∇' is referred to as the *projective 1-form* associated with ∇ and ∇'. Since $\psi_a = \frac{1}{n+1}P^b_{ab} = \frac{1}{n+1}(\Gamma'^b_{ab} - \Gamma^b_{ab})$ and since ∇ and ∇' are the respective Levi-Civita connections for the metrics g and g' on M one may use the expression for Γ^a_{bc} in chapter 3 to find Γ^b_{ab} (see, e.g., [37]) and hence to get $\psi_a = \frac{1}{2(n+1)}(\log |\frac{\det g'}{\det g}|)_{,a}$, so that ψ is a global gradient (an exact 1-form) on M. [It is remarked that if one, or both, of the symmetric connections ∇ and ∇' is not metric then, although (7.9) still holds, ψ need not be exact. In fact, given a smooth, symmetric *metric* connection ∇ on M and a smooth global 1-form ψ on M which is *not* exact, one may construct a symmetric connection ∇' on M according to (7.9) which is then *not* metric but shares the same unparametrised geodesics as ∇. Thus knowledge of all such unparametrised geodesics on M does not disclose whether a symmetric connection yielding those geodesics is metric or not (cf [98]).]

Equation (7.9) for connections ∇ and ∇' gives a geometrical interpretation of the 1-form ψ. Suppose a path $c(t)$ is an affinely parametrised geodesic for ∇ so that, in the above notation, $k^a{}_{;b}k^b = 0$ along $c(I)$. One then finds $k^a{}_{|b}k^b = \mu(t)k^a$ with $\mu = 2\psi_a k^a$ on $c(I)$. Thus ψ controls the deviation of the parameter on c from being affine for ∇'. (cf section 6.4 and note that if ∇ and ∇' are metric connections one has $\psi = d\chi$ for some smooth function $\chi : M \to \mathbb{R}$ and so $\mu = 2\frac{d\chi}{dt}$.) If ψ is identically zero on M, $\nabla' = \nabla$, and ∇ and ∇' agree as to their affine parameters. In this case ∇ and ∇' (or g and g') are said to be *affinely related* or *affinely equivalent* on M. If ψ is not identically zero on M, that is, $\nabla' \neq \nabla$, (M, g') and (M, g), or ∇' and ∇, or g' and g are said to be *properly projectively related*, on M.

If ∇ and ∇' are smooth symmetric metric connections on M they are projectively related if and only if their respective Christoffel symbols satisfy (7.9) on M for some global, smooth, exact 1-form on M. This equivalence can also be described in terms of the ∇-covariant derivative of the metric g', $\nabla g'$. To see this one first writes out the identity $\nabla' g' = 0$ in any coordinate domain in M and then replaces the ∇' Christoffel symbols with those from ∇ and the components of ψ, using (7.9), to get

$$g'_{ab;c} = 2g'_{ab}\psi_c + g'_{ac}\psi_b + g'_{bc}\psi_a. \tag{7.10}$$

This is the desired expression for $\nabla g'$. Conversely, if (7.10) holds in each coordinate domain of M (and noting that the identity $\nabla' g' = 0$ is just $g'_{ab|c} = 0$) write $g'_{ab;c} - g'_{ab|c} = g'_{ab;c}$ on any coordinate domain. The left hand side of this equation can be written out in terms of the components g'_{ab}, Γ'^a_{bc} and Γ^a_{bc} and the right hand side may be replaced by (7.10) to give

$$g'_{ad}P^d_{bc} + g'_{bd}P^d_{ac} = 2g'_{ab}\psi_c + g'_{ac}\psi_b + g'_{bc}\psi_a. \tag{7.11}$$

Regarding (7.11) as the "abc" equation, compute the equation "abc"-"bca"+"cab" to get (7.9). Thus the following theorem has been established.

Theorem 7.1 *Let g and g' be smooth metrics on M with respective Levi-Civita connections ∇ and ∇'. Then the following are equivalent.*

(i) *∇ and ∇' (that is, g and g') are projectively related.*

(ii) *There exists a smooth, global, exact 1-form ψ on M such that the Christoffel symbols arising from ∇ and ∇' satisfy (7.9) in any coordinate domain.*

(*iii*) There exists a smooth, global, exact 1-form ψ on M such that g' and ∇ satisfy (7.10) in any coordinate domain on M.

Given that the above (equivalent) conditions hold the 1-forms ψ in parts (*ii*) and (*iii*) are equal. [It is remarked here that if $0 \neq \rho \in \mathbb{R}$, g and ρg are always projectively related (in fact, affinely related) since then $\nabla = \nabla'$.] The condition (*ii*) (equation (7.9)) above allows a neat comparison between the curvature and Ricci tensors arising from (the projectively related) ∇ and ∇' [37]. To achieve this write the curvature tensor components $R'^a{}_{bcd}$ for ∇' as in (2.30) and eliminate the terms Γ'^a_{bc} using (7.9). A tedious but straightforward calculation gives

$$R'^a{}_{bcd} = R^a{}_{bcd} + \delta^a_d \psi_{bc} - \delta^a_c \psi_{bd} \qquad (7.12)$$

where ψ_{ab} is the symmetric tensor with components $\psi_{ab} \equiv \psi_{a;b} - \psi_a \psi_b$. A contraction over the indices a and c then relates the Ricci tensors $Ricc$ and $Ricc'$ from ∇ and ∇', respectively, with components R_{ab} and R'_{ab}, as

$$R'_{ab} = R_{ab} - (n-1)\psi_{ab}. \qquad (7.13)$$

One can then construct a smooth, global, type $(1,3)$ tensor W from the tensors $Riem$ and $Ricc$ for ∇ on M as

$$W^a{}_{bcd} \equiv R^a{}_{bcd} + \frac{1}{(n-1)}(\delta^a_d R_{bc} - \delta^a_c R_{bd}). \qquad (7.14)$$

If one computes the tensors W and W' as in (7.14) but associated with the projectively related connections ∇ and ∇' so that equations (7.12) and (7.13) hold, one finds that $W' = W$, that is, the tensors W and W' agree on M if the symmetric connections ∇ and ∇' are projectively related on M. In this sense, W is a "projective invariant". This tensor and the consequence $W' = W$ were discovered by Weyl [30] and W is called the *Weyl projective tensor*. [It should be compared with the Weyl conformal tensor C introduced earlier for conformally related metrics, but *not* confused with it. It is noted that C depends on a metric for its existence whereas W depends only on the connection ∇ (through the Ricci and curvature tensors).] It is pointed out here that the converse of this result by Weyl (in the sense that two metrics with equal (and nowhere zero) Weyl projective tensors are projectively related) is not true [88]. To see this let g be the Lorentz metric given in coordinates u, v, x, y in the general form (4.49) with the coordinate restriction $u > 0$ and with the chart chosen so that $Riem$ is nowhere zero. Choose the function H to satisfy $\partial^2 H/\partial x^2 + \partial^2 H/\partial y^2 = 0$ which can be shown to give the vacuum condition $Ricc \equiv 0$. Such a metric g is called a *pp-wave* [64]. Then consider the metric $g' = u^{-2}g$ which is conformally related to g. In the usual notation it can be shown that g' is also vacuum and so, since the Weyl conformal tensors C and C' are equal and $Ricc = Ricc' \equiv 0$, one gets $Riem = Riem'$. It follows that $W' = W$ and is nowhere zero. However a simple calculation from (7.10) shows that, quite generally, two conformally related metrics cannot be projectively related unless the conformal factor is constant (see, e.g. [100]). Thus g and g' are not projectively related and the converse is thus false.

It should be remarked here that the computations in (7.12), (7.13) and (7.14) depended on the symmetry of the tensor ψ_{ab}. This, in turn, relied on the fact that ∇ and ∇' were each metric and symmetric connections (and hence their associated Ricci tensors are symmetric). In fact, if one makes no assumptions on ∇ and ∇', one must drop the symmetry of ψ_{ab} and also of $Ricc$ and $Ricc'$. However, the expression for $Riem$ and definition of $Ricc$ given in chapter 2 still hold but there occur extra terms $\delta_b^a(\psi_{d;c}-\psi_{c;d})$ on the right hand side of (7.12) and $\psi_{b;a} - \psi_{a;b}$ on the right hand side of (7.13). Then it easily follows from this new version of (7.13) that if $Ricc'$ and $Ricc$ are symmetric, so is ψ_{ab} and hence also $\psi_{a;b}$. [This latter remark was noted in [97]]. Then the original equations (7.12), (7.13) and (7.14) hold and, further, if one assumes that ∇ is a symmetric (but not necessarily metric) connection, the symmetry of $\psi_{a;b}$ leads to the symmetry of $\psi_{a,b}$ and hence to the fact that ψ is now a closed (but not necessarily exact) 1-form, that is, it is locally a gradient. That ψ may not be exact can be seen by simply choosing ψ in (7.9) to be closed but not exact [98]. However, even without these assumptions one may still construct a tensor with the above projective invariant properties [97] but this will not be done here. The remainder of this book will be concerned only with the case when ∇ and ∇' are symmetric, metric connections with metrics g and g'. In this case one has the smooth type $(0,4)$ Weyl projective tensor with components $W_{abcd} = g_{ae}W^e{}_{bcd}$ and with $W^a{}_{bcd}$ as in (7.14). However, it is noted that the above projective invariance property applies only to the type $(1,3)$ tensor W.

Lemma 7.2 *Let (M,g) give rise to the smooth tensors $Riem$, $Ricc$, R, E, C and W, as in chapter 3. Then at any $p \in M$,*

(i) $W^a{}_{acd} = 0$, $W^c{}_{acb} = 0$, $W^a{}_{bcd} = -W^a{}_{bdc}$, $W^a{}_{[bcd]} = 0$.

(ii) $W(p) = 0 \Leftrightarrow Riem$ *takes the constant curvature form at p.*

(iii) $W(p) = Riem(p) \Leftrightarrow Ricc(p) = 0$.

(iv) $W(p) = C(p) \Leftrightarrow E(p) = 0 \Leftrightarrow$ *the Einstein space condition holds at $p \Leftrightarrow W_{abcd} = -W_{bacd}$ at p. If these equivalent conditions hold, $W_{abcd} = W_{cdab}$.*

(v) *If g' is another smooth metric on M which is projectively related to g, (M,g) is of constant curvature $\Leftrightarrow (M,g')$ is of constant curvature.*

Proof Part (i) follows easily from (7.14). For (ii) the condition $W(p) = 0$ gives $R_{abcd} = \frac{1}{n-1}(g_{ac}R_{bd} - g_{ad}R_{bc})$ at p which, since $R_{abcd} = R_{cdab}$, implies the Einstein space condition on $Ricc(p)$ and which, on substitution into (7.14), gives the constant curvature condition at p, which for $\dim M = n$ is $R_{abcd} = \frac{R}{n(n-1)}(g_{ac}g_{bd}-g_{ad}g_{bc})$. Conversely, the constant curvature condition at p gives the Einstein space condition at p and then (7.14) reveals $W(p) = 0$. Part (iii) is immediate from (7.14). For (iv) it is convenient to define a type $(0, 4)$ tensor Q at p by

$$Q_{abcd} = W_{abcd} + W_{bacd} = \frac{1}{n-1}(g_{ad}R_{bc} - g_{ac}R_{bd} + g_{bd}R_{ac} - g_{bc}R_{ad}). \quad (7.15)$$

Thus $Q = 0 \Rightarrow Q^c{}_{acb} = 0 \Rightarrow \widetilde{R}_{ab} = 0$ at p where $\widetilde{Ricc}$ is the tracefree Ricci tensor at p, the latter equation here being equivalent to the Einstein space condition at p and to the vanishing of the tensor E at p (chapter 3). It then follows that $R_{abcd} = C_{abcd} + \frac{R}{n(n-1)}(g_{ac}g_{bd} - g_{ad}g_{bc})$ at p. From this the Einstein space condition at p follows and (7.14) reveals that $W(p) = C(p)$. Finally the index symmetries of C show that if $W(p) = C(p)$ then $Q(p) = 0$ and (iv) is established. For part (v) if g and g' are projectively related, $W'^a{}_{bcd} = W^a{}_{bcd}$ on M and the result follows from part (ii). $\square$

Part (iv) appears in [88] (but is mentioned in [99]) and part (v) is given in [37].

In the case when $\dim M = 4$ the following results hold.

Lemma 7.3 (i) *For (M, g), if there exists $p \in M$ at which Riem takes the constant curvature form with Ricci scalar non-zero, the infinitesimal holonomy algebra and hence the holonomy algeba is $6-$dimensional and is hence $o(4)$, $o(1, 3)$ or $o(2, 2)$, depending on signature.*

(ii) *If (M, g) and (M, g') are projectively related and if each of their respective holonomy algebras has dimension < 6 then Riem and Riem' agree as to their zeros, that is, the sets $\{p \in M : Riem(p) = 0\}$ and $\{p \in M : Riem'(p) = 0\}$ are the same.*

Proof For (i) one simply notes that R_{abcd} is then a multiple of the bivector metric P (chapter 3) and so the range space of the curvature map is $6-$dimensional. For (ii) suppose $Riem(p) = 0$ for some $p \in M$. Then $Ricc(p) = 0$ and so $W(p) = 0$ (lemma 7.2). Hence $W'(p) = 0$ and part (ii) of the previous lemma shows that $Riem'$ takes the constant curvature form at p and the first part of this lemma reveals that the associated Ricci scalar, and hence $Riem'$, vanish at p. The reverse proof is identical. $\square$

7.3 The Sinjukov Transformation

In this section $\dim M = n > 2$. For a given geometry (M, g), where g has Levi-Civita connection ∇, the formal procedure for finding another smooth metric g' on M with Levi-Civita connection ∇' and which is projectively related to g is thus to solve (7.9) for ∇' and ψ, or to solve (7.10) for g' and

ψ. Thus, with (M, g) given, one seeks pairs (g', ψ) satisfying (7.10). There is a little redundancy in (7.10) which can be removed by using the following rather neat transformation due to Sinjukov [101] and which, starting with (M, g), instead of attempting to solve (7.10) for the pair (g', ψ) tries to find a certain pair (a, λ), where a is a non-degenerate, smooth type $(0, 2)$ symmetric tensor on M and λ a smooth 1-form on M and which are constrained by a differential relation, the Sinjukov equation, and which corresponds to (7.10). Having found a and λ the procedure provides a means of recovering the desired solution pair (g', ψ).

To achieve this one starts with (M, g) and a metric g' and *exact* 1-form ψ on M satisfying (7.10) and writes $\psi = d\chi$ for some global smooth function $\chi : M \to \mathbb{R}$. In any coordinate domain $U \subset M$ one then defines a smooth, symmetric, non-degenerate tensor a and a smooth 1-form λ, on U, by

$$a_{ab} = e^{2\chi} g'^{cd} g_{ac} g_{bd}, \qquad \lambda_a = -e^{2\chi} \psi_b g'^{bc} g_{ac}, \tag{7.16}$$

from which it follows that $\lambda_a = -a_{ab}\psi^b$ where $\psi^a = g^{ab}\psi_b$. One may then invert (7.16) to get

$$g'^{ab} = e^{-2\chi} a_{cd} g^{ac} g^{bd}, \qquad \psi_a = -e^{-2\chi} \lambda_b g^{bc} g'_{ac}, \tag{7.17}$$

where the g'^{ab} are the contravariant components of g', so that $g'^{ab} g'_{bc} = \delta^a_c$. Assuming that g and g' are projectively related, so that (7.10) holds, one may rewrite (7.10) by first noting that $(g'^{ab} g'_{bc})_{;d} = 0$, where a semi-colon denotes a $\nabla-$covariant derivative. Expanding this result and using (7.10) gives

$$g'^{ab}{}_{;c} = -2g'^{ab}\psi_c - \delta^a_c g'^{db}\psi_d - \delta^b_c g'^{ad}\psi_d. \tag{7.18}$$

Then using the first equation in (7.16) one finds

$$a_{ab;c} = 2e^{2\chi} \chi_c g'^{ed} g_{ae} g_{bd} + e^{2\chi} g'^{ed}{}_{;c} g_{ae} g_{bd} \tag{7.19}$$

into which (7.18) is substituted. The result, using the second of (7.16), is then

$$a_{ab;c} = g_{ac}\lambda_b + g_{bc}\lambda_a. \tag{7.20}$$

This the required *Sinjukov equation* [101] and, given (M, g), is to be solved for the global, smooth, symmetric, type $(0, 2)$ *Sinjukov tensor* a and the global, smooth 1-form λ, on M. It immediately follows by contracting (7.20) with g^{ab} that λ is exact, being the global gradient of the global smooth function $\frac{1}{2} a_{ab} g^{ab}$. Thus from the pair (g', ψ) satisfying (7.10) one has computed the pair (a, λ), as in (7.16), each uniquely up to the same multiplicative, non-zero constant which arises from the freedom in the choice of χ. The pair (a, λ) then satisfies (7.20).

Now suppose (M, g) is given and one has a smooth solution pair (a, λ) to (7.20) on M with a symmetric and non-degenerate and λ smooth (and necessarily exact). One must then show how to construct the pair (g', ψ) with

g' a smooth metric and ψ a smooth, exact 1-form, on M which, together, satisfy (7.10). To do this define a unique, smooth, symmetric type $(2,0)$ tensor b on M by stating that at any point in any coordinate domain of M the components b^{ab} constitute the inverse matrix to a_{ab} at that point, that is, $a_{ac}b^{cb} = \delta_a^b$. The b so defined is easily checked to be a tensor and to be smooth and non-degenerate. Of course, one still raises and lowers indices associated with a and b using the original metric g, so that $b_{ac}a^{cb} = \delta_a^b$. Now define a global, smooth 1-form ψ on M by the component expression $\psi_a = -b_{ab}\lambda^b$ in any coordinate domain. Then index manipulation (using g) gives $\lambda_a = -a_{ab}\psi^b$. To see that the 1-form ψ is exact one takes a $\nabla-$covariant derivative of the relation $a_{ac}b^{cb} = \delta_a^b$ to get $(a_{ac}b^{cb})_{;d} = 0$ and which, on expansion and use of (7.20), gives

$$(g_{ad}\lambda_c + g_{cd}\lambda_a)b^{cb} + a_{ac}b^{cb}{}_{;d} = 0. \qquad (7.21)$$

On multiplying (7.21) by b^{ae} and using some index manipulation, one gets

$$b_d{}^e b^{cb}\lambda_c + b_d{}^b b^{ae}\lambda_a + b^{eb}{}_{;d} = 0 \qquad (7.22)$$

where $b_d{}^e = g_{ad}b^{ae}$ and which, since $\psi_a = -b_{ab}\lambda^b$, finally gives

$$b_{ab;c} = b_{ac}\psi_b + b_{bc}\psi_a. \qquad (7.23)$$

Next define a smooth symmetric connection ∇'' on M by taking, in any coordinate domain in M, its Christoffel symbols as $\Gamma''^a_{bc} \equiv \Gamma^a_{bc} - \psi^a g_{bc}$. Using a double vertical stroke for $\nabla''-$covariant derivatives one finds

$$a_{ab||c} = a_{ab,c} - a_{ad}\Gamma''^d_{bc} - a_{bd}\Gamma''^d_{ac} = a_{ab;c} + a_{ad}\psi^d g_{bc} + a_{bd}\psi^d g_{ac} = 0 \quad (7.24)$$

where the final step is obtained using $\lambda_a = -a_{ab}\psi^b$ and (7.20). Thus $\nabla''a = 0$ and so ∇'' is a metric connection with a as a compatible metric (chapter 3). The definition $\Gamma''^a_{bc} = \Gamma^a_{bc} - \psi^a g_{bc}$, on contraction over the indices a and b then gives

$$\psi_a = \Gamma^b_{ab} - \Gamma''^b_{ab} = (\frac{1}{2}\log(\frac{|det\,g|}{|det\,a|}))_{,a} \qquad (7.25)$$

which confirms that ψ is exact on M, being the gradient of $\chi = \frac{1}{2}\log(\frac{|detg|}{|deta|})$, $\psi = d\chi$. Finally define a global, smooth metric g' on M by $g' = e^{2\chi}b$. It then immediately follows from (7.23) that g' and ψ satisfy (7.10) on M. Further, $g'_{ab} = e^{2\chi}b_{ab}$, $g'^{ab} = e^{-2\chi}a^{ab}$ and so $a_{ab} \equiv g_{ac}g_{bd}a^{cd} = g_{ac}g_{bd}e^{2\chi}g'^{cd}$ and $-e^{2\chi}\psi_b g'^{bc}g_{ac} = -e^{2\chi}\psi_b e^{-2\chi}a^{bc}g_{ac} = -\psi_b a^{bc}g_{ac} = \lambda_a$ and one recovers the original relations (7.16) with g' determined up to a multiplicative constant.

Thus the problem of solving (7.10) for the metric g' projectively related to the original metric g, and the exact 1-form ψ, has been expressed in the equivalent form of solving (7.20) for a and λ using (7.16) and (7.17) to convert back to g' and ψ. In practice, whilst (7.10) is useful, it is usually more convenient to deal with the Sinjukov form (7.20).

7.4 Introduction of the Curvature Tensor

The curvature tensor may be introduced into the problem by applying the Ricci identity (chapter 2) to the Sinjukov tensor using the Sinjukov equation (7.20). One gets, in any coordinate system of M

$$a_{ae}R^e{}_{bcd} + a_{be}R^e{}_{acd}(= a_{ab;cd} - a_{ab;dc}) = g_{ac}\lambda_{bd} + g_{bc}\lambda_{ad} - g_{ad}\lambda_{bc} - g_{bd}\lambda_{ac} \tag{7.26}$$

where $\lambda_{ab} = \lambda_{a;b} = \lambda_{ba}$ since λ is exact. On contracting (7.26) with g^{ac} one finds

$$n\lambda_{bd} = \Psi g_{bd} + a^{ec}R_{ebcd} - a_{be}R^e{}_d \tag{7.27}$$

where $\Psi \equiv \lambda^a_{;a} = \lambda^a_a$ and where the Ricci tensor is introduced. Since the left hand side and the first two terms on the right hand side of (7.27 are symmetric in b and d it follows that $a_{be}R^e{}_d = a_{de}R^e{}_b$. A $\nabla-$ covariant differentiation of (7.27) then gives

$$n\lambda_{b;df} = \Psi_{,f}g_{bd} + a^{ce}{}_{;f}R_{ebcd} + a^{ec}R_{ebcd;f} - a_{be;f}R^e{}_d - a_{be}R^e{}_{d;f} \tag{7.28}$$

and the $\nabla-$covariant derivative of a may be removed using (7.20) to give

$$n\lambda_{b;df} = \Psi_{,f}g_{bd} + \lambda^e R_{ebfd} + \lambda^e R_{fbed} + a^{ae}R_{ebad;f} - g_{bf}R_{ed}\lambda^e - R_{fd}\lambda_b - a_{be}R^e{}_{d;f}. \tag{7.29}$$

Next the Ricci identity for λ gives $\lambda_{b;df} - \lambda_{b;fd} = \lambda^e R_{ebdf}$ and using this in the previous equation one finds

$$n\lambda_{b;fd} = \Psi_{,f}g_{bd} + (n+1)\lambda^e R_{ebfd} + \lambda^e R_{fbed} + a^{ae}R_{ebad;f} - g_{bf}R_{ed}\lambda^e - R_{fd}\lambda_b - a_{be}R^e{}_{d;f}. \tag{7.30}$$

Then a contraction of (7.30) with g^{bf} gives

$$(n-1)\Psi_{,d} = -2(n+1)R_{ed}\lambda^e - a^{ae}R^b{}_{ead;b} - a^{be}R_{ed;b}. \tag{7.31}$$

Finally the Bianchi identity $R_{be[ca;d]} = 0$, on contraction with g^{bc}, gives $R^b{}_{ead;b} = R_{ed;a} - R_{ea;d}$ and on substitution into (7.31) gives

$$(n-1)\Psi_{,d} = -2(n+1)R_{ed}\lambda^e + a^{ec}(R_{ec;d} - 2R_{ed;c}). \tag{7.32}$$

Next consider equations (7.20), (7.27) and (7.32). These give a system of first order differential equations for the global quantities a_{ab}, λ_a and Ψ in the sense described in the previous chapter [101, 97, 104, 102]. It follows that the values of the members of the triple (a, λ, Ψ) at each point of M are uniquely determined by the $\frac{1}{2}(n+1)(n+2)$ quantities $a(p)$, $\lambda(p)$ and $\Psi(p)$ at any point $p \in M$. Thus one has the following theorem [101] (see also [97], p 150).

Theorem 7.2 *Let (M, g) be an $n-$dimensional manifold with smooth metric g of arbitrary signature. Then (7.20), (7.27) and (7.32) represent a first order system of differential equations for the global objects a_{ab}, λ_a and Ψ on M.*

7.5 Einstein Spaces

Continuing the argument of the previous sections let (M, g) be an n−dimensional Einstein space $(n > 2)$ (which is not flat) so that the Ricci tensor satisfies $Ricc = \frac{R}{n}g$ with constant Ricci scalar R and, as a consequence, the Bianchi identity (Chapter 2) gives $R^a{}_{bcd;a} = 0$ on M. Then an obvious covariant differentiation of (7.26) using this last identity gives, on any coordinate domain of M,

$$a_{ae;d}R^e{}_{bc}{}^d + a_{be;d}R^e{}_{ac}{}^d = g_{ac}\lambda_b{}^d{}_d + g_{bc}\lambda_a{}^d{}_d - \delta_a^d\lambda_{bcd} - \delta_b^d\lambda_{acd} \qquad (7.33)$$

where $\lambda_{abc} \equiv \lambda_{a;bc} = \lambda_{bac}$ and this can be simplified using (7.20) to get

$$\lambda^e R_{ebca} + \lambda^e R_{eacb} - \lambda_a R_{bc} - \lambda_b R_{ac} = g_{ac}k_b + g_{bc}k_a - \lambda_{bca} - \lambda_{acb} \qquad (7.34)$$

where $k_a \equiv \lambda_a{}^d{}_d$. Taking into account the symmetry $R^a{}_{[bcd]} = 0$, the Ricci identity for λ, $\lambda_{bac} - \lambda_{bca} = \lambda^d R_{dbac}$ and the identity $\lambda_{acb} = \lambda_{cab}$ one can take the skew part of (7.34) over the indices a and c to get (see [104])

$$4\lambda^e R_{ebca} = R_{bc}\lambda_a - R_{ba}\lambda_c + g_{bc}k_a - g_{ab}k_c. \qquad (7.35)$$

Then use the Einstein space condition $R_{ab} = \frac{R}{n}g_{ab}$ to get

$$\lambda^a R_{abcd} = \frac{1}{4}[g_{bc}(\frac{R}{n}\lambda_d + k_d) - g_{bd}(\frac{R}{n}\lambda_c + k_c)] \qquad (7.36)$$

followed by a contraction with g^{bd} to find

$$\lambda^a R_{ab} = -\frac{(n-1)}{4}(\frac{R}{n}\lambda_b + k_b). \qquad (7.37)$$

But $\lambda^a R_{ab} = \frac{R}{n}\lambda_b$ and so $k_a = -\frac{(3+n)R}{n(n-1)}\lambda_a$ Now since (M, g) is an Einstein space, $E = 0$ in chapter 3 and so one achieves the following expression for the Weyl conformal tensor,

$$C_{abcd} = R_{abcd} - \frac{R}{n(n-1)}(g_{ac}g_{bd} - g_{ad}g_{bc}), \qquad (7.38)$$

and then (7.36) and (7.38) give the important result [103]

$$\lambda^a C_{abcd} = 0. \qquad (7.39)$$

A back substitution into (7.38) yields

$$\lambda^a R_{abcd} = \frac{R}{n(n-1)}(g_{bd}\lambda_c - g_{bc}\lambda_d). \qquad (7.40)$$

Next use the Ricci identity for λ^a recalling that $\lambda_{abc} = \lambda_{a;bc}$ to get

$$\lambda_{bca} + \lambda_{acb} = \lambda_{cba} + \lambda_{cab} = \lambda_{cba} - \lambda_{cab} + 2\lambda_{cab} = 2\lambda_{c;ab} + \lambda^d R_{dcba}. \quad (7.41)$$

Then using (7.40) and the above expression for k the Einstein space condition in (7.34) gives

$$\frac{R}{n(n-1)}(2g_{ab}\lambda_c + 3g_{ac}\lambda_b + 3g_{bc}\lambda_a) = -\lambda_{bca} - \lambda_{acb} \quad (7.42)$$

and use of (7.41) (and (7.40)) finally gives

$$\lambda_{a;bc} = 2g_{ab}\mu_c + g_{ac}\mu_b + g_{bc}\mu_a \quad (7.43)$$

where the 1-form $\mu \equiv -\frac{R}{n(n-1)}\lambda$. Since these calculations hold in any coordinate domain of M, (7.43) reveals, from (6.21) that, because $\lambda_{a;b} = \lambda_{b;a}$, the tensor $h_{ab} \equiv \lambda_{a;b} + \lambda_{b;a}$ satisfies the conditions for the associated global, smooth vector field λ^a to be a *projective vector field* on M with projective 1-form 2μ and zero projective bivector (since $\lambda_{a;b} = \lambda_{b;a}$).

Suppose now that (M,g), in addition to being an Einstein space, (is not flat and) is *properly* projectively related to (M,g') as above (that is, λ is not identically zero on M), and define the open subset $U \subset M$ by $U = \{p \in M : Riem(p) \neq 0\}$ so that $U \neq \emptyset$. Then define the open subset $V \subset M$ by $V = \{p \in M : \lambda(p) \neq 0\}$. Since λ is a projective vector field on M it vanishes on M if it vanishes on some non-empty open subset of M (chapter 6). Thus the subset V is open and dense in M and $U \cap V$ is non-empty and open in M. Then for $p \in U \cap V$ let $F^{ab} = u^a v^b - v^a u^b$ be a simple 2−form at p for independent $u, v \in T_p M$. Suppose T is a symmetric tensor at p and that $T_{ae}F^e{}_b + T_{be}F^e{}_a = 0$, where $F^e{}_a = F^{eb}g_{ba}$. Then, defining $u'^a = T^a{}_b u^b$ and $v'^a = T^a{}_b v^b$ (and $u'_a = g_{ab}u'^b$, $v'_a = g_{ab}v'^b$) one finds $u'_a v_b - v'_a u_b + u'_b v_a - v'_b u_a = 0$ and selecting $0 \neq w \in T_p M$ with $w \cdot u = 0 \neq w \cdot v$ (as one always can since $\dim M > 2$ and where $\cdot$ denotes an inner product from g) an obvious contraction with w^b shows that u' is a linear combination of u and v, and similarly for v'. It follows that the 2−space $Sp(u,v)$ is invariant for (the linear map represented by) T (cf a similar result in the 4−dimensional case discussed in earlier chapters). Writing $(u'_a =)T_{ab}u^b = au_a + bv_a$ and $(v'_a =)T_{ab}v^b = cu_a + dv_a$ and substituting back then easily gives $b = c = 0$ and $a = d$ and so $Sp(u,v)$ is an eigenspace of T at p. Now, for $p \in U \cap V$, extend $\lambda^a(p)$ to a basis $\lambda^a(p), q_1^a, ..., q_{n-1}^a$ for $T_p M$ and note, from (7.40) that each 2−form $F^{ab}_{(m)} \equiv \lambda^a q_m^b - q_m^a \lambda^b$ satisfies $R_{abcd}F^{cd}_{(m)} = \frac{2R}{n(n-1)}F_{(m)ab}$ $(1 \leq m \leq n-1)$. A contraction of (7.26) with $F^{cd}_{(m)}$ then reveals that $T_{ae}F^e{}_{(m)b} + T_{be}F^e{}_{(m)a} = 0$ for each m and with $T_{ab} = \frac{2R}{n(n-1)}a_{ab} + 2\lambda_{a;b}$ giving the components of a global, smooth, symmetric tensor on M and hence, from the above, that $T_p M$ is an eigenspace of T for each $p \in M$. It follows that, at each $p \in U \cap V$, T is a multiple of g and hence that (cf [97])

$$\lambda_{a,b} = \sigma g_{ab} - \frac{R}{n(n-1)}a_{ab} \quad (7.44)$$

holds on $U \cap V$ for some smooth function $\sigma : (U \cap V) \to \mathbb{R}$. A back substitution of (7.44) into (7.26) can be used to eliminate λ from the latter equation and gives

$$a_{ae}R^e{}_{bcd} + a_{be}R^e{}_{acd} = \frac{-R}{n(n-1)}(g_{ac}a_{bd} + g_{bc}a_{ad} - g_{ad}a_{bc} - g_{bd}a_{ac}). \quad (7.45)$$

Combining this with (7.38) finally gives, on $U \cap V$

$$a_{ae}C^e{}_{bcd} + a_{be}C^e{}_{acd} = 0. \quad (7.46)$$

Now *Riem*, and hence C, vanish on $M \setminus U$ and so (7.46) holds on $(U \cap V) \cup (M \setminus U)$ and hence on V and since V is open and dense in M, (7.46) holds on M.

Now specialise to the case when dim$M = 4$ and in which (7.39) and (7.46) will play important roles. Define the subsets P and Q of M by $P \equiv \{p \in M : C(p) \neq 0\}$ and $Q \equiv \{p \in M : C(p) = 0\}$ so that P is open and $Q = M \setminus P$ is closed, in M. Whichever signature is chosen, if $P = \emptyset$ (and since (M, g) is an Einstein space) (M, g) has constant curvature (chapter 3). So suppose (M, g) *is not of constant curvature* so that the open set $P \neq \emptyset$. In the positive definite case (7.39) shows that $\lambda = 0$ on P (chapter 3) and since λ is projective, one gets the contradiction that $\lambda = 0$ on M (chapter 6). Hence (from (7.16)) the 1-form ψ vanishes on M and so $\nabla' = \nabla$.

Now suppose g has Lorentz signature and consider the subset P above. By assumption λ cannot vanish on P (otherwise it vanishes on M and $\nabla' = \nabla$). So suppose $p \in P$ with $\lambda(p) \neq 0$. Then there exists an open subset $W \subset P$ on which λ does not vanish. Then (7.39) shows that λ is *null* and C is of Petrov type **N** on W with (repeated) pnd spanned by λ (chapter 4). Thus, on W, the range space of the Weyl map f_C is spanned by a dual pair of null bivectors F and $\overset{*}{F}$ which, writing l for λ, can be written at any $p \in W$ as $F = l \wedge x$ and $\overset{*}{F} = -l \wedge y$ with l, n, x, y a null basis at p. Then (7.46) shows that $a_{ae}F^e{}_b + a_{be}F^e{}_a = 0$ at p and similarly for $\overset{*}{F}$ and so the blades of F and $\overset{*}{F}$ are each eigenspaces of a at each $p \in P$. It follows that $a_{ab} = \alpha g_{ab} + \beta l_a l_b$ in any coordinate domain in W and for smooth functions α and β and smooth, null vector field l, on W. Since $l_a l^a = 0$ on W, $l^a l_{a;b} = 0$ on W and a substitution of this expression for a into (7.20) and a contraction with g^{ab} gives $2\alpha_{,a} = l_a$ whilst a contraction with l^a gives $\alpha_{,a} = l_a$. This contradiction leads, in turn, to the contradiction that $\lambda = 0$ on W. Thus λ vanishes on M and again $\nabla' = \nabla$.

Finally suppose that g has neutral signature. A similar argument to that given above reveals the open subset W on which λ and C are nowhere-zero and on which (7.39) and (7.46) hold. Thus, from the classification of C given in chapter 5, λ is null on W and the algebraic type of C at each $p \in P$ is either $(\mathbf{N}, \mathbf{N})$ or $(\mathbf{N}, \mathbf{0})$ with λ a pnd for C in each case. In the first case the Weyl map f_C is of rank 2 and its range is spanned by bivectors $l \wedge N$ and $l \wedge L$ whilst in the second case it is of rank 1 and is spanned by a bivector

$l \wedge N'$ where (l, n, L, N) and (l, n', L', N') are null bases for this signature and with $l^a = \lambda^a$. Thus, in the first case $l \wedge N$ and $l \wedge L$ are eigenspaces for a and in the second case $l \wedge N'$ is an eigenspace of a, from (7.46). It is clear that one may disjointly decompose W into the (open in W and M) subset W_2 of points where rank$f_C = 2$ and the subset W_1 of points where the rank$f_C = 1$. If $W_2 \neq \emptyset$ the above argument in the Lorentz case gives the contradiction that $\lambda = 0$ on W_2 and, as before, $\nabla' = \nabla$. If, however, $W_2 = \emptyset$, $W_1 = W$ is open and so, dropping primes for convenience, a smooth null basis (l, n, L, N) may be chosen locally smooth with $l \wedge N$ and eigenspace for a from (7.46) and one gets

$$a_{ab} = \alpha g_{ab} + \beta l_a l_b + \gamma N_a N_b + \delta(l_a N_b + N_a l_b) \qquad (7.47)$$

for smooth functions α, β, γ and δ on some appropriate open neighbourhood $W' \subset W$ of any $p \in W$ (and, as before, $l = \lambda$ and $l^a l_{a;b} = 0$). A substitution into (7.20) followed by successive contractions with g^{ab} and N^a (noting that $N^a N_a = 0 \Rightarrow N^a N_{a;b} = 0$ and $l^a N_a = 0 \Rightarrow l^a N_{a;c} + N^a l_{a;c} = 0$) then give the equations

$$2\alpha_{,c} = l_c, \qquad \alpha_{,c} + \delta N^a l_{a;c} = 0, \qquad \beta N^a l_{a;c} = N_c. \qquad (7.48)$$

Now since $\lambda(= l)$ and N are nowhere zero on W' one sees that $\alpha_{,a}, \delta$ and β are nowhere zero on W' and hence that l and N are proportional on W'. This contradiction leads, in turn, to the contradiction that λ vanishes on W'. Thus, as before, λ vanishes on M and so $\nabla' = \nabla$. It is recalled that, independently of signature (lemma 7.2), (M, g') is of constant curvature if and only if (M, g) is of constant curvature. Thus one has the following theorem which appears to have been discovered independently in [49, 103, 104] (see also [98]).

Theorem 7.3 *Let g and g' be smooth metrics of arbitrary signature on the 4-dimensional manifold M with Levi-Civita connections ∇ and ∇', respectively, and with (M, g) an Einstein space. If ∇' and ∇ are projectively related then either $\nabla' = \nabla$ or (M, g) and (M, g') are each of constant curvature.*

Still in the 4-dimensional case and with (M, g) not of constant curvature and projectively related to (M, g') one has $\nabla' = \nabla$ and hence $Riem' = Riem$ and $Ricc' = Ricc$. In the case when (M, g) is Ricci-flat, that is, $Ricc = 0$ on M, (M, g') is also Ricci-flat, $Ricc' = 0$. Suppose (M, g) is of Lorentz signature and is vacuum. The holonomy types of ∇ and ∇' are the same and are either R_8, R_{14} or R_{15} (theorem 4.11). The general techniques of chapter 4 show that $g' = cg$ for $0 \neq c \in \mathbb{R}$ unless this type is R_8. In this latter case there exists an open neighbourhood W about any $p \in M$ and a smooth vector field l on W such that $g'_{ab} = ag_{ab} + bl_a l_b$ for $a, b \in \mathbb{R}$ with $a \neq 0$ and with l a smooth (g- and g'-) null vector field satisfying $\nabla l = \nabla' l = 0$ on W (since $\nabla' = \nabla$). Thus g and g' are pp-waves (chapter 4). However, if (M, g) is a proper Einstein space ($R \neq 0$) the (equal) holonomy types of (M, g) and (M, g') are R_7, R_{14} or R_{15}. This gives $g' = cg$ for $0 \neq c \in \mathbb{R}$ unless this type is R_7 and it can be checked that in this case, locally as above, $g'_{ab} = ag_{ab} + b(l_a n_b + n_a l_b)$ with

l and n (∇ and ∇') recurrent ($g-$ and $g'-$) null vector fields on W. In this case, (M, g') may not be an Einstein space. [103, 105] Similar remarks may be made regarding the positive definite and neutral signature cases.

As a final remark and with $\dim M = 4$ and (M, g) an Einstein space of any signature and which is not of constant curvature suppose X is a projective vector field on X satisfying (6.20) and (6.21) with $\psi_a = \chi_{,a}$. It has been pointed out in [97], quite generally, that if one defines a global tensor field a on M in component form by $a_{ab} \equiv h_{ab} - 2\chi g_{ab}$ then a and the $1-$ form ψ satisfy Sinjukov's equation (7.20) on M. Thus (M, g) is projectively related to some manifold (M, g'). Theorem 7.3 then shows that $\psi = 0$ on M and so X is affine. Thus *any projective vector field on M is affine*. This strengthens a result in [113, 13] where this result was shown only for Lorentz signature *and* in the case when the Weyl conformal tensor was nowhere-zero over some open dense subset of M.

7.6 Projective Relatedness and Geometrical Symmetry

Returning to the general case with $\dim M = n$, let g and g' be smooth metrics on M with respective Levi-Civita connections ∇ and ∇' and suppose ∇ and ∇' are projectively related. Let X be a projective vector field for (M, g). Then (chapter 6) if ϕ_t is any local flow for X it maps a geodesic of ∇ into a geodesic of ∇ and hence, since ∇ and ∇' are projectively related, a geodesic of ∇' into a geodesic of ∇' (and vice versa) and so X is a projective vector field for (M, g'). It follows that the projective algebras of ∇ and ∇' are the same. Further, with ϕ_t as above and U an open subset of M on which ϕ_t is defined, it is easily seen that the metric g (restricted to U) and the metric $\phi_t^* g$ on U (pulled back from the metric g restricted to $\phi_t(U)$) are projectively related. The corresponding Weyl projective tensors W (from g restricted to U) and the one from the metric $\phi_t^* g$ are thus equal (section 7.2). But this latter tensor is just the pullback $\phi_t^* W$ of the Weyl projective tensor from g restricted to $\phi_t(U)$ and so $\phi_t^* W = W$ for each such ϕ_t. It follows that $\mathcal{L}_X W = 0$ for each projective vector field on M. Global smooth vector fields X satisfying $\mathcal{L}_X W = 0$ are referred to as *Weyl projective vector fields* and the collection of such vector fields on M, labelled $WP(M)$, is a Lie algebra under the usual bracket operation called the *Weyl projective algebra*. It easily follows that $P(M) \subset WP(M)$. However, one need not have equality here. To see this, briefly, one returns to the $pp-$waves of general relativity (chapter 4) which satisfy the vacuum condition $Ricc = 0$ and hence $W = Riem$ (section 7.2). It is known that for such metrics the solutions of $\mathcal{L}_X Riem(= \mathcal{L}_X W) = 0$ give an infinite-dimensional Lie algebra [106] whereas (chapter 6) $P(M)$ is finite-dimensional. Hence $P(M)$ is a *proper* subset of $WP(M)$ in this case.

Staying with the general case when $\dim M = n > 2$ and with projectively related metrics g and g' on M and with respective Levi-Civita connections

∇ and ∇', let X be a Killing vector field on M with respect to g so that $\mathcal{L}_X g = 0$ or, equivalently in any chart domain of M, $X_{a;b} = F_{ab}(= -F_{ba})$ where $X_a = g_{ab}X^b$, F is the Killing bivector and a semi-colon means a ∇–covariant derivative. The metrics g and g' satisfy (7.9) and (7.10) where the global smooth 1-form $\psi = d\chi$ for a global, smooth function χ on M. Then define a global, smooth vector field Y on M by the component relations in any chart of M as $Y^a = e^{2\chi}g'^{ab}g_{bc}X^c$ and then define $Y_a = g'_{ab}Y^b$ (noting the use of g' to lower the index on Y). Then following [107] (taking the opportunity to correct an error in that reference) and using a vertical stroke to denote a ∇'–covariant derivative one finds from (7.9)

$$X_{a;b} = X_{a|b} + P^c_{ab}X_c = X_{a|b} + X_a\psi_b + \psi_a X_b \qquad (7.49)$$

and so

$$(\mathcal{L}_X g)_{ab} = X_{a;b} + X_{b;a} = X_{a|b} + X_{b|a} + 2(X_a\psi_b + \psi_a X_b). \qquad (7.50)$$

Next compute

$$e^{-2\chi}(e^{2\chi}X_a)_{|b} = X_{a|b} + e^{-2\chi}(2e^{2\chi}\chi_{,b}X_a) = X_{a|b} + 2X_a\psi_b. \qquad (7.51)$$

Now, from the definition of Y, $Y_a = g'_{ab}Y^b = e^{2\chi}g'_{ab}g'^{bc}g_{cd}X^d = e^{2\chi}X_a$ and so, from (7.51), $e^{-2\chi}Y_{a|b} = X_{a|b} + 2X_a\psi_b$. Then using (7.50)

$$(\mathcal{L}_X g)_{ab} = e^{-2\chi}(Y_{a|b} + Y_{b|a}) = e^{-2\chi}(\mathcal{L}_Y g')_{ab}. \qquad (7.52)$$

Thus *if X is a Killing vector field for (M,g), Y is a Killing vector field for (M,g')*. Further, with the above definitions of X and Y, the map $X \to Y$ is linear and injective and inverts to give $X^a = e^{-2\chi}g^{ae}g'_{be}Y^b$. Hence the Killing algebras of (M,g) and (M,g') are isomorphic as vector spaces and so have the same dimension. (It has been pointed out to the author that this result was known much earlier [92]). Thus for $p \in M$, $X(p) = 0 \Leftrightarrow Y(p) = 0$ and then, at p, the general relation $Y_a = e^{2\chi}X_a$ and (7.49) show that the Killing bivectors $F_{ab} = X_{a;b}$ and $G_{ab} = Y_{a|b}$ for X and Y, respectively, satisfy $G_{ab}(p) = e^{2\chi}(p)F_{ab}(p)$. It follows that the isotropy algebras for g and g' are of the same dimension at any $p \in M$ and hence so are the Killing orbits through p for g and g'. The following result has been established [107].

Theorem 7.4 *If g and g' are projectively related on M then (M,g) and (M,g') have the same projective algebras, their Killing algebras are isomorphic as vector spaces and the Killing orbits for g and g' through any $p \in M$ have the same dimension.*

[It is noted that the above Killing vector field X for g is not necessarily Killing for g' but it is a projective vector field for g'. Examples of this feature will be given later.]

It is remarked briefly here that one may now establish theorem 7.3 in another, shorter, way as follows. Having established the existence of the projective vector field with components λ^a on M as in (7.43) let X represent this global vector field on M so that (6.20) and (6.21) hold and also, from (7.39), $C_{abcd}X^d = 0$. Because of the Einstein space condition, one has $W = C$ from lemma 7.2 and since $\mathcal{L}_X W = 0$, one gets $\mathcal{L}_X C = 0$ and $\mathcal{L}_X g \equiv h$ from (6.20) and (6.21). Now $g_{ae}C^e{}_{bcd} + g_{be}C^e{}_{acd} = 0$ and Lie differentiating this last equation reveals $h_{ae}C^e{}_{bcd} + h_{be}C^e{}_{acd} = 0$ [113]. Now suppose that $\dim M = 4$ and that the set P of section 7.5 is non-empty. In the event that g is positive definite one sees immediately from (7.39) that X, that is, λ, vanishes on P and hence on M. If g is of Lorentz signature (7.39) shows that if X does not vanish on P one may choose P so that (M, g) is of Petrov type $\mathbf{N}$ on P (chapter 4) whilst if g is of neutral signature the algebraic type of C is $(\mathbf{N}, \mathbf{N})$ or $(\mathbf{N}, \mathbf{O})$ on P (chapter 5). In each case the result follows as in section 7.5 above using (6.21) instead of 7.20.

7.7 The 1-form ψ

Returning to the case when $\dim M = 4$ and with g and g' projectively related metrics on M one can further investigate the role of the curvature tensor. Consider the vector space of bivectors F at p satisfying $R_{abcd}F^{cd} = 0$, that is, the kernel, $ker f(p)$, of the curvature map f at p. Then, with $F \in ker f(p)$, contract (7.26) with F^{cd} (noting that $F^a{}_b = F^{ac}g_{bc} = -F_b{}^a$) to get

$$\lambda_{ac}F^c{}_b + \lambda_{bc}F^c{}_a = 0 \tag{7.53}$$

where $\lambda_{ab} = \lambda_{a;b} = \lambda_{ba}$, since λ is exact on M. One now sees from lemma 3.6, lemma 4.7 and lemma 5.10 how each F satisfying (7.53) supplies information on the algebraic structure of λ_{ab} and, in particular, if such an F is simple its blade is an eigenspace of λ_{ab}. Suppose that, at each $p \in M$, $ker f(p)$ supplies sufficient information to ensure that $T_p M$ is an eigenspace of λ. Then $\lambda_{ab}(p)$ is a multiple of $g(p)$ at each $p \in M$ and one gets $\lambda_{ab} = \sigma g_{ab}$ in any coordinate domain of M for a smooth function $\sigma : M \to \mathbb{R}$. The Ricci identity on λ then gives on M

$$\lambda_d R^d{}_{abc} = \lambda_{a;bc} - \lambda_{a;cb} = g_{ab}\sigma_{,c} - g_{ac}\sigma_{,b} \tag{7.54}$$

and then a contraction with F^{bc} any non-zero member of $ker f(p)$ gives, on M,

$$0 = F^{bc}g_{ab}\sigma_{,c} - F^{bc}g_{ac}\sigma_{,b} = 2F_a{}^b\sigma_{,b}. \tag{7.55}$$

Further (7.26) then gives on M

$$a_{ae}R^e{}_{bcd} + a_{be}R^e{}_{acd} = 0. \tag{7.56}$$

(Thus far one could have stayed in the general case $\dim M > 2$ but from now on the imposition $\dim M = 4$ is made). Suppose now that for $p \in M$,

$\sigma_a(p) \neq 0$, where $\sigma_a \equiv \sigma_{,a}$. Then (7.55) shows that every bivector in $ker f(p)$ is simple and that they possess a common annihilator and so, from lemma 3.2, $\dim ker f(p) \leq 3$ and this lemma gives the possibilities for $ker f(p)$. These are (*i*) $\dim ker f(p) = 1$, (*ii*) $\dim ker f(p) = 2$ with $ker f(p)$ spanned by two simple bivectors whose blades intersect in a 1−dimensional subspace of $T_p M$, (*iii*) $\dim ker f(p) = 3$ with $ker f(p)$ spanned by three simple bivectors whose blades intersect in a 1−dimensional subspace of $T_p M$, or (*iv*) $\dim ker f(p) = 3$ with $ker f(p)$ spanned by three simple bivectors the blades of whose duals intersect in a 1−dimensional subspace of $T_p M$. It is now easily checked that the conditions required of $ker f(p)$ in order to obtain $\nabla \lambda = \sigma g$ are not satisfied in cases (*i*), (*ii*) and (*iv*) but are satisfied in case (*iii*) since, in this case, the blades of the members of $ker f(p)$ span $T_p M$. (For case (*iv*) $ker f(p) =$ Sp(P, Q, R) for bivectors P, Q and R the blades of whose duals intersect in Sp(k) for $0 \neq k \in T_p M$. Then k annihilates each member of the blades of P, Q and R and so the span of the union of these blades cannot be $T_p M$.) Then (7.55) implies that $\sigma_{,a}(p) = 0$ contradicting the assumption that $\sigma_{,a}(p) \neq 0$. Thus $\sigma_{,a}$ vanishes on M and, since M is connected, σ is constant on M, say $\sigma = c$ for $c \in \mathbb{R}$. Then (7.54) shows that $\lambda_d R^d{}_{abc} = 0$ on M. The following has been proved [108].

Lemma 7.4 *Let M be a 4−dimensional manifold and g and g' projectively related, smooth metrics on M so that the results (and notation) of section 7.3 hold. Let ∇ be the Levi-Civita connection arising from g and f the associated curvature map. Suppose $ker f(p)$ is such that it forces $T_p M$ to be an eigenspace of $\nabla \lambda(p)$, as above, for each $p \in M$. Then the following hold in any coordinate domain of M, for $c \in \mathbb{R}$.*

$$(a) \; \lambda_{a;b} = c g_{ab}, \qquad (b) \; \lambda_d R^d{}_{abc} = 0, \qquad (c) \; a_{ae} R^e{}_{bcd} + a_{be} R^e{}_{acd} = 0. \quad (7.57)$$

Suppose now that the conditions of lemma 7.4 hold. Then the vector field with components λ^a is a *global, homothetic vector field on M with zero homothetic bivector*. [If $ker f(p)$ is such that it forces $T_p M$ to be an eigenspace of $\nabla \lambda(p)$, as above, over some non-empty, connected, open subset U of M, then (7.57) holds on U and λ^a is homothetic on U, but may not be homothetic on M.] Then if λ can be shown to vanish over some non-empty, open subset of M it vanishes on M, as follows from the general theory of chapter 6, and then $\nabla = \nabla'$. [In fact one can say a little more here. If $c \neq 0$, λ is a proper homothetic vector field on M. If (M, g) is of positive definite signature the zeros of $\lambda(= \{p' \in M : \lambda(p') = 0\})$ are isolated (that is, if $\lambda(p) = 0$ there exists an open neighbourhood V of p such that λ does not vanish at any point of $V \setminus \{p\}$ and, further, V may be chosen so that *Riem* vanishes on V [10]. Continuing with this case if (M, g) is of Lorentz or neutral signature the zeros of λ need not be isolated but they are in this case since the homothetic bivector of λ vanishes on M. (See [13] for the Lorentz case; the neutral case is similar). Then, if $\lambda(p) = 0$, the equations $\mathcal{L}_\lambda Riem = 0$, $\mathcal{L}_\lambda Ricc = 0$ and $\mathcal{L}_\lambda C = 0$ for the curvature, Ricci and Weyl conformal tensor, respectively, which are easily seen to hold since λ is homothetic, give $Riem(p) = 0$, $Ricc(p) = 0$ and $C(p) = 0$. These follow from the vanishing of the homothetic bivector and (6.10). These results can be used to get weaker conditions on λ to ensure it vanishes on M and hence that $\nabla' = \nabla$.] Continuing with the conditions of lemma 7.4 assumed to hold, and with A and B denoting, as usual, the subsets of M of curvature classes A and B, if the necessarily open subset $A \cup B$ is non-empty, (7.57(b)) shows that λ vanishes on $A \cup B$ and hence on M and again $\nabla' = \nabla$ (see section 3.5, section 4.5 and section 5.6). Finally if the range of the curvature map f is known over M, (7.57c) can be used to to find algebraic expressions for the Sinjukov tensor a (lemma 3.6, lemma 4.7 and lemma 5.10) since then $a_{ae}F^e{}_b + a_{be}F^e{}_a = 0$ for each $F \in rgf(p)$ and for each $p \in M$. On substituting these into (7.20) one achieves information on λ.

Thus to study projective relatedness for $\dim M = 4$ and for all three signatures the general idea is to start with (M, g) and to assume some holonomy group Φ and algebra ϕ for (M, g). Thus with ϕ'_p the infinitesimal holonomy algebra at p one has $rgf(p) \subset \phi'_p \subset \phi$. Then one makes an assumption (that is, a choice) about the holonomy algebra for (M, g) and writes down a possible disjoint decomposition of M into its curvature classes. From ϕ one may also check from lemma 7.1 if any local parallel or recurrent vector fields are admitted. Some general consequences of lemma 7.4 can now be given and which will reduce repetition in the subsequent arguments.

Lemma 7.5 *Let M be a 4−dimensional manifold and g and g' projectively related, smooth metrics on M so that the results (and notation) of section 7.3 hold. Let ∇ and ∇' be the Levi-Civita connections arising from g and g' and let f be the curvature map associated with (M, g). Suppose that (M, g) is not flat.*

(i) *If $\dim rgf(p) \leq 2$ for those p in some non-empty open subset $U \subset M$ (and note that this automatically holds for each such subset U if the holonomy group Φ of (M, g) has dimension ≤ 2) the conditions of lemma 7.4 are satisfied on U (that is, (7.57) holds and $\nabla\lambda = cg$ on U with $c \in \mathbb{R}$).*

(ii) *Suppose (M, g) satisfies the conditions of lemma 7.4 and that one may choose a non-empty open subset $U \subset M$ on which $Riem$ is nowhere zero and which admits a smooth, properly recurrent vector field. Then $\lambda = 0$ on M and $\nabla' = \nabla$.*

(iii) *Suppose (M, g) satisfies the conditions of lemma 7.4 and that one may choose a non-empty open subset $U \subset M$ on which $Riem$ is nowhere zero and which admits a smooth, nowhere-zero, parallel vector field k. Suppose $\dim rgf(p) \geq 2$ at each point of U. Then $\lambda = 0$ on M and so $\nabla' = \nabla$.*

(iv) *Suppose (M, g), U and k are as in part (iii) and where k lies in the blade of some simple member G of $rgf(p)$ at each $p \in U$. Then k is null and $\lambda = 0$ on M, hence $\nabla' = \nabla$.*

(v) *Suppose (M, g) satisfies the conditions of lemma 7.4 and that one may choose a non-empty open subset $U \subset M$ on which Riem is nowhere-zero. Suppose $rgf(p)$ is 1-dimensional and spanned by a null bivector F at each $p \in U$. Suppose also that there exists a parallel, nowhere-zero and nowhere-null vector field z on U which is orthogonal to the blade of F at each $p \in U$. Then either $\nabla' = \nabla$ or (M, g) admits a global, parallel null vector field.*

(vi) *Suppose (M, g) satisfies the conditions of lemma 7.4 and that one may choose a non-empty open subset $U \subset M$ on which Riem is nowhere-zero and which admits two independent, smooth, nowhere-zero parallel vector fields. Then $\lambda = 0$ on U and hence on M and $\nabla' = \nabla$.*

Proof

For part (i) one has $\dim rgf(p) \leq 2$ for each $p \in U$ and so $\dim ker f(p) \geq 4$ for each $p \in U$. Thus the bivectors F satisfying (7.53) constitute (at least a) 4-dimensional subspace of the appropriate orthogonal algebra. But if G and H are bivectors satisfying (7.53) then, as seen before, $[G, H]$ also satisfies (7.53) and so the solutions of (7.53) for F constitute (at least a) 4-dimensional subalgebra of the appropriate orthogonal algebra and it is then easily checked from chapters 3, 4 and 5 that $\nabla \lambda$ is a multiple of g at each $p \in U$. But then (7.55) shows that if $\sigma_{,a} \neq 0$ at any $p \in U$, $\ker f(p)$ is a 4-dimensional subspace of simple members of $\Lambda_p M$ contradicting lemma 3.2. The result follows.

For part (ii) let k be the properly recurrent vector field on the open subset U the latter chosen so that $k_{a;b} = k_a P_b$ for a smooth 1-form field P on U. Then (7.1) gives $k^d R_{dabc} = k_a Q_{bc}$ where Q is the 2-form $Q_{ab} = P_{a;b} - P_{b;a}$. So assume U is chosen so that Q is nowhere zero on U (as it can be). Thus $k_{[a} Q_{bc]} = 0$ and so Q is simple with k in its blade (chapter 3) and clearly lies in $rgf(p)$ at each $p \in U$. It now follows from (7.57)(c) that k is an eigenvector of a over U, $a_{ab} k^b = \alpha k_a$, for smooth $\alpha : U \to \mathbb{R}$. Differentiating this last equation and using (7.20) one finds

$$(a_{ab} k^b)_{;c} = (\alpha k_a)_{;c} \Rightarrow (g_{ac} \lambda_b + g_{bc} \lambda_a) k^b + \alpha k_a P_c = k_a \alpha_{,c} + \alpha k_a P_c. \quad (7.58)$$

Now (7.57)(b) shows that $\lambda^a k_a = 0$. Then (7.58) gives $\lambda_a k_c = k_a \alpha_{,c}$ on U. Since k is properly recurrent it is null on U and if λ does not vanish on U one may choose λ to be nowhere zero on U and the last equation then shows that λ is proportional to k, and hence null, on U. But $\lambda_{a;b} = c g_{ab}$ and $\lambda^a \lambda_{a;b} = 0$ on U and so $c = 0$ on U, that is, λ is parallel on U. But this forces k to be proportional to the parallel vector field λ on U contradicting the assumption that it is properly recurrent on U. It follows that λ vanishes on U and hence on M and $\nabla' = \nabla$.

For part (iii) the Ricci identity gives $k^d R_{dabc} = 0$ and then (7.57)(b) and the assumption on $\dim rgf$ shows that $\lambda = \gamma k$ on U for smooth γ. If λ is not identically zero on U one may choose U so that λ (and hence k) are nowhere zero on U and so γ is nowhere zero on U. Then $\lambda_{a;b} = k_a \gamma_{,b}$ on U and (57)(a) shows that $c = 0$. Since k and λ are each parallel one may write $\lambda^a = k^a$ on U. Then $rgf(p)$ is spanned by (two or three) simple bivectors having k as their common annihilator and so k and the 3−space orthogonal to it give rise to eigenspaces of a, from $(7.57)(c)$. Thus, on U

$$a_{ab} = \alpha g_{ab} + \beta k_a k_b \qquad (7.59)$$

for smooth functions α and β on U. Then (7.20) gives (since $k^a = \lambda^a$)

$$\alpha_{,c} g_{ab} + \beta_{,c} k_a k_b = g_{ac} k_b + g_{bc} k_a. \qquad (7.60)$$

If k is null at any $p \in U$ it is null on U and contractions of (7.60) first with g^{ab} and then with k^a reveal the contradiction that $k = 0$ on U. So k is non-null on U and then a contraction of (7.60) with k^a and use of the non-degeneracy of g again gives the contradiction that $k = 0$ on U. It follows that $\lambda = 0$ on U and hence on M and that $\nabla' = \nabla$.

For part (iv) since k lies in the blade of G (and G is in rgf) over U the Ricci identity on k immediately shows that k is null on U and one may choose U so that $G = k \wedge r$ on U for a smooth, nowhere zero vector field r on U and, from $(7.57)(b)$, in an obvious notation, $\lambda \cdot k = \lambda \cdot r = 0$ on U. Then $k \wedge r$ is an eigenspace of a on U and so $a_{ab}k^b = \alpha k_a$ and $a_{ab}r^b = \alpha r_a$ for a smooth function α. Differentiating these on U, using (7.20), one gets

$$\lambda_a k_c = k_a \alpha_{,c} \qquad (7.61)$$

and

$$\lambda_a r_c + a_{ab} r^b{}_{;c} = r_a \alpha_{,c} + \alpha r_{a;c}. \qquad (7.62)$$

If at some $p \in U$ r is *not* null, one may assume U chosen so that $r_a r^a \neq 0$ on U and that r is scaled so that $r_a r^a =$ constant $\neq 0$ and hence that $r^a r_{a;b} = 0$ on U. A contraction of (7.62) with r^a then gives $\alpha_{,a} = 0$ on U and then (7.61) shows that λ vanishes on U. Otherwise, r is null on U (and note that (7.61) then gives $(r \cdot k)\alpha_{,a} = 0$ on U since $r \cdot \lambda = 0$ there) and so one may choose U so that either $\alpha_{,a} = 0$ on U which, as before, leads to $\lambda = 0$ on U, or $\alpha_{,a}$ is nowhere zero on U, which, from (7.61), implies that $k \cdot r = 0$ on U. In this latter case G is totally null on U and one is forced into neutral signature. Relabelling k by l and r by N one sees that $G = l \wedge N$ and is an eigenspace of a on U. Thus, extending to a smooth null basis on U (which gives $l^a N_{a;b} = 0$ since $k = l$ is parallel) and using the completeness relation for neutral signature, one finds

$$a_{ab} = \alpha g_{ab} + \beta l_a l_b + \gamma N_a N_b + \delta(l_a N_b + N_a l_b) \qquad (7.63)$$

for smooth functions α, β, γ and δ on U. Then (7.61) becomes $\lambda_a l_c = l_a \alpha_{,c}$ and a substitution of (7.63) into (7.20) and a contraction with g^{ab} gives $2\alpha_{,a} = \lambda_a$ on U. Thus λ_a and l_a are proportional and these equations can only be consistent if $\lambda = 0$ on U and hence on M. It follows that $\nabla' = \nabla$.

For part (v) one may choose U, smooth vector fields l and x on U with l null, x spacelike (the case x timelike, if applicable, is similar) and $F = l \wedge x$ with $l \cdot x = 0$ on U. Suppose λ is not identically zero on U. Then one may choose U so that λ is nowhere-zero on U and z is smooth on U. Then $l \cdot z = x \cdot z = 0$ on U and, from $(7.57)(b)$, $\lambda \cdot l = \lambda \cdot x = 0$ on U. Also the blade of F is an eigenspace of a on U and so its orthogonal complement $l \wedge z$ is invariant for a on U. Thus, on U, $a_{ab}z^b = \rho l_a + \sigma z_a$ for smooth ρ and σ and (7.20) contracted with z^b gives

$$(g_{ac}\lambda_b + g_{bc}\lambda_a)z^b = l_a\rho_{,c} + \rho l_{a;c} + z_a\sigma_{,c}. \tag{7.64}$$

A contraction with l^a then shows that $\lambda \cdot z = 0$ on U which reveals, since l, x and z are independent at each $p \in U$, that $\lambda = \mu l$ on U with μ smooth on U. Then $l^a\lambda_{a;b} = 0$ and so from $(7.57)(a)$, $c = 0$. Thus $\nabla\lambda = 0$ on M and either $\lambda = 0$ $(\Leftrightarrow \nabla' = \nabla)$ on M or λ gives rise to a global, parallel, null vector field on M.

Finally for part (vi) let the parallel vector fields on U be r and s chosen, as one always can, so that $r \cdot s = 0$ on U with *Riem* nowhere zero on U. Then the Ricci identities give $R_{abcd}r^d = 0$ and $R_{abcd}s^d = 0$ on U and *Riem* is clearly of curvature class D on U and $R_{abcd} = AQ_{ab}Q_{cd}$ for a smooth function A and smooth bivector Q with $\overset{*}{Q} = r \wedge s$. Then from $(7.57)(b)$ and $(7.57)(c)$, the blade of Q is an eigenspace of a and λ, r and s lie in the blade of $\overset{*}{Q}$, $\lambda^a = \alpha r^a + \beta s^a$, for smooth α and $\beta : U \to \mathbb{R}$. Next $(7.57)(a)$ gives

$$cg_{ab} = r_a\alpha_{,b} + s_a\beta_{,b} \, (\Rightarrow cr_a = (r \cdot r)\alpha_{,a}, \, cs_a = (s \cdot s)\beta_{,a}). \tag{7.65}$$

Since the blade of Q is an eigenspace of a the blade of $\overset{*}{Q}$ is invariant for a, that is, $a_{ab}r^b = \gamma r_a + \delta s_a$ and $a_{ab}s^b = \rho r_a + \sigma s_a$ for smooth functions γ, δ, ρ and σ. A differentiation of the first of these and use of (7.20) give

$$(\lambda \cdot r)g_{ac} + \lambda_a r_c = r_a\gamma_{,c} + s_a\delta_{,c}. \tag{7.66}$$

For $p \in U$ and $0 \neq k \in T_pM$ with $k \cdot r = k \cdot s = 0$ $(\Rightarrow k^a\lambda_a = 0)$ a contraction of (7.66) with k^a gives $\lambda \cdot r = 0$ and similarly $\lambda \cdot s = 0$, on U. If $\lambda(p) \neq 0$ for some $p \in U$ one may arrange U so that λ is nowhere zero on U and then λ lies in the blades of both Q and $\overset{*}{Q}$ and is hence null. Thus in the positive definite case λ vanishes on U and so $\nabla' = \nabla$. For Lorentz or neutral signature if Q is timelike or spacelike at some $p \in U$ it remains so in some open neighbourhood of p (and which may be chosen as U) and again $\nabla' = \nabla$. For Lorentz signature the only remaining case is when Q and $\overset{*}{Q}$ are null on U. One may then adjust U so that l, n, x, y form a smooth, real null basis on U with $Q = l \wedge x$, $\overset{*}{Q} = -l \wedge y$ and with λ proportional to l on U. Then $l \wedge x$ gives an eigenspace of a over U, $a_{ab}l^b = \mu l_a$, $a_{ab}x^b = \mu x_a$ on U for smooth $\mu : U \to \mathbb{R}$ and with $l_{a;b} = y_{a;b} = 0$ on U. Then using the completeness relation for this basis

$$a_{ab} = \mu g_{ab} + \alpha l_a l_b + \beta y_a y_b + \gamma(l_a y_b + y_a l_b) \tag{7.67}$$

for $\alpha, \beta, \gamma : U \to \mathbb{R}$ which are easily seen to be smooth. On substituting (7.67) into (7.20) one finds

$$g_{ac}\lambda_b + g_{bc}\lambda_a = \mu_{,c}g_{ab} + \alpha_{,c}l_a l_b + \beta_{,c}y_a y_b + \gamma_{,c}(l_a y_b + y_a l_b). \qquad (7.68)$$

Now $\lambda^a l_a = \lambda^a x_a = \lambda^a y_a = 0$ and a contraction of (7.68) with $n^a x^b$ yields $\lambda^a n_a = 0$. It follows that $\lambda = 0$ on U, hence on M and so $\nabla' = \nabla$. For neutral signature one has that Q is null or totally null over U and if there exists $p \in U$ where Q is null then Q differs from $\overset{*}{Q}$ at p and hence on some open neighbourhood of p which may be chosen to be U. The proof that $\lambda \equiv 0$ on U then follows in a similar way to the Lorentz case. Otherwise Q is totally null over U and one may choose a smooth null basis l, n, L, N on U with $Q = l \wedge N = \overset{*}{Q}$, l and N parallel and the blade of Q spanning an eigenspace of a. One then proceeds, as above, to obtain

$$a_{ab} = \mu g_{ab} + \alpha l_a l_b + \beta N_a N_b + \gamma(l_a N_b + N_a l_b) \qquad (7.69)$$

for smooth α, β, γ and with λ^a in the blade $l \wedge N$ at each point of U. A substitution of (7.69) into (7.20) and contractions with $l^a L^b$ and $n^a N^b$ yield, successively, $\lambda^a L_a = \lambda^a n_a = 0$ and so λ vanishes on U. It follows that $\nabla' = \nabla$. $\square$

7.8   Projective-Relatedness in 4-dimensional Manifolds

In considering projective relatedness in a 4−dimensional manifold (M, g) with g of arbitrary signature, a significant amount of information may be obtained by investigating those cases where (M, g) satisfies the conditions of lemma 7.4. This is achieved by first assuming some holonomy group Φ for (M, g) with associated holonomy algebra ϕ. Then one takes the decomposition of M into its curvature classes as in lemma 3.1 and checks for each non-empty open subset U of constant curvature class the consistency of the inclusion $rgf(p) \subset \phi'_p \subset \phi$ for each $p \in U$, where f is the curvature map and ϕ'_p the infinitesimal holonomy algebra at p, recalling that ϕ'_p is a subalgebra of ϕ. If the resulting $rgf(p)$ leads to $\ker f(p)$ satisfying the conditions of lemma 7.4 one can check if Φ leads to locally properly recurrent or parallel vector fields on M using lemma 7.1. Then lemma 7.5 may be used to investigate the cases arising. This procedure will be followed for each signature using the appropriate subalgebra tables in chapters 3, 4 and 5.

7.8.1   The positive definite case

Suppose (M, g) has positive definite signature. Then the holonomy algebra ϕ arising from Φ is, from section 3.6, of type S_0, S_1, S_2, S_3, $\overset{+}{S_3}$, $\overset{-}{S_3}$, $\overset{+}{S_4}$, $\overset{-}{S_4}$

or S_6. Now suppose g' is another smooth metric on M of arbitrary signature such that g and g', with respective Levi-Civita connections ∇ and ∇', are projectively related. If (M, g) is of constant curvature so also is (M, g') (lemma 7.2) and so, in particular, if (M, g) is flat (holonomy type S_0), (M, g') is either flat or of (non-zero) constant curvature (holonomy type S_6) (lemma 7.3). Now suppose (M, g) has holonomy type S_1 so that $\dim rgf \leq 1$ on M but for some $p \in M$ and in some open neighbourhood U of p where *Riem* does not vanish, $\dim rgf = 1$. Then the curvature class is O where *Riem* vanishes and D elsewhere. Also lemma 7.5(i) shows that the conditions of lemma 7.4 and thus (7.57) hold on M. Then ϕ is spanned by a simple bivector F and so, from lemma 7.1, one may choose U so that it admits two independent, parallel vector fields u and v spanning the blade of $\overset{*}{F}$ on U ($\nabla u = 0 = \nabla v$). Then lemma 7.5(vi) shows that $\nabla' = \nabla$. It follows that $Riem' = Riem$ on M and that, on U, the blade of F is an eigenspace of g' so that (section 3.8), g' takes the form

$$g'_{ab} = ag_{ab} + bu_a u_b + cv_a v_b + d(u_a v_b + v_a u_b) \tag{7.70}$$

where a, b, c and d are smooth functions on U, $u_a = g_{ac} u^c$ and $v_a = g_{ac} v^c$. Since $\nabla g = \nabla g' = 0$, a, b, c and d are constant. Judicious choices of a, b, c, d with $a \neq 0$ (and consistent with g' being non-degenerate), can be used to make g' any signature on U. Equation (7.57)(c) reveals a similar expression for a.

Now suppose (M, g) has holonomy type S_2 so the conditions of lemma 7.4 are satisfied by lemma 7.5(i). Then one sees that for $p \in M$ $\dim rgf(p)$ equals 0, 1 or 2 and where, in the 2$-$dimensional case, $rgf(p)$ is spanned by two orthogonal, simple bivectors. Thus the curvature class at p is O, D or B and so M decomposes as $M = B \cup \mathrm{int} D \cup \mathrm{int} O \cup Z$, with $\mathrm{int} Z = \emptyset$ (theorem 3.1). The rank theorem shows that B is open (and possibly empty). The holonomy algebra ϕ at $p \in M$ is $\mathrm{Sp}(x \wedge y, z \wedge w)$ where x, y, z, w is an orthonormal basis at p. Thus $x + iy$ is a complex eigenvector of each member of ϕ at p and leads to a local complex recurrent vector field $X + iY$ on some open neighbourhood U of p for real, smooth vector fields X and Y on U (lemma 7.1). Similarly, $z + iw$ is a complex eigenvector of each member of ϕ and one obtains smooth vector fields Z and W on U with $Z + iW$ recurrent. So $(X_a + iY_a)_{;b} = (X_a + iY_a)(P_b + Q_b)$ for a complex recurrence 1-form $P + iQ$, and similarly for $Z + iW$. It then follows from the properties of recurrent vector fields in section 7.1 that the parallel transport of $X + iY$ along any path is a (complex) multiple of $X + iY$ at each point (and similarly for $Z + iW$ and so there exists a pair of smooth, orthogonal, 2$-$dimensional, holonomy invariant distributions which, at p, are $\mathrm{Sp}(x \wedge y)$ and $\mathrm{Sp}(z \wedge w)$ and in the neighbourhood U are spanned by the vector field pairs (X, Y) and (Z, W) chosen so that $|X| = |Y| = |Z| = |W| = 1$. The above recurrence relations and a simple contraction then give $X^a{}_{;b} = -Y^a Q'_b$ and $Y^a{}_{;b} = X^a Q'_b$ for some smooth 1-form Q'. Similar results arise for Z and W. It follows that the bivectors $X \wedge Y$ and $Z \wedge W$ are parallel on U. Now if the subset $B \neq \emptyset$ there are no non-trivial solutions for λ of (7.57)(b) and so

$\lambda = 0$ on B which gives $\nabla' = \nabla$. Since this gives $Riem' = Riem$, the local relationship between g and g' on B can be found using techniques described in chapter 3. If $B = \emptyset$ then D is open and non-empty and one may choose a non-empty open subset $U \subset D$ such that rgf is spanned by a simple bivector which is a linear combination of the holonomy spanning members $X \wedge Y$ and $Z \wedge W$ on U and so must be a multiple of one or the other, say of $Z \wedge W$. Then $Z \wedge W$ is an eigenspace of a at each point of U from (7.57)(c). This gives, on U

$$a_{ab} = a' g_{ab} + b' X_a X_b + c' Y_a Y_b + d'(X_a Y_b + Y_a X_b) \qquad (7.71)$$

for functions a', b', c', d' which are easily seen to be smooth on U. A substitution into (7.20) and contractions with $Z^a W^b$, $X^a Z^b$ and $Y^a Z^b$ then show that $\lambda \cdot X = \lambda \cdot Y = \lambda \cdot Z = \lambda \cdot W = 0$ and so λ vanishes on U. So $\nabla' = \nabla$. As in the case for the set B one can relate g and g' locally on D.

Next suppose (M, g) has holonomy type S_3. The holonomy structure here shows that at each $p \in M$ there exists $0 \neq k \in T_p M$ which annihilates each member of $rgf(p)$ and hence annihilates $Riem(p)$. Thus three independent simple bivectors whose blades contain k lie in $kerf$ and it easily follows that $kerf$ satisfies the conditions of lemma 7.4 and that, from lemma 7.1, each $p \in M$ admits a non-empty open neighbourhood on which a non-trivial parallel vector field exists. Thus one has the decomposition $M = C \cup \text{int} D \cup \text{int} O \cup Z$ with C open (possibly empty) and $\text{int} Z = \emptyset$. It then follows from lemma 7.5(iii) that if $C \neq \emptyset$ then $\lambda = 0$ on M and $\nabla' = \nabla$. The situation when $C = \emptyset$ will be discussed in detail later.

If (M, g) has holonomy type $\overset{+}{S_3}$ then for each $p \in M$ $rgf(p)$ is a subspace of the 3–dimensional subspace $\overset{+}{S_3}$. Hence $kerf(p)$ contains a subspace isomorphic to the orthogonal complement $\overset{-}{S_3}$ of $\overset{+}{S_3}$. Similar ideas to those used in theorem 3.4 show the conditions of lemma 7.4 are satisfied. Further since no member of $\overset{+}{S_3}$ is simple, M admits no points of curvature class B, C or D and hence decomposes as $M = A \cup \text{int} O \cup Z$ with A open and $\text{int} Z = \emptyset$ and clearly $A \neq \emptyset$. Thus each $p \in A$ admits an open neighbourhood $U \subset A$ on which (7.57)(b) has only the solution $\lambda = 0$ and so $\nabla' = \nabla$. The type $\overset{-}{S_3}$ is similar. Theorem 3.4 then shows that $g' = \kappa g$ on M with $0 \neq \kappa \in \mathbb{R}$.

If (M, g) has holonomy type $\overset{+}{S_4}$ (the type $\overset{-}{S_4}$ is similar) it follows from section 3.6 that M decomposes as $M = A \cup B \cup D \cup O$ with A and $A \cup B$ open, possibly empty. That the subset C is empty follows since in this case the holonomy algebra $\phi = \text{Sp}(\overset{+}{S_3}, G)$ for $G \in \overset{-}{S_3}$ and any simple member may be written as $F + \alpha G$ for $F \in \overset{+}{S_3}$ and then $(F + \alpha G) \cdot (F - \alpha G) = 0$ from lemma 3.1. Thus there are only two independent such simple members and these are essentially the pair $x \wedge y$ and $z \wedge w$ in table 3.1 and these have no common annihilator. Also since $\dim \overset{+}{S_4} = 4$, $kerf$ contains $(\overset{+}{S_4})^\perp$ and has dimension ≥ 2 and so contains two independent members of $\overset{-}{S_3}$. Then from lemmas 3.5(v) and 3.6(iv), T_pM is an eigenspace of $\nabla\lambda$ and so the conditions of lemma 7.4 are satisfied on M. So if $A \cup B \neq \emptyset$, (7.57)(b) has only trivial solutions for λ on $A \cup B$, and so $\nabla' = \nabla$. Thus either $\nabla' = \nabla$ and (see chapter 3) $g' = \kappa g$ on M ($0 \neq \kappa \in \mathbb{R}$), or $A = B = \emptyset$ and so $M = D \cup O$. This case will be explored later. The following has been proved.

Theorem 7.5 *Suppose g and g' are projectively related metrics on M with respective Levi-Civita connections ∇ and ∇' and with g positive definite. Suppose the holonomy type of (M, g) is S_1, S_2, $\overset{+}{S_3}$, $\overset{-}{S_3}$, or one of the types S_3, $\overset{+}{S_4}$ or $\overset{-}{S_4}$ and for which $\dim \mathrm{rg} f(p) \geq 2$ for some $p \in M$, then $\nabla' = \nabla$ (and so (M, g) and (M, g') have the same holonomy type). For the types $\overset{+}{S_3}$ and $\overset{-}{S_3}$ and for the types $\overset{+}{S_4}$ and $\overset{-}{S_4}$ with the above restrictive clause, $g' = \kappa g$ for $0 \neq \kappa \in \mathbb{R}$. For the other types g and g' are related as described following theorem 3.3 and g' need not be of positive definite signature.*

Theorem 7.6 *Suppose (M, g) and (M, g') are projectively related with g positive definite. Then, in the above notation,*

(i) *If (M, g) has holonomy type $\overset{+}{S_4}$ either $\nabla' = \nabla$ and $g' = \kappa g$ for $0 \neq \kappa \in \mathbb{R}$ or $(M, g)'$ has holonomy type S_6 (and similarly with $\overset{+}{S_4}$ replaced by $\overset{-}{S_4}$)*

(ii) *If (M, g) has holonomy type S_3 either $\nabla' = \nabla$ or (M, g) satisfies $\dim \mathrm{rg} f \leq 1$ on M and then (M, g') has holonomy type S_3 or S_6.*

Proof For part (i) suppose (M, g) has holonomy type $\overset{+}{S_4}$ and $\nabla' \neq \nabla$. Then from theorem 7.5 (M, g') cannot have holonomy type S_1, S_2, $\overset{+}{S_3}$ $\overset{-}{S_3}$, nor type S_3, $\overset{+}{S_4}$ or $\overset{-}{S_4}$ if there exists $p \in M$ with $\dim \mathrm{rg} f'(p) \geq 2$ where f' is the curvature map for (M, g') otherwise $\nabla' = \nabla$. Also (M, g') cannot be of type S_0 since then (M, g) would have holonomy type S_0 or S_6 (lemma 7.3). So if (M, g') does not have holonomy type S_6 it follows (theorem 7.5) that either $\nabla' = \nabla$ or $\nabla' \neq \nabla$ *and* $\dim \mathrm{rg} f \leq 1 \geq \dim \mathrm{rg} f'$ over M and so the curvature class decompositions of M with respect to $Riem$ and $Riem'$ are $M = D \cup O = D' \cup O'$ with D and D' open and non-empty and (lemma 7.3) $O = O'$ (and so $D = D'$). Then for any $p \in M$ there exists $r, s, r', s' \in T_pM$ with r and s independent and r' and s' independent and which satisfy $R^a{}_{bcd}r^d = R^a{}_{bcd}s^d = 0$ and $R'^a{}_{bcd}r'^d = R'^a{}_{bcd}s'^d = 0$. Then a contraction of (7.12) with $r^b r'^c$ gives $(\psi_{bc}r^b r'^c)\delta^a_d - (\psi_{bd}r^b)r'^a = 0$ from which it easily follows that $\psi_{bc}r^b r'^c = 0$ and then that $\psi_{bd}r^b = 0$ at p. Then contractions of (7.12) with r^b and with r^c give $R'^a{}_{bcd}r^b = 0$ and finally $\psi_{ab} = 0$, at p. Thus the tensor with components $\psi_{ab} \equiv \psi_{a;b} - \psi_a \psi_b$ in any coordinate domain of M vanishes on M and so $Riem' = Riem$ on M. But it is assumed that $\nabla' \neq \nabla$ and so the global 1-form $\psi(= d\chi)$ is not identically zero on M. However, the vanishing of the tensor ψ_{ab} reveals that $\nabla(e^{-\chi}\psi) = 0$ so that $e^{-\chi}\psi$ is a non-trivial, global parallel 1-form on M. [In fact, (7.9) then gives $(e^\chi\psi_a)_{|b} - (e^\chi\psi_a)_{;b} =$

$-e^\chi \psi_c(\delta^c_a \psi_b + \delta^c_b \psi_a)$ and, as a consequence, $\nabla'(e^\chi \psi) = 0$ on M. Thus each of (M, g) and (M, g') admit global, non-trivial, parallel 1-forms and the obvious associated non-trivial parallel vector fields with components $e^{-\chi} g^{ab}\psi_b$ and $e^\chi g'^{ab}\psi_b$.] This contradicts the holonomy type $\overset{+}{S}_4$ of (M, g) since then no such global, parallel vector fields exist. It follows that either (M, g') has holonomy type S_6 or $\nabla' = \nabla$ and then theorem 3.4 completes the proof.

For (ii) if (M, g) has holonomy type S_3 the previous theorem and part (i) of this theorem show, using similar techniques, that if $\nabla' \neq \nabla$, (M, g') has holonomy type S_3, $\overset{+}{S}_4$, $\overset{-}{S}_4$ or S_6 and, if $\overset{+}{S}_4$ or $\overset{-}{S}_4$, one again has the curvature decompositions $M = D \cup O = D' \cup O'$ with $O' = O$ and with $D' = D$ open and non-empty and the contradiction that (M, g) admits a global, non-trivial, parallel vector field. The result now follows. If (M, g') is of holonomy type S_3 and $\nabla' \neq \nabla$ the argument of part (i) reveals the stronger result that each of (M, g) and (M, g') admits a non-trivial, global, parallel vector field (whereas this holonomy type leads, in general, only to local such vector fields; see lemma 7.1). □

In the event that (M, g) has holonomy type S_3 or $\overset{+}{S}_4$ (or $\overset{-}{S}_4$) it need not be the case that $\nabla' = \nabla$ and counterexamples are available. These will be given in the next section which discusses the Lorentz situation.

7.8.2   The Lorentz case

As a preamble to this section and motivation for this chapter a brief, elementary, somewhat simplistic discussion and comparison of the principles of equivalence in classical Newtonian physics and Einstein's general relativity theory is now given. Following [69] consider (any) two gravitationally attracting (freely falling, spherical, uncharged, etc—such conditions will always be assumed) particles in Newtonian theory labelled M and m. One may identify three types of mass parameter with each of them: the *active* gravitational mass M_{AG} (m_{AG}) (the power a mass has to attract another mass), the *passive* gravitational mass M_{PG} (m_{PG}) (the susceptability of a mass to be attracted by another mass) and the *inertial* mass $M_I(m_I)$ (a particle's resistance to acceleration under the application of a force). In Newton's theory one can use his third law combined with his inverse square law of gravitational attraction to write, in some reference frame,

$$GM_{AG}m_{PG}r^{-2} = GM_{PG}m_{AG}r^{-2} \tag{7.72}$$

where G is Newton's gravitational constant and r the distance from M to m in this frame. One immediately gets $\frac{M_{AG}}{M_{PG}} = \frac{m_{AG}}{m_{PG}}$, and so, since M and m were arbitrary, one may choose units of active and gravitational mass so that for any such particle its active and passive gravitational masses are the same (and, say, labelled with the suffix g). Now assume that m is a "test particle" in that it is sufficiently small that the gravitational acceleration of M due to

m can be ignored. Then Newton's second law gives in an inertial frame in which M is at rest (by the test particle assumption)

$$GM_g m_g r^{-2} = m_I a \qquad (7.73)$$

where a is the acceleration of m in that frame. Then considering the gravitational field of (the assumed fixed particle) M equation (7.73) above shows that the quantity $aG^{-1}\frac{m_I}{m_g}(=\frac{M_g}{r^2})$ is the same for all such particles m at a fixed point of space and a fixed instant in time (a fixed event). In Newtonian theory one accepts, of course, the constancy of G and, in addition, the remarkable experimental result of Eötvös (and more recently of Dicke [70]) that a is the same for all such particles at a fixed event in this, and hence by arbitrariness, any gravitational field. It follows that the ratio $\frac{m_I}{m_g}$ is particle independent and thus on choosing appropriate units of measurement of inertial mass one may take $m_g = m_I$ for any such particle m and so conclude that only one mass parameter is required for each particle. It follows that a given gravitational field leads, through Newton's second law, his inverse square law and an elementary cancellation of this common mass parameter, to a well-defined *gravitational acceleration* being prescribed to any such particle at that event, independently of the particle and its make-up, and so, from the theory of second order differential equations, the path of such a particle passing through some event depends, at least locally, only on that event and its velocity at that event. This is the conclusion of what might be called (one form of) Newton's principle of equivalence. It is a consequence of the result $m_g = m_I$. (No attempt will be made here to get into the thorny problem of what "freely falling", spherical, etc mean!)

The above work displays the *indiscriminate* nature of the gravitational force in Newton's theory in that the acceleration it imparts to a body at some fixed event in a gravitational field is independent of that body. This feature, unusual for a force, is shared with the indiscriminate behaviour of the inertial acceleration imposed on bodies in a frame of reference undergoing acceleration with respect to some inertial frame. Hence one is tempted to postulate that these two types of forces are indistinguishable. This leads, through the well-known, so-called *Einstein lift experiment* [72] (at the risk of gross oversimplification), to the main idea of Einstein's general relativity theory. This is that the gravitational field is a geometrical phenomenon described in the setting of a 4-dimensional manifold M (space-time) admitting a metric of Lorentz signature and which is restricted by certain equations (Einstein's field equations for the determination of this metric and which then represents the "gravitational field").

The (weak) principle of equivalence in Einstein's theory then uses the classical Newtonian result above to make the assumption about the path c of a freely falling, etc, particle passing through an event $p \in M$ to be a "natural" path in M determined, at least locally, by the event p and the tangent vector to c at p (and c is assumed timelike for reasons to do with the particle's local speed not exceeding the speed of light). This is one way of arguing for the assumption of (timelike) *geodesic* motion for freely falling, etc, particles in Einstein's theory, in that such particles are assumed to follow timelike geodesics in the geometry of M and which are locally determined at any $p \in M$ by p and the tangent to the path at p (see chapter 2), independently of the particle's make-up. The remainder of this subsection may thus be interpreted as an attempt to find how much information on M and its metric (and its associated Levi-Civita connection) can be achieved from the knowledge of the geodesics of this geometry. Further details on general relativity theory may be found in [71, 13].

In the case when (M, g) is of Lorentz signature the possible holonomy algebras for (M, g) are R_1, R_2, R_3,...,R_{15} (chapter 4–and note that, because metric connections are being dealt with, type R_5 is excluded since it is spanned by a single, non-simple bivector) and techniques similar to those of the last subsection, using lemmas 7.4 and 7.5, may be used. It is noted that if (M, g) has holonomy type R_1, that is, (M, g) is flat, then (M, g') is of constant curvature and is hence either flat or of non-zero constant curvature (holonomy type R_{15}–see lemma 7.2(v)).

Theorem 7.7 *[108] Suppose g and g' are smooth projectively related metrics on M with Levi-Civita connections ∇ and ∇', respectively, and with g of Lorentz signature. If the holonomy type of (M, g) is R_2, R_3, R_4, R_6, R_7, R_8 or R_{12}, or if it is of holonomy type R_{10}, R_{11} or R_{13} and there exists $p \in M$ with $\dim rgf(p) \geq 2$ then $\nabla' = \nabla$. The signature of g' may not be Lorentz.*

Proof If (M, g) has any of these holonomy types the conditions of lemma 7.4 are satisfied everywhere and hence (7.57) holds. For types R_2,...,R_8 this follows from lemma 7.5(i). For holonomy type R_{12}, the holonomy algebra takes the form $Sp(l \wedge x, l \wedge y, l \wedge n + wx \wedge y)$ for a real null basis l, n, x, y and $0 \neq w \in \mathbb{R}$. In this case the kernel of the curvature map, $ker f$, contains the bivectors $l \wedge x, l \wedge y$ and $l \wedge n + w^{-1}x \wedge y$ and it easily follows from lemma 4.7 that the conditions of lemma 7.4 are satisfied on M.

If (M, g) has holonomy type R_2, R_3 or R_4, one has the decomposition $M = D \cup O$ with D non-empty and open in M and for $p \in D$, $rgf(p) = Sp(G)$ for a non-simple bivector G. It is now clear that lemma 7.5(vi) applies since two independent, local, parallel vector fields are admitted (lemma 7.1) and so $\nabla' = \nabla$.

If (M, g) has holonomy type R_6 with holonomy algebra $\phi = Sp(l \wedge n, l \wedge x)$ in some real null tetrad l, n, x, y, M decomposes as $M = C \cup D \cup O$ and M admits, locally, a recurrent null vector field and a parallel spacelike vector field. If $C \neq \emptyset$, then C is open and the recurrent null vector field is properly recurrent (section 7.1) so that lemma 7.5(ii) completes the proof. If $C = \emptyset$, D is non-empty and open and if there exists $p \in D$ with $rgf(p) = Sp(l \wedge n + \mu l \wedge x)$ ($\mu \in \mathbb{R}$) this is true of some open neighbourhood U of p and again a null, properly recurrent vector field exists on U so that lemma 7.5(ii) again applies and $\nabla' = \nabla$. Otherwise rgf is spanned by a smooth bivector $l \wedge x$ over U and the recurrent null vector field l satisfies $R^a{}_{bcd}l^d = 0$ from the Ricci identity and so l may be scaled so that it is parallel on U (section 7.1) and so two independent parallel null vector fields arise on some non-empty, open subset of M and lemma 7.5(vi) completes the proof that $\nabla' = \nabla$.

If (M, g) has holonomy type R_7 with holonomy algebra $\mathrm{Sp}(l \wedge n, x \wedge y)$, one has the decomposition $M = B \cup D \cup O$ with B open in M. If $B \neq \emptyset$ (7.57)(b) shows that $\lambda = 0$ on B and hence on M. If $B = \emptyset \neq D$ with D necessarily open then, on D, rgf must be spanned by a simple bivector (since $R_{a[bcd]} = 0$) and so D contains either an open subset on which rgf is spanned by a timelike bivector or an open subset where rgf is spanned by a spacelike bivector. Thus one either has two independent, properly recurrent, null vector fields or two parallel, null vector fields on some non-empty subset of D and lemmas 7.5(ii) and 7.5(vi) show that $\nabla' = \nabla$. If (M, g) has holonomy type R_8, $M = C \cup D \cup O$ then a local parallel null vector field is admitted and lemma 7.5(iv) shows that $\nabla' = \nabla$. If the holonomy type is R_{12} with $\phi =\mathrm{Sp}(l \wedge x, l \wedge y, l \wedge n + w(x \wedge y))$ $(0 \neq w \in \mathbb{R})$ one has $M = A \cup C \cup D \cup O$. (Here $B = \emptyset$ since the only simple members of this span can be checked to be $l \wedge x$ and $l \wedge y$.) If the open subset $A \neq \emptyset$ the only solution of (7.57)(b) on A is $\lambda = 0$ and so $\nabla' = \nabla$. If $A = \emptyset$ arguments similar to those above reveal that $\nabla' = \nabla$.

If (M, g) has holonomy type R_{10}, R_{11} or R_{13} one has, in terms of an orthonormal basis x, y, z, t and an associated real null basis l, n, x, y at p with $\sqrt{2}z = l + n$, $\sqrt{2}t = l - n$, respective holonomy algebras $\mathrm{Sp}(l \wedge x, n \wedge x, l \wedge n)$ $(=\mathrm{Sp}(t \wedge x, t \wedge z, z \wedge x))$, $\mathrm{Sp}(l \wedge x, l \wedge y, x \wedge y)$ and $\mathrm{Sp}(x \wedge y, x \wedge z, y \wedge z)$ and in each case a decomposition $M = C \cup D \cup O$. Again $B = \emptyset$ since each of these subalgebras has a common annihilator. The restriction $rgf \geq 2$ forces C to be non-empty and open. Thus a local parallel vector field is admitted in each case, being spacelike for R_{10}, null for R_{11} and timelike for R_{13}. lemma 7.5(iii) then shows that $\nabla' = \nabla$. Without this restriction on rgf the conclusion $\nabla' = \nabla$ need not follow and this will be explored later.   □

In the cases of holonomy type R_{10}, R_{11} and R_{13} and with $\dim rgf \leq 1$ on M, the subset $D \subset M$ is non-empty and open. Writing D_N (respectively, D_S, D_T) for those subsets of D where rgf is spanned by a null (respectively, spacelike or timelike) bivector one has, for R_{10}, $M = D_N \cup D_S \cup D_T \cup O$, for type R_{11}, $M = D_N \cup D_S \cup O$ and for R_{13}, $M = D_S \cup O$. If $\mathrm{int} D_N \neq \emptyset$ in cases R_{10} and R_{11} one still gets $\nabla' = \nabla$ using lemma 7.5(iv) for R_{11} and lemma 7.5(v) for R_{10}, the last result following from the fact that no global parallel *null* vector field is admitted in this case.

Now suppose (M, g) has holonomy type R_9 $(=\mathrm{Sp}(l \wedge n, l \wedge x, l \wedge y))$ with M decomposing as $M = A \cup C \cup D \cup O$. That the subset of curvature class B is empty is clear since each member of the holonomy is simple with l in its blade. Also a local, null, recurrent (not necessarily properly recurrent) vector field arising from l is admitted. In this case it is straightforward to check that the conditions required on $ker f$ in lemma 7.4 may *not* be satisfied at certain points since then $ker f$ contains $\mathrm{Sp}(l \wedge x, l \wedge y, x \wedge y)$. It is also clear that given $p \in M$, $\dim rgf(p) = 3 \Leftrightarrow p \in A$ and $\dim rgf(p) = 2 \Leftrightarrow$

$p \in C$. In this latter case the common annihilator of $rgf(p)$ is either spacelike, or null and, if null, it is proportional to the local recurrent vector field l. Write $C = C_S \cup C_N$ where C_S (respectively, C_N) are those subsets where this annihilator is spacelike (respectively, null). Similarly, using the notation of the previous paragraph, write $D = D_N \cup D_T$ since clearly $D_S = \emptyset$. Thus $M = A \cup C_S \cup C_N \cup D_T \cup D_N \cup O$. Problems arise here when $A \neq \emptyset$ and so assume that $A = \emptyset$. Then the conditions of lemma 7.4 are satisfied on M from lemma 7.5(i) and if any of intC_S, intC_N, intD_T and intD_N is not empty (and clearly one of them must be), λ vanishes on that open subset of M and so $\nabla' = \nabla$. This follows (briefly) for these respective interiors in the following way. For intC_S and intD_T rgf is easily checked to contain a timelike bivector whose blade contains l and so l gives rise to a local, properly recurrent null vector field and the result follows from lemma 7.5(ii) and for int C_N and intD_N the recurrent null vector l annihilates $Riem$ and hence may be scaled to be parallel over the relevant open subset. The result then follows from lemma 7.5 parts (iii) and (iv), respectively. The situation when $A \neq \emptyset$ will be considered later.

If (M, g) has holonomy type R_{14} with holonomy algebra $\phi = \mathrm{Sp}(l \wedge x, l \wedge y, l \wedge n, x \wedge y)$ one gets the decomposition $M = A \cup B \cup C \cup D \cup O$ and a local, null, recurrent vector field arising from l is admitted. The conditions of lemma 7.4 are not necessarily satisfied (since problems arise if $A \neq \emptyset$) and so it is assumed that $A = \emptyset$. Then B is open and in this case the conditions of lemma 7.4 are satisfied. To see this examine rgf over the subsets B, C and D noting that l spans the unique common eigendirection of the members of the holonomy algebra. If $B \neq \emptyset$ certainly $ker f$ satisfies the conditions of lemma 7.4 over B by lemma 7.5(i) and if $p \in B$ $rgf(p) = \mathrm{Sp}(G, \overset{*}{G})$ where $\overset{*}{G}$ is a spacelike bivector. Then $\overset{*}{G}$ is a linear combination of the members of ϕ and is annihilated by a timelike $t \in T_p M$. Performing this contraction immediately shows that $\overset{*}{G}$ is a linear combination of $l \wedge x$, $l \wedge y$ and $x \wedge y$ only and hence that $\overset{*}{G}_{ab} l^b = 0$. Thus l lies in the blade of G (and hence does not annihilate $Riem$ at p) and so gives rise to a local properly recurrent null vector field in some neighbourhood of p. If $C \neq \emptyset$ let $p \in C$. If dim$rgf(p) = 3$ the common annihilator of $rgf(p)$ cannot be timelike or spacelike otherwise there would exist an orthonormal basis x', y', z', t' at p with $rgf(p) = \mathrm{Sp}(x' \wedge y', x' \wedge z', y' \wedge z')$ or $rgf(p) = \mathrm{Sp}(y' \wedge t', z' \wedge t', y' \wedge z')$ and in neither case could there be a common null eigendirection for the spanning members of $rgf(p)$. Hence the common annihilator is null and so l must be a null eigenvector for each member of $rgf(p)$. It follows that $rgf(p) = \mathrm{Sp}(l \wedge x, l \wedge y, x \wedge y)$. If C_3 is the (necessarily open) subset of such points then, for $p \in C_3$, $ker f(p) = \mathrm{Sp}(l \wedge x, l \wedge y, l \wedge n)$ and the conditions of lemma 7.4 are satisfied on C_3 and then they are satisfied everywhere else on M by lemma 7.5(i). Clearly a local recurrent vector field exists on C_3 which annihilates $Riem$ everywhere on C_3 and so can be scaled to a local parallel null vector field. On the rest of the subset C, dim$rgf = 2$

and with a similar notation to that used in the previous case one decomposes this latter set as $C_N \cup C_S$ since $C_T = \emptyset$. [This follows as before since any linear combination Q of members of the R_{14} holonomy algebra which is orthogonal to a timelike vector cannot contain a non-zero contribution from $l \wedge n$. But then the remaining members have l as an annihilator and so Q is annihilated by two independent vectors and is thus unique up to a scaling and is, in fact, $x \wedge y$.] Similarly one decomposes D as $D = D_S \cup D_T \cup D_N$. So suppose $A = \emptyset$. One thus has the open subset of M given by $B \cup C_3 \cup \mathrm{int} C_S \cup \mathrm{int} C_N \cup \mathrm{int} D_T \cup \mathrm{int} D_N$ (note: $\mathrm{int} D_S$ is excluded here). It is then clear from the above and from parts (ii), (iii) and (iv) of lemma 7.5 that if this union of open subsets (equivalently, some member of this union) is non-empty then $\lambda = 0$ and $\nabla' = \nabla$. One thus has the following theorem [102, 109].

Theorem 7.8 *Suppose g and g' are smooth projectively related metrics of M with Levi-Civita connections ∇ and ∇' and with g of Lorentz signature. If the holonomy type of (M,g) is R_9 with $A = \emptyset$, or R_{14} with $A = \emptyset$ and with at least one of B, C_3, $\mathrm{int} C_S$, $\mathrm{int} C_N$, $\mathrm{int} D_T$ or $\mathrm{int} D_N$ non-empty, $\nabla' = \nabla$.*

It follows that the interesting cases arising from the previous theorems can be encapsulated by considering holonomy types R_{10} when $M = D_S$ or $M = D_T$, R_{11} when $M = D_S$, R_{13} when $M = D_S$, R_9 when $M = A$ and R_{14} when $M = A$ or $M = D_S$.

Let (M,g) have holonomy type R_{11} with $M = D_S$ [108] and for $p \in M$ let U be an connected, open coordinate neighbourhood of p such that U admits a smooth, parallel, null vector field l and a smooth null vector field n satisfying $l \cdot n = 1$ on U and with $R_{abcd} = \delta F_{ab} F_{cd}$ on U for smooth nowhere-zero function δ and smooth, nowhere-zero, spacelike bivector F, on U. Suppose also that g' is a smooth metric on M projectively related to g but with $\nabla' \neq \nabla$ so that one may choose U so that λ is nowhere zero on U. As mentioned earlier the conditions of lemma 7.4 are satisfied and so (7.57) holds on M and it will be assumed that δ and F are chosen so that $\overset{*}{F} = l \wedge n$ and so l and n are orthogonal to the blade of F and so annihilate *Riem*. Then (7.57) gives, on U,

$$\lambda_a = \sigma l_a + \rho n_a, \qquad \lambda_{a;b} = c g_{ab}, \qquad (c \in \mathbb{R}) \qquad (7.74)$$

for smooth functions σ and ρ on U and which combine to give

$$\rho n_{a;b} + n_a \rho_{,b} + l_a \sigma_{,b} = c g_{ab}. \qquad (7.75)$$

One may assume U chosen so that $F = x \wedge y$ where x, y are smooth unit spacelike vector fields on U satisfying $x \cdot y = 0$ on U and which, together with l and n comprise a real null basis on U. The (7.57)(c) shows that the blade $x \wedge y$ of F is an eigenspace of a on U and so

$$a_{ab} = w g_{ab} + \alpha l_a l_b + \beta n_a n_b + \gamma (l_a n_b + n_a l_b) \qquad (7.76)$$

where the functions w, α, β and γ are smooth on U (since g, a and the basis members are). Now choose a smooth vector field q in the blade of F on U ($\Rightarrow l \cdot q = n \cdot q = \lambda \cdot q = 0$) so that $q \cdot q$ is a constant (necessarily non-zero) on U ($\Rightarrow q^a q_{a;b} = 0$ on U where a semi-colon denotes a covariant derivative with respect to ∇). Then (7.76) gives $a_{ab} q^b = w q_a$ and so $a_{ab;c} q^b + a_{ab} q^b_{;c} = w_{,c} q_a + w q_{a;c}$. Then, use of (7.20) together with a contraction with q^a shows that w is constant on U (and non-zero since a is non-degenerate). Since n is null, $l_{a;b} = 0$ and $l \cdot n = 1$, one gets $n^a n_{a;b} = l^a n_{a,b} = 0$ and so a contraction of (7.75) with l^a and n^a gives

$$cl_a = \rho_{,a}, \qquad cn_a = \sigma_{,a}, \tag{7.77}$$

on U and then (7.75) reveals that

$$\rho n_{a;b} = cT_{ab}, \qquad (T_{ab} = g_{ab} - l_a n_b - n_a l_b = T_{ba}), \tag{7.78}$$

for a nowhere-zero, symmetric tensor field T on U. A covariant differentiation of (7.76) and use of (7.20) and (7.74) give

$$g_{ac}(\sigma l_b + \rho n_b) + g_{bc}(\sigma l_a + \rho n_a) = \alpha_{,c} l_a l_b + \beta_{,c} n_a n_b \tag{7.79}$$
$$+\beta(n_{a;c} n_b + n_a n_{b;c}) + \gamma_{,c}(l_a n_b + n_a l_b) + \gamma(l_a n_{b;c} + n_{a;c} l_b)$$

and successive contractions of (7.79) with $l^a l^b$, $n^a n^b$ and $l^a n^b$ reveal

$$2\rho l_a = \beta_{,a}, \qquad 2\sigma n_a = \alpha_{,a}, \qquad \sigma l_a + \rho n_a = \gamma_{,a}. \tag{7.80}$$

Finally a contraction of (7.79) first with l^a, then with n^a and use of (7.80) give

$$\beta n_{a;b} = \rho T_{ab}, \qquad \gamma n_{a;b} = \sigma T_{ab}. \tag{7.81}$$

Now suppose $\nabla n = 0$ on U. Then (7.78) and (7.81) show that c, σ and ρ are each zero on U. Then (7.74) gives the contradiction that λ vanishes on U. Thus one may take ∇n to be nowhere-zero on U. Now consider the constant c. If $c \neq 0$ (7.77) shows that l_a and n_a are gradients on U whilst if $c = 0$, $\nabla \lambda = 0$ and (7.77) shows that σ and ρ are constant on U with $\rho = 0$ from (7.78) and $\sigma \neq 0$ from (7.74), so that λ_a is a non-zero constant multiple of l_a. Finally (7.80) reveals that l_a and n_a are gradients on U. (Conversely, if λ_a is proportional to l_a on U (7.74) shows that ρ vanishes on U and $c = 0$ from (7.78)). Thus whether $c = 0$ or $c \neq 0$ one may write $l_a = u_{,a}$ and $n_a = v_{,a}$ for smooth functions u and v on U and then $l_{a;b} n^b - n_{a;b} l^b = -n_{b;a} l^b = 0$ so that l and n have zero Lie bracket on U. Then U may be chosen so that l and n span a 2–dimensional distribution on U. So choosing coordinates x^3 and x^4 on the relevant part of the 2–dimensional submanifold $u = v = 0$ (where the functions u and v are assumed adjusted with additive constants to make this possible) and the other two coordinates as parameters along the integral curves of l and n these parameter coordinates may be chosen as u and v and $l = \partial/\partial v$ and $n = \partial/\partial u$. With the coordinates u, v, x^3, x^4 the metric g is given by

$$ds^2 = 2du\,dv + g_{\alpha\beta} dx^\alpha dx^\beta \tag{7.82}$$

where $\alpha, \beta = 3, 4$, $l^a = (0, 1, 0, 0)$ and $n^a = (1, 0, 0, 0)$, $l_a = (1, 0, 0, 0)$, $n_a = (0, 1, 0, 0)$ and with summation over $\alpha, \beta = 3, 4$. Then (7.78) shows that $g_{\alpha\beta} = T_{\alpha\beta}$.

If $c \neq 0$ (7.77) shows that one may use the translational freedom in the coordinate u to arrange that $\rho = cu$ whilst if $c = 0$ (7.78) and the nowhere vanishing of ∇n on U gives $\rho = 0$ on U and then (7.74) and (7.77) show that σ is a non-zero constant on U (because, since $\nabla' \neq \nabla$, λ is not identically zero on U). In this latter case another use of the translational freedom in u allows one to write $\gamma = \sigma u$. Thus whether c is zero or not one has $un_{a;b} = T_{ab}$ on U. Next, since $\nabla l = 0$, l is a Killing vector field for g on U and so the components g_{ab} are independent of v (chapter 6). The equation $un_{a;b} = T_{ab}$ then gives $un_{\alpha;\beta} = T_{\alpha\beta} = g_{\alpha\beta}$ and so $-u\Gamma^2_{\alpha\beta} = g_{\alpha\beta}$ and finally $\frac{u}{2}\partial g_{\alpha\beta}/\partial u = g_{\alpha\beta}$. It follows that $g_{\alpha\beta} = u^2 h_{\alpha\beta}$ for a non-degenerate 2×2 matrix function h which is independent of u and v. Thus (7.82) becomes

$$ds^2 = 2dudv + u^2 h_{\alpha\beta} dx^\alpha dx^\beta. \qquad (7.83)$$

The equations (7.77) together with $l_a = u_{,a}$ and $n_a = v_{,a}$ give, after another use of the translational freedom in u, $\rho = cu$ and $\sigma = cv + e_1$ and then (7.80) gives $\alpha = cv^2 + 2e_1 v + e_2$, $\beta = cu^2 + e_3$ and $\gamma = cuv + e_1 u + e_4$ for constants e_1, e_2, e_3 and e_4. Finally (7.81) together with $n_{\alpha;\beta} = \frac{1}{2}\partial g_{\alpha\beta}/\partial u$ implies $e_3 = e_4 = 0$ and so $\alpha = cv^2 + 2e_1 v + e_2$, $\beta = cu^2$ and $\gamma = cuv + e_1 u$ and so (7.76) gives

$$a_{ab} = w\{g_{ab} + (cv^2 + 2e_1 v + e_2)l_a l_b + cu^2 n_a n_b + (cuv + e_1 u)(l_a n_b + n_a l_b)\}. \qquad (7.84)$$

The work and notation of section 7.3 can now be introduced. First one finds λ from (7.74), then calculates $\psi_a = -b_{ab}\lambda^b$ and then finds the function χ ($\psi = d\chi$). Then the required metric g' is, in the above coordinates, given by $g'_{ab} = e^{2\chi}b_{ab}$. This calculation is done using MAPLE and yields [108]

$$\psi = d\chi, \qquad \chi = \frac{1}{2}\log F,$$
$$F = \kappa^4[1 + 2cuv + 2e_1 u + (e_1^2 - ce_2)u^2]^{-1}, \qquad (7.85)$$

and finally for the metric g' with line element ds'^2

$$ds'^2 = \kappa F g - \kappa^{-3}F^2[(cv^2 + 2e_1 v + e_2)du^2 + cu^2 dv^2$$
$$+ 2u(cv + e_1 + (e_1^2 - ce_2)u)dudv] \qquad (7.86)$$

for $\kappa = w^{-1} > 0$. The special case $c = 0$ implies $F = F(u)$, $\chi = \chi(u)$, $\rho = 0$ and $\sigma = $ constant $\neq 0$ and so λ_a is a constant multiple of l_a. Conversely, the assumption that λ_a is a constant multiple of l_a forces $\rho = 0$ and hence $c = 0$. For this special case (7.85) shows that $\psi_a = r(u)l_a$ for some smooth function r and then, since $\nabla l = 0$, (7.9) reveals, using a vertical stroke to denote a ∇' covariant derivative, that $l_{a|b} = l_{a|b} - l_{a;b} = -l_c P^c_{ab} = -2r(u)l_a l_b$ and hence that $l' \equiv e^{2\int r du}l$ is parallel with respect to ∇' and from (7.86), null with respect to g'. Then (7.85) shows that $F = \kappa^4(1 + e_1u)^{-2}$ and, restricting the coordinate domain by $1 + e_1u \neq 0$, that $\psi_a = \chi_{,a} = -e_1(1 + e_1u)^{-1}l_a$ and so $\psi_{a;b} = \psi_a\psi_b$ and (section 7.2) $\psi_{ab} = \psi_{a;b} - \psi_a\psi_b = 0$. It follows from (7.12) that $R'^a{}_{bcd} = R^a{}_{bcd}$ in this (restricted) coordinate domain and hence that (M, g') has curvature rank at most 1 with l' being g'−null and $\nabla'-$ parallel.

The metric (7.86) with $c \neq 0$ can be simplified with the coordinate transformation $(u, v, x^3, x^4) \rightarrow (u', v', x^3, x^4)$ given by

$$u = u'(1 - cu'v')^{-1}, \qquad (7.87)$$
$$v = [ce_2u' + 2cv' - 2e_1 - u'(e_1 - cv')^2][2c(1 - cu'v')]^{-1}$$

and becomes (up to a constant scaling)

$$ds'^2 = 2du'dv' + cu'^2dv'^2 + u'^2h_{\alpha\beta}(x^3, x^4)dx^\alpha dx^\beta. \qquad (7.88)$$

The metric (7.86) with $c = 0$ is tidied up by the coordinate transformation

$$u = u'(1 - e_1u')^{-1}, \qquad v = (v' + \frac{e_2u'}{2})(1 - e_1u')^{-1} \qquad (7.89)$$

and becomes

$$ds'^2 = 2du'dv' + u'^2h_{\alpha\beta}(x^3, x^4)dx^\alpha dx^\beta. \qquad (7.90)$$

It is observed that (7.90) is of the same form as the original metric (7.83) and so, in any component of the intersection of these coordinate domains, g' is also of Lorentz signature and ∇' is also of holonomy type R_{11}. Thus one can find a local form for the projectively related metrics g and g'.

The cases of holonomy type R_{10} with $M = D_S$ or with $M = D_T$ or for type R_{13} with $M = D_S$ (and also the holonomy type S_3 when g has positive definite signature) are similar and will be handled together. With the notation as above let r be a nowhere-zero, smooth, parallel vector field on U ($\nabla r = 0$ on U) and let s be another nowhere-zero, smooth vector field on U with $|r|$, $|s|$ constant, $r \cdot s = 0$ and $R_{abcd}r^d = R_{abcd}s^d = 0$ on U. Then lemma 7.4 is satisfied on M and so on U one has

$$\lambda_{a;b} = cg_{ab}, \qquad \lambda_a = \sigma r_a + \rho s_a, \qquad cg_{ab} = r_a\sigma_{,b} + s_a\rho_{,b} + \rho s_{a;b} \qquad (7.91)$$

and

$$a_{ab} = wg_{ab} + \alpha r_a r_b + \beta s_a s_b + \gamma(r_a s_b + s_a r_b), \qquad (7.92)$$

where, as before, $r^a r_{a;b} = s^a s_{a;b} = r^a s_{a;b} = 0$. To accommodate each of the above cases one arranges $r^a r_a = \epsilon_1$ and $s^a s_a = \epsilon_2$ so that for R_{10} ($M = D_T$) and S_3, $\epsilon_1 = \epsilon_2 = 1$ whilst for R_{10} ($M = D_S$) $\epsilon_1 = 1$, $\epsilon_2 = -1$ (that is the parallel vector field r is spacelike) and for R_{13}, $\epsilon_1 = -1$, $\epsilon_2 = 1$ (so that r is timelike).

Contractions of the last equation in (7.91) with r^a and s^a give

$$cr_a = \epsilon_1\sigma_{,a}, \qquad cs_a = \epsilon_2\rho_{,a}, \qquad \rho s_{a;b} = cT_{ab}, \qquad (7.93)$$

where $T_{ab} = g_{ab} - \epsilon_1 r_a r_b - \epsilon_2 s_a s_b$. A substitution of (7.92) into (7.20) then gives $w =$ constant $\neq 0$ as before and contractions with $r^a r^b$, $s^a s^b$, $r^a s^b$, r^a and s^a give

$$\alpha_{,a} = 2\sigma\epsilon_1 r_a, \qquad \beta_{,a} = 2\rho\epsilon_2 s_a, \qquad \epsilon_1\epsilon_2\gamma_{,a} = \rho\epsilon_2 r_a + \sigma\epsilon_1 s_a, \qquad (7.94)$$
$$\beta s_{a;b} = \rho T_{ab}, \qquad \gamma s_{a;b} = \sigma T_{ab}.$$

Then with the assumption that λ is nowhere zero on U one finds that ∇s is nowhere zero on U and that in a coordinate domain on some (possibly reduced) U with coordinates $t, z, x^3, x^4)$ one has $r = \partial/\partial t$, $s = \partial/\partial z$, $r_a = \epsilon_1 dt$, $s_a = \epsilon_2 dz$ and the metric g is

$$ds^2 = \epsilon_1 dt^2 + \epsilon_2 dz^2 + z^2 h_{\alpha\beta} dx^\alpha dx^\beta \qquad (7.95)$$

with the $h_{\alpha\beta}$ a non-degenerate 2×2 matrix whose components are independent of t and z. Then

$$a_{ab} = w\{g_{ab} + (ct^2 + 2c_1 t + c_2)r_a r_b + cz^2 s_a s_b + (ctz + c_1 z)(r_a s_b + s_a r_b)\} \quad (7.96)$$

for constants c_1 and c_2. MAPLE calculations then reveal that

$$\psi = d\chi, \qquad \chi = \frac{1}{2}\log F, \qquad F = \kappa^4\{1 + \epsilon_2 q z^2 + \epsilon_1(ct^2 + 2c_1 t + c_2)\}^{-1}, \quad (7.97)$$

where $q = c + \epsilon_1(c_2 c - c_1^2)$ and the metric g' is given by

$$ds'^2 = \kappa F g - \kappa^{-3} F^2 \{(ct^2 + 2c_1 t + c_2 + \epsilon_2(cc_2 - c_1^2)dt^2$$
$$+ qz^2 dz^2 + 2\epsilon_1\epsilon_2(ct + c_1)z dz dt\}. \qquad (7.98)$$

A coordinate change of the form $(t, z, x^3, x^4) \to (t', z', x^3, x^4)$ with $t' = t'(t)$ and $z' = z'(t, z)$ can be used to simplify (7.98) up to a constant scaling as

$$ds'^2 = \epsilon_1(1 - \epsilon_2 q z'^2)dt'^2 + \epsilon_2(1 - \epsilon_2 q z'^2)^{-1}dz'^2 + z'^2 h_{\alpha\beta} dx^\alpha dx^\beta. \quad (7.99)$$

The holonomy type R_{10} requires $\epsilon_1 = 1$ and $\epsilon_2 = \pm 1$, the type R_{13} requires $\epsilon_1 = -1$ and $\epsilon_2 = 1$ and the positive definite type S_3 requires $\epsilon_1 = \epsilon_2 = 1$. It is noted that, in (7.95) $\partial/\partial t$ is a Killing vector field on U.

Theorem 7.9 *Suppose g and g' are projectively related on M with $\nabla' \neq \nabla$.*
(i) Suppose (M, g) has holonomy type R_{11} (respectively, R_{10} or R_{13}). Then either (M, g') has the same holonomy type as (M, g) or the type R_{15}.
(ii) Suppose (M, g) has holonomy type which is one of the types R_9 or R_{14}. Then (M, g') has holonomy type which is one of R_9 or R_{14}, or type R_{15}.

Proof For part (i) suppose (M, g) has type R_{11}. Then (M, g') cannot be of type R_2, R_3, R_4, R_6, R_7, R_8 or R_{12} by theorem 7.7. So assume (M, g') is one of the types R_{11}, R_{10} and R_{13}. Then theorem 7.7 shows that, since $\nabla' \neq \nabla$, the curvature maps f and f' for g and g' must satisfy dim$rgf \leq 1 \geq$ dimrgf' on M.

Thus, as in the proof of theorem 7.6, M decomposes as $M = D \cup O = D' \cup O'$ with $D' = D$ and $O' = O$ and with (M, g) (respectively, (M, g')) admitting a global nowhere-zero parallel vector field $e^{-\chi}\psi$ (respectively, $e^{\chi}\psi$). Then any $p \in D$ admits a connected open coordinate neighbourhood U on which g takes the form (7.82) and in which it is easily checked that, up to a constant scaling, the null (co)vector field $u_{,a}$ is the only g-parallel covector field. Since $e^{-\chi}\psi$ is g-parallel on U, χ is a function of u only in (7.85) and so $c = 0$. But then g' takes the form (7.90) on U and so (recalling (7.89)) $e^{\chi}\psi$ is g'−null on U and hence on M. It follows that (M, g') has holonomy type R_{11}. Similar results follow for the types R_{10} and R_{13} (using the condition $q = 0$ instead of $c = 0$). The remainder of the proof of part (i) will arise after the next part of the argument.

Now suppose that (M, g) has holonomy type R_9 and that $\nabla' \neq \nabla$. Then [102] from theorems 7.7 and 7.8 one has, in the decomposition of M for *Riem*, that A is open and non-empty and that (M, g') has either holonomy type R_9 (with, in an obvious notation, $A' \neq \emptyset$ for *Riem'*), R_{14} (with A' and/or intD'_S the only non-empty open subset(s) of M), holonomy type R_{10}, R_{11}, or R_{13} (each with dim$rgf \leq 1$ on M) or R_{15}. Choose $p \in A$ and an open neighbourhood U of p with $U \subset A$ and a smooth properly recurrent null vector field l on U satisfying $l_{a;b} = l_a q_b$ for some smooth 1-form q on U ($l_a = g_{ab}l^b$). It is then easily checked that $l_{[a;b}l_{c]} = 0$. But this can be shown to imply (after a possible reduction in U) that $l_a = \alpha u_{,a}$ on U for smooth functions α and u on U. Thus by rescaling l (but retaining the same symbol for it) one may write $l_a = u_{,a}$ on U so that l is a gradient and hence $l_{a;b} = l_{b;a}$ on U. Thus $l_{a;b} = \beta l_a l_b$ for some smooth function β on U. The Ricci identity then gives (see the proof of lemma 7.5)

$$l_d R^d{}_{abc} = l_a G_{bc} \qquad (7.100)$$

for some smooth bivector G on U with U assumed chosen so that G is nowhere-zero on U (since l is properly recurrent). Then consider the smooth orthogonal complement of the smooth 1−dimensional distribution on U spanned by l and which may be spanned by l and smooth spacelike vector fields x and y on (a possibly reduced) U. Then although the conditions of lemma 7.4 do not hold, contractions of (7.26) on the indices c and d with $F = l \wedge x$ and $F' = l \wedge y$ show that (7.53) holds for F and F' Thus the blades of F and F' are each eigenspaces of $\nabla\lambda$ at each point of U and hence

$$\lambda_{a;b} = \rho g_{ab} + \sigma l_a l_b \qquad (7.101)$$

on U for smooth functions ρ and σ A back substitution then gives

$$a_{ae}R^e{}_{bcd} + a_{be}R^e{}_{acd} = \sigma(g_{ac}l_b l_d + g_{bc}l_a l_d - g_{ad}l_b l_c - g_{bd}l_a l_c). \qquad (7.102)$$

Now define a nowhere-zero vector field s on U by $s_a = a_{ab}l^b$ and contract (7.102) with l^a using (7.100) to get

$$s_e R^e{}_{bcd} = s_b G_{cd}. \qquad (7.103)$$

Contractions of (7.103) with s^b and with l^b reveal that s is null and (using the symmetries of $Riem$) that l and s are orthogonal. Thus l and s are proportional on U and it follows that $a_{ab}l^b = \kappa l_a$ for some smooth, nowhere-zero (since a is non-degenerate) function κ. Then since $l_a = u_{,a}$, (7.20) and the equation $l_{a;b} = \beta l_a l_b$ give

$$(g_{ac}\lambda_b + g_{bc}\lambda_a)l^b = l_a\kappa_{,c} \Rightarrow (\lambda_b l^b)g_{ac} + \lambda_a l_c = l_a\kappa_{,c} \qquad (7.104)$$

and so, by a rank argument, $\lambda_a l^a = 0$ on U. It then follows that $\lambda_a = \kappa' l_a$ for some smooth function κ' on U (and hence $\rho = 0$ in (7.101) and then, since $\lambda_a = -a_{ab}\psi^b$ (see after (7.16)) and a is non-degenerate, ψ_a is proportional to l_a on U. It then follows from (7.9) that

$$l^a{}_{|b} - l^a{}_{;b} = P^a_{bc}l^c = l^a\psi_b \qquad (7.105)$$

and so l is recurrent on U for ∇' also, with recurrence 1-form $q_a + \psi_a$ and since ψ is a global gradient it follows, since l is properly recurrent for ∇, that it is also properly recurrent for ∇' (section 7.1). Then, as above for $Riem$, one may choose U so that $l^d R'_{dabc} = l'_a G'_{bc}$ where $l'_a = g'_{ab}l^d$ and G' is nowhere-zero on U (and so one may choose U so that $Riem'$ is nowhere-zero on U.) Next for (M, g), since ψ_a is proportional to l_a on U and since $\psi_{a;b}$ is symmetric, $\psi_{ab} \equiv \psi_{a;b} - \psi_a\psi_b$ is proportional to $l_a l_b$ on U. So if (M, g') has holonomy type R_{10}, R_{11} or R_{13} with $\dim rgf \leq 1$ on M in each case (recalling theorem 7.7 and the assumption $\nabla' \neq \nabla$) then for each $p \in U$ consider the 2$-$dimensional subspace of T_pM consisting of annihilators of $Riem'$ (since $Riem'$ is nowhere-zero on U and since $\dim rgf \leq 1$ on M) and the 3$-$dimensional subspace $(\mathrm{Sp}(l))^\perp$ of T_pM. They intersect in (at least a) 1$-$dimensional subspace of T_pM containing, say, $0 \neq k \in T_pM$. A contraction of (7.12) with k^b then shows that $R^a{}_{bcd}k^d = 0$ and this contradicts the choice $p \in U \subset A$ for $Riem$ since no such non-zero vectors k exist at $p \in A$. It follows that (M, g') is either of holonomy type R_9 with $A' \neq \emptyset$ or type R_{14} with $A' \cup \mathrm{int} D'_S \neq \emptyset$ or R_{15}.

If (M, g) has holonomy type R_{14} with A' and/or $\mathrm{int} D_S$ the only non-empty open subsets of M then if $A' \neq \emptyset$ similar arguments lead to the same contradiction whilst if $A' = \emptyset \neq \mathrm{int} D_S$ then each of (M, g) and (M, g') has curvature rank at most one on M and the proof of theorem 7.6 reveals the contradiction of a global, parallel vector field for (M, g). This finishes the proof of part (ii) and completes the proof of part (i). $\square$

To obtain a metric for holonomy types R_9 and R_{14} is a little more complicated. However, one does have the following results [102, 109], and which were largely the work of Dr David Lonie. Suppose (M, g) is of holonomy type R_{14} and is such that there exists an open coordinate neighbourhood $U \subset M$ such that U, together with its induced metric from g on M, is *also of holonomy type R_{14}*. Suppose also that (M, g) is projectively related to a metric g' on U with $\nabla' \neq \nabla$. Suppose also that U is contained in the curvature class A subset for (M, g). Then with a mild restriction on U one may choose it as a coordinate domain with coordinates u, z, x, y such that g takes the form

$$ds^2 = 2dudz + \sqrt{z}a(u)du^2 + u^2e^{2w(x,y)}(dx^2 + dy^2) \tag{7.106}$$

for an arbitrary smooth function a and smooth function w satisfying the condition that $\partial^2 w/\partial x^2 + \partial^2 w/\partial y^2$ is nowhere-zero on U. The metric g' is such that, after a coordinate transformation $(u, z, x, y) \to (u', z', x', y')$ with $u = u(u')$, $z = z(u', z')$, $x = x'$ $y = y'$, it takes exactly the same form in terms of u', z', x', y' as g does in terms of u, z, x, y in (7.106) and with $a(u)$ replaced by a function $a'(u')$ but with w unchanged, that is, $w(x', y') = w(x, y)$. So g' has holonomy type R_{14} on U.

If the above conditions on U are retained but with g now of holonomy type R_9 one can achieve g either in the form (7.106) but now with w harmonic on U, $\partial^2 w/\partial x^2 + \partial^2 w/\partial y^2 = 0$ on U, or in the form

$$ds^2 = 2dudz + \sqrt{z}b(u)du^2 + u^2dx^2 + (u - \alpha)^2dy^2 \tag{7.107}$$

with α constant and b an arbitrary smooth function. The metric g' is then such that a coordinate transformation $(u, z, x, y) \to (u', z', x', y')$ given by $u = u(u')$, $z = z(u', z')$, $x = x'$ and $y = \frac{y}{\beta}$ for some constant β casts g' into exactly the same form in terms of (u', z', x', y') as g was in terms of (u, z, x, y) with $b(u)$ replaced by a function $b'(u')$ and α replaced by some constant α'. Thus g' has holonomy type R_9 on U. In each of the above cases the 1-form ψ_a is a nowhere-zero multiple of $u_{,a}$ on U and so $\nabla' \neq \nabla$. It is noted that (7.107) admits $\partial/\partial x$ and $\partial/\partial y$ as Killing vector fields on U.

To end this section a few remarks can be made regarding projective structure and symmetry. First, consider the metric (7.83) and let the coordinate domain for which (7.83) holds be taken as M. Then (M, g) admits the global nowhere-zero Killing (in fact, parallel) vector field $X = \partial/\partial v$ with components in this chart given by $X^a = (0, 1, 0, 0)$ and then $X_a = g_{ab}X^b = (1, 0, 0, 0)(= u_{,a})$ and $X_{a;b} = 0$. Then consider the vector field $Z = u\partial/\partial v$ with components $Z^a = (0, u, 0, 0)$ and $Z_a = (u, 0, 0, 0)$ so that $Z_a = uX_a$ and hence $Z_{a;b} = X_au_b = u_{,a}u_{,b}$. It follows that $Z_{a;bc} = 0$, that is, if $Z_{a;b} \equiv h_{ab} = h_{ba}$, $h_{ab;c} = 0$ and so Z is a proper affine vector field on M (chapter 6). Then define a vector field R by $R = u\partial/\partial u + v\partial/\partial v$ with components $R^a = (u, v, 0, 0)$ and $R_a = (v, u, 0, 0)$. Thus $R_a = vu_{,a} + uv_{,a}$ so that $R_{a;b} = u_{,a}v_{,b} + v_{,a}u_{,b} - u\Gamma^2_{ab} = u_{,a}v_{,b} + v_{,a}u_{,b} + \frac{1}{2}\partial g_{\alpha\beta}/\partial u$ and so $R_{a;b} = g_{ab}$. Hence R is a proper homothetic vector field (chapter 6). Finally define the vector field S by $S = uR$ with components $S^a = (u^2, uv, 0, 0)$ and so $S_a = uR_a$ and one finds $S_{a;b} = R_au_{,b} + ug_{ab}$. It follows that

$$\mathcal{L}_Sg_{ab} = S_{a;b} + S_{b;a} = h_{ab}, \quad (h_{ab} = R_au_{,b} + u_{,a}R_b + 2ug_{ab}),$$
$$h_{ab;c} = 2g_{ab}u_{,c} + g_{ac}u_{,b} + g_{bc}u_{,a} \tag{7.108}$$

and so S is a proper projective vector field with projective 1-form $u_{,a}$. The Killing bivector field for X, the homothetic bivector field for R and the affine bivector field for Z (but not the projective bivector field for S) are each zero. No new symmetry vector fields arise from the Lie bracket operations between X, Z, R and S.

Now return to the metric g given in (7.83) in the coordinates u, v, x^3 and x^4 together with its projectively related metric g' in the case when $c = 0$ so that g' is given by (7.90). Now regard U chosen so that the above coordinates and their primed counterparts each make sense on U and that U is connected and as before take $M = U$. Thus the projective algebras for (M, g) and (M, g') are equal (section 7.6) and their Killing algebras are isomorphic as vector spaces. To elaborate a little on this last point consider the Killing vector field $X' = \partial/\partial v'$ for g' on M arising from the ignorable coordinate v' in (7.90). In coordinates u', v', x^3, x^4, X' has components $(0, 1, 0, 0)$ whereas in the coordinates u, v, x^3, x^4, X' has components $(0, \partial v/\partial v', 0, 0) = (0, (1 - e_1 u')^{-1}, 0, 0) = (0, 1 + e_1 u, 0, 0)$ (from (7.89)) and so $X' = (0, 1, 0, 0) + e_1(0, u, 0, 0) = X + e_1 Z$ and, being a linear combination of a Killing vector field X for g and a proper affine vector field Z for g, is proper affine but *not* Killing for g (c.f. the remarks following theorem 7.4.) However, considering the Killing vector field X for g with components $(0, 1, 0, 0)$ in coordinates (u, v, x^3, x^4) the associated Killing vector field Y for g' (section 7.6 and using the notation of that section) has components in this coordinate system given by $Y^a = e^{2\chi} g'^{ab} g_{bc} X^c = e^{2\chi} g'^{ab} X_b = e^{2\chi} g'^{a1}$. Then, choosing the arbitrary constant in χ to be zero and recalling the choice $c = 0$ in (7.85) and (7.86) one has $\kappa = 1$ and $e^{2\chi} = F = (1 + e_1 u)^{-2}$ and, after a straightforward inverting, $g'^{a1} = (0, g_{12}^{-1}, 0, 0) = (0, (1 + e_1 u)^3, 0, 0)$. Thus, up to a constant multiple, $Y^a = (0, 1 + e_1 u, 0, 0)$ which, as shown above, is the Killing vector field $\partial/\partial v'$ for g'.

As a final example in Lorentz signature let $M = I \times M'$ where I is an open interval in $\mathbb{R}$, M' is a 3$-$dimensional, smooth, connected manifold admitting a smooth positive definite metric γ of constant curvature and admitting a local chart with coordinates r, θ, ϕ. Thus M admits a Lorentz metric g given in the obvious chart with coordinates t, r, θ, ϕ by

$$ds^2 = -dt^2 + R(t)^2 \{dr^2 + f(r)^2 (d\theta^2 + \sin^2\theta d\phi^2)\} \tag{7.109}$$

where γ is represented by the curly bracketed term and R is a smooth function on M (the *scaling function*). This metric is the well-known Friedmann-Robertson-Walker-Lemaître (F.R.W.L.) metric of relativistic cosmology. The coordinate t is the cosmic time function and $f(r)$ equals $\sin r$, $\sinh r$ or r according as M' has positive, negative or zero (constant) curvature. The chart given above will be regarded as a global chart, that is, its domain will be taken as being equal to M. There are a number of special cases of (7.109) which are less interesting both mathematically and physically. They are the cases when (M, g) itself has constant curvature and which, if positive (respectively, negative or zero), give (open submanifolds of) the well-known *de Sitter* (respectively the *anti-de Sitter* and *Minkowski* metrics), or when R is constant $\neq 0$ when it is an *Einstein static type metric*. {The Einstein static type metric [115] has much historical interest. It admits an extra independent, global, Killing (in fact, parallel) vector field given by $u = \partial/\partial t$ and a proper affine vector field $t\partial/\partial t$. There are no proper projective or proper homothetic vector fields and so $\dim K(M) = 7$, $\dim A(M) = \dim P(M) = 8$ and there is a single Killing orbit equal to M. It has holonomy type R_{13} with $\dim rgf = 3$ everywhere and any metric g' projectively related to it satisfies $\nabla' = \nabla$ with $g'_{ab} = ag_{ab} + u_a u_b$ $(a, b \in \mathbb{R})$ and is also of the Einstein static type (see theorem 7.7).}

The idea is, in some sense, to exclude these special cases by assuming that (M, g) does not contain any non-empty open subsets which, with the induced metric from g, is diffeomorphic to any of the above special types. In this case (M, g) will be referred to as a *generic* F.R.W.L metric and admits a 6−dimensional Killing algebra whose orbits are the spacelike 3−dimensional submanifolds of constant t (copies of M') with positive definite metric γ. The Killing vector fields are tangent to these orbits and give rise to a 6−dimensional Killing algebra in these orbits with respect to the metric γ also. The vector field $\partial/\partial t$ represents a fluid (galactic) flow in the application to cosmology. The Killing isotropy subalgebra is $o(3)$ at each $p \in M$ and in general (M, g) has a 6−dimensional projective algebra. There are, however, certain special choices of the function R which lead to the admission of a single independent proper homothetic or a single independent proper projective vector field (but not both) and, in addition, no proper affine vector fields are admitted [110]. Thus $\dim P(M) = 6$ or 7. The holonomy type is R_{15} since it is known that $\dim rgf(p) = 6$ for each $p \in M$.

Now suppose g' is a Lorentz metric on M projectively related to g. Then the projective algebra for g', $P'(M)$, equals that for g and it has a Killing algebra $K'(M)$ of dimension 6 whose orbits are 3−dimensional (theorem 7.4). Since $K(M)$ and $K'(M)$ are subalgebras of $P(M)$ and $\dim P(M) \le 7$ it follows that either $K(M) = K'(M)$ or $K \equiv K(M) \cap K'(M)$ is a 5−dimensional subalgebra of $K(M)$. In this latter case any orbit of (M, g) arising from K must lie within a necessarily g−spacelike orbit of $K(M)$ (and is hence g−spacelike) and so cannot have dimension 1 or 2 since then it would have to be g−null at each of its points [13] (Stictly speaking the theorem alluded to here applies to the Killing algebra of M whereas here $K \ne K(M)$. However K is a Lie algebra of Killing vector fields and this is sufficient for this result to hold.) Hence the orbits arising from K coincide with those of $K(M)$. But then the isotropy subalgebra arising from K would be a $(5-3=2)$-dimensional subalgebra of $o(3)$ which does not exist. It follows that $K(M) = K'(M)$ and the isotropy at p is $o(3)$ and if $X \in K(M)$, $\mathcal{L}_X g' = 0$.

Now suppose ϕ_t is a local flow for some member of the isotropy algebra K_p^* arising from $K(M)(= K'(M))$ at $p \in M$, so that $\phi_t(p) = p$, then the vector $u = \partial/\partial t$ in (7.109) is the unique (up to a scaling) vector in $T_p M$ satisfying $\phi_{t*} u = u$ and so u is both $g-$ and $g'-$ orthogonal to the subspace $H_p \subset T_p M$ tangent to the orbit at p. So, for any $x \in H_p$ one has $g'(x, u) = g(x, u) = 0$. Now let I_p be the subset of $T_p M$ of $g-$unit vectors at p. If $v \in I_p$, suppose

$g'(v, v) = \alpha g(v, v) (= \alpha)$ for $\alpha \in \mathbb{R}$. If $v' \in I_p$ there exists a local flow ϕ_t of $X \in K_p^*$ such that $\phi_{t*}(v) = v'$ and so $g'(v', v') = g'(\phi_{t*}v, \phi_{t*}v) = \phi_t^* g'(v, v) = g'(v, v)$ and similarly $g(v', v') = g(v, v)$. It follows that $g'(v', v') = \alpha g(v', v')$ for each $v' \in I_p$. Then by linearity it follows that $g'(v, v) = \alpha g(v, v)$ for each $v \in H_p$. Next if $x, y \in H_p$, $g'(x+y, x+y) = \alpha g(x+y, x+y)$ and on expanding this out one finds that $g'(x, y) = \alpha g(x, y)$ for each $x, y \in H_p$. Then if $X \in H_p$ is fixed and $k \in T_pM$ it follows from the above that by decomposing k as a member of H_p and a multiple of u, $k = \mu y + \nu u$ for $y \in H_p$ and $\mu, \nu \in \mathbb{R}$, one finds that $(g' - \alpha g)(X, k) = 0$ for any fixed $X \in H_p$ and for any $k \in T_pM$. Thus, in components and with $h = g' - \alpha g$, $(h_{ab}X^b)$ is orthogonal to each member of T_pM and is hence zero. Thus each member of H_p is an eigenvector of g' with respect to g with the same eigenvalue α. It follows that either g' is a multiple of g or H_p is the α-eigenspace of g' and $\mathrm{Sp}(u)$ its orthogonal complement and so, in either case, u is also an eigenvector of g'. So, at $p \in M$, $g'_{ab} = \alpha g_{ab} + \beta u_a u_b$ for $\beta \in \mathbb{R}$ and, since g' is non-degenerate and $|u| = -1$, $\alpha \neq \beta$. Now, regarding α and β as functions on M and if q also lies in the Killing orbit through p, say with $\phi_t(p) = q$ for some local flow of a member of $K(M)$, then for $x, y \in H_p$, $\alpha(p)g_p(x, y) = g'_p(x, y) = (\phi_t^* g')_p(x, y) = g'_q(\phi_{t*}x, \phi_{t*}y) = \alpha(q)g_q(\phi_{t*}x, \phi_{t*}y) = \alpha(q)(\phi_t^* g)_p(x, y) = \alpha(q)g_p(x, y)$ for $\phi_{t*}x, \phi_{t*}y \in H_q$, (where $g_p = g(p)$, etc) and since $\phi_t^* g' = g'$, $\phi_t^* g = g$ and $\phi_{t*}u = u$. This, and a similar calculation for u shows that the functions α and β are constants on the orbits and hence (smooth) functions of t only. Thus

$$g'_{ab} = \alpha(t)g_{ab} + \beta(t)u_a u_b \qquad (7.110)$$

where $\alpha > 0 < \alpha - \beta$ for g' non-degenerate and of Lorentz signature. It follows that g' is also an F.R.W.L metric whose submanifolds of constant t (orbits) are the same (constant) curvature as those for g.

One may now impose the remainder of the conditions (7.10) on the functions α and β to compute the form for g' up to a constant conformal factor as [107]. The calculation was done by Dr David Lonie using MAPLE.

$$ds'^2 = \frac{-dt^2}{(1 + eR^2(t))^2} + \frac{R^2(t)}{1 + eR^2(t)}[dr^2 + f^2(r)(d\theta^2 + \sin^2\theta d\phi^2)]. \qquad (7.111)$$

The metric g' is thus controlled by the single constant $e \in \mathbb{R}$ and the single function $1 + eR^2(t)$. Any metric g' of the form (7.111) is projective related to the metric (7.109). The following is thus proved [107].

Theorem 7.10 *If (M, g) is a generic F.R.W.L. metric as in (7.109) and g' is also a Lorentz metric on M projectively related to g, then g' is also a generic F.R.W.L. metric given by (7.111) and is of the same general form as (7.109). Any metric of the form (7.111) is projectively related to (7.109). The situation for the Einstein static type metrics is as described earlier.*

It is noted here that examples of projectively related F.R.W.L. metrics g and g' on M exist for which the common projective algebra is 7-dimensional and with M admitting a vector field X which is proper homothetic for g (and hence no proper projective vector fields are admitted by (M,g)—see earlier) and proper projective for g' (but with (M,g') admitting no proper homothetic vector fields) [107]. This should be noted in connection with theorem 7.4. It is remarked that [111] quotes a result without proof but which is attributed to Sinjukov [112] and which can be shown to be equivalent to theorem 7.10. Further results may be found in [107, 110, 113, 114] and the references therein.

Finally one may return briefly to the positive definite case. It turns out that using similar techniques as above, but this time for positive definite signature, a version of theorem 7.10 can be found (see [34]). One can also consider the situation when (M,g) has holonomy type $\overset{+}{S}_4$. In this latter case, and with g' projectively related to g, in order to get an example when $\nabla' \neq \nabla$ one needs $\dim rgf(p) \leq 1$ for each $p \in M$ (theorems 7.5 and 7.6). Consider the following metric in a global coordinate system x, y, z, w with $w > 0$

$$ds^2 = w^2\{dx^2 + dz^2 + (4x^2 + 1)dy^2 + 4x\,dy\,dz\} + dw^2. \tag{7.112}$$

This metric has holonomy type $\overset{+}{S}_4$ and the only non-vanishing component of *Riem* (up to index symmetries) is R_{1212}. Thus *Riem* is of curvature class D on M and g admits Killing vector fields $\partial/\partial y$ and $\partial/\partial z$ and a parallel bivector $\frac{1}{w}\{\partial/\partial w \wedge \partial/\partial z + \partial/\partial x \wedge \frac{1}{w}(\partial/\partial y - 2x\partial/\partial z)\}$. The metric g' is found using the inversion procedure [34, 67] and is given up to a constant scaling by

$$ds'^2 = (1 + bw^2)^{-1}\{w^2[dx^2 + dz^2 + 4x\,dy\,dz + (4x^2 + 1)dy^2] + (1 + bw^2)^{-1}dw^2\} \tag{7.113}$$

where b is a positive constant. It has positive definite signature and holonomy type S_6 (see theorem 7.6). It admits two Killing vector fields $\partial/\partial y$ and $\partial/\partial z$.

7.8.3 The neutral signature case

Suppose now that $\dim M = 4$ and that g and g' are smooth metrics on M of neutral signature. The techniques used earlier for the other signatures may be used here also but the general situation is more complicated. The main ideas will be presented together with references where some examples can be found but the treatment will otherwise be brief. It is noted that since a metric exists the holonomy type of (M,g), if 1–dimensional, must be $1a$, $1b$, $1c$ or $1d$ and cannot be $2l$, $3e$, $3f$ or $4d$ (see earlier). The uncertain type $2j$ will be included.

Theorem 7.11 *Let $\dim M = 4$ with g and g' projectively related metrics of neutral signature on M. Suppose g has holonomy algebra of dimension 1 or 2. Then in the usual notation $\nabla' = \nabla$.*

Proof Let ϕ be the holonomy algebra associated with g and use the corresponding bases l, n, L, N and x, y, s, t associated with this signature. Lemma

7.5 shows that the conditions of lemma 7.4 hold in all cases and U will be used to denote the open subset of M used in lemma 7.5 and in which *Riem* is nowhere zero. If ϕ is of type $1a$, $1b$, $1c$ or $1d$ then U may be chosen to admit two independent parallel vector fields and lemma 7.5(vi) applies in each case to give $\lambda = 0$ on M and hence that $\nabla' = \nabla$. For ϕ of type $2a = \text{Sp}(l \wedge N, l \wedge n - L \wedge N)$, (M, g) admits two independent, local, recurrent, null vector fields on U arising from l and N. If one may take U so that $\dim rgf(p) = 1$ for each $p \in U$ then rgf is spanned by $l \wedge N$ on U and $R^a{}_{bcd}l^d = R^a{}_{bcd}N^d = 0$ on U and so U may be chosen to admit two independent, local, parallel vector fields (section 7.1) and lemma 7.5(vi) shows that $\nabla' = \nabla$. Otherwise, for some $p \in U$ and hence on some open neighbourhood of p, chosen as U, $\dim rgf = 2$, and for each $p \in U$, $rgf(p)$ contains a non-simple member. Then $(7.57)(b)$ gives $\nabla' = \nabla$. For holonomy types $2b = \text{Sp}(l \wedge n, L \wedge N)$ and $2d = \text{Sp}(l \wedge n - L \wedge N, l \wedge L)$ the proof is similar. For type $2c = \text{Sp}(l \wedge n - L \wedge N, x \wedge y - s \wedge t = \text{Sp}(l \wedge n - L \wedge N, l \wedge L + n \wedge N)$ it is easy to check that all non-zero members of $rgf(p)$ are non-simple. (To see this note that if $G \equiv l \wedge n - L \wedge N \in \overset{+}{S}$ and $H \equiv x \wedge y - s \wedge t \in \overset{-}{S}$ then $(G + \mu H) \cdot (\overset{*}{G} + \mu \overset{*}{H}) = |G| - \mu^2 |H|$ $(\mu \in \mathbb{R})$ is never zero since $|G| < 0 < |H|$.) Then $(7.57)(b)$ shows that $\lambda = 0$ and the result follows. The case when ϕ is of type $2e = \text{Sp}(x \wedge y, s \wedge t)$ is similar to the type S_2 case in the positive definite situation. For type $2f = \text{Sp}(l \wedge N + n \wedge L, l \wedge L)$ one has local, complex recurrent vector fields $l \pm iL$ on U and the only simple members of ϕ are multiples of $l \wedge L$. So either one may choose U so that $rgf(p)$ is spanned by $l \wedge L$ on U or $rgf(p)$ has a non-simple member at each $p \in U$. In the former case, $l \pm iL$ give rise to complex conjugate parallel vector fields on U and hence to two real null parallel vector fields on U lying in the blade of $l \wedge L$ and lemma 7.5(vi) gives $\lambda = 0$ on U and hence on M. In the latter case rgf has a non-simple member at each $p \in U$ and again $\lambda = 0$ on U and hence M. Thus $\nabla' = \nabla$. For type $2g = \text{Sp}(l \wedge N, l \wedge L)$ there is a local null parallel vector field on U arising from l and one may choose U so that either $\dim rgf = 1$ on U or $\dim rgf(p) = 2$ on U. In these cases lemma 7.5(iv), respectively, lemma 7.5(iii), completes the proof. If the type is $2h = \text{Sp}(l \wedge N, \alpha(l \wedge n) + \beta(L \wedge N)$ $(\alpha \neq \pm\beta)$ or type $2j = \text{Sp}(l \wedge N, \alpha(l \wedge n - L \wedge N) + \beta(l \wedge L))$ $(\alpha\beta \neq 0)$ then, in each case, a local recurrent vector field is admitted arising from l. For $2j$ the only simple member in rgf (up to scaling) is $l \wedge N$. (To see this let $F = l \wedge N, G = l \wedge n - L \wedge N$ and $H = l \wedge L$. Putting $Q = F + \mu(\alpha G + \beta H)$ one finds $Q \cdot \overset{*}{Q} = \mu^2 \alpha^2 |G|$ and so Q is simple if and only if $\mu = 0$.) The result now follows for this type from lemma 7.5. For type $2h$ if $\alpha\beta \neq 0$ a similar argument shows that $l \wedge N$ is the only simple member and the conclusion that $\nabla' = \nabla$ follows as above. If $\alpha = 0 \neq \beta$ or if $\alpha \neq 0 = \beta$ all members of rgf are simple. If one can choose U so that $\dim rgf = 2$ on U a properly recurrent null vector field is admitted. Otherwise one may choose U so that $\dim rgf = 1$ on U and either a null parallel vector field is admitted on U which lies in the blade of the spanning member of rgf, or a null properly recurrent vector field is admitted on U. In either case appropriate parts of lemma 7.5 shows that $\nabla' = \nabla$. For type $2k =\mathrm{Sp}(l \wedge n, l \wedge y)$ or $\mathrm{Sp}(l \wedge n, l \wedge s)$ the result again follows from lemma $7.5(ii)$ or $5(iv)$. This completes the proof of the theorem. □

Now suppose the holonomy type is $3a =\mathrm{Sp}(l \wedge N, l \wedge n, L \wedge N)$. Then M admits two local recurrent null vector fields arising from l and N in some open neighbourhood of any $p \in M$. The kernel of f consists, at any $p \in M$, of at least $\mathrm{Sp}(l \wedge N, l \wedge L, n \wedge N)$ and lemma 5.10 shows that the conditions of lemma 7.4 hold. If there exists $p \in M$ where $rgf(p)$ contains a non-simple member (and this is necessarily the case if $\dim rgf(p) = 3$) then this is true over some open neighbourhood U of p and $\lambda = 0$ on U. Thus $\nabla' = \nabla$. Otherwise, suppose there exists $p \in M$ where $\dim rgf(p) \leq 2$ with $rgf(p)$ containing only simple members. Suppose that either $F_{ab}l^b \neq 0$ or $G_{ab}N^b \neq 0$ for some (not necessarily independent) members $F, G \in rgf(p)$. Then this supposition holds over some open neighbourhood U of p and a properly recurrent null vector field exists on U. Thus $\nabla' = \nabla$ on M from lemma $7.5(ii)$. Otherwise rgf is spanned by $l \wedge N$ on some open subset of M and $\nabla' = \nabla$ on M from lemma $7.5(iv)$.

If ϕ has holonomy type $3b =\mathrm{Sp}(l \wedge n - L \wedge N, l \wedge N, l \wedge L)$, then at any $p \in M$, $kerf$ contains $\mathrm{Sp}(l \wedge N, l \wedge L, l \wedge n + L \wedge N)$ and, recalling the proof of lemma 7.4, lemma 5.10 shows that $\mathrm{Sp}(l, N, L)$ forms (part of) an eigenspace of $\nabla \lambda$ whilst $l \wedge L$ and $n \wedge N$ are invariant 2-spaces for $\nabla \lambda$. The invariance of $n \wedge N$ and the symmetry of $\nabla \lambda (= \lambda_{a;b})$ then shows that n is also in this eigenspace and thus $\nabla \lambda$ is a multiple of g on M and the conditions of lemma 7.4 hold. Also a null local recurrent vector field arises from l. If for $p \in M$ $rgf(p)$ contains a non-simple member (and this is necessarily the case if $\dim rgf = 3$) it does so in some open neighbourhood of p and $\lambda = 0$ on M, that is $\nabla' = \nabla$. Otherwise, either there exists a non-empty open subset on which $\dim rgf = 2$ ($\Rightarrow rgf =\mathrm{Sp}(l \wedge N, l \wedge L)$) or $\dim rgf = 1 (\Rightarrow rgf=\mathrm{Sp}(l \wedge (aL + bN))$ for $a, b \in \mathbb{R}$. In each case lemma 7.5 shows that $\nabla' = \nabla$.

For holonomy type $3c =\mathrm{Sp}(x \wedge y, x \wedge t, y \wedge t)$ or $\mathrm{Sp}(x \wedge s, x \wedge t, s \wedge t)$ these case are similar to each other and to the case R_{10} in Lorentz signature and the conditions of lemma 7.4 are again satisfied. One finds that $\nabla' = \nabla$ except when $\dim rgf \leq 1$ on M. In this latter case one again finds $\nabla' = \nabla$ if on some open subset $U \subset M$, rgf is null (lemma $7.5(v)$), but otherwise, as in the R_{10} case, ∇' and ∇ are not necessarily equal and counterexamples exist [66, 67].

For holonomy type $3d =\mathrm{Sp}(l \wedge N, l \wedge L, \alpha(l \wedge n) + \beta(L \wedge N))$ with $\alpha, \beta \in \mathbb{R}$ and $\alpha \neq \pm \beta$ a local recurrent vector field arises from l and which may be chosen locally parallel if $\alpha = 0$ ($\Rightarrow \beta \neq 0$). The kernel of f contains $\mathrm{Sp}(l \wedge N, l \wedge L, \beta(l \wedge n) - \alpha(L \wedge N))$ and the conditions of lemma 7.4 are seen to be satisfied provided $\beta \neq 0$ (lemma 5.10).For most of this type the discussion is similar to that given above and so it will be dealt with briefly. The following breakdown of cases simplifies the approach. First suppose $\alpha \neq 0 \neq \beta$ (so that the conditions if lemma 7.4 are satisfied). Then if there exists $p \in M$ such that $rgf(p)$ contains a non-simple member (and this is always true if $\dim rgf(p) = 3$) then $\nabla' = \nabla$.

Otherwise, one considers the existence of $p \in M$ at which $\dim rgf(p) = 2$ and then at which $\dim rgf(p) = 1$ with $rgf(p) \subset \mathrm{Sp}(l \wedge N, l \wedge L)$ in each case. Each leads to a local parallel null vector field arising from l and lemma 7.5(iv) gives $\nabla' = \nabla$. If $\alpha = 0 \neq \beta$ the conditions of lemma 7.4 are satisfied, all members of rgf are simple at each $p \in M$ and l gives rise to a local parallel null vector field. One finds (lemma 7.5) $\nabla' = \nabla$ except possibly when $\dim rgf \leq 1$ on M. In this case if there exists a non-empty open subset U of M on which rgf is spanned by a null or totally null bivector one again finds $\nabla' = \nabla$ from lemma 7.5(iv). Otherwise, for such open subsets rgf is spanned by a timelike bivector and one may achieve $\nabla' \neq \nabla$ with g taking a form similar to (7.83) [66, 67] and from which g' may be calculated. If $\alpha \neq 0 = \beta$ the conditions of lemma 7.4 are not satisfied and further details may be found in [66, 67].

For the holonomy type $4a = \mathrm{Sp}(\overset{+}{S}, l \wedge n + L \wedge N) = \mathrm{Sp}(l \wedge n, L \wedge N, l \wedge N, n \wedge L)$ at each $p \in M$ $kerf$ contains at least $\mathrm{Sp}(l \wedge L, n \wedge N)$ and so, from (7.54), $l \wedge L$ and $n \wedge N$ are eigenspaces of the symmetric tensor $h = \nabla \lambda$. Thus at any $p \in M$ $h_{ab}l^b = \alpha l_a$, $h_{ab}n^b = \beta n_a$ $(\alpha, \beta \in \mathbb{R})$. But then $h(l, n) = h(n, l)$ shows that $\alpha = \beta$ and hence that $\nabla \lambda$ is a multiple of g. It follows that the conditions of lemma 7.4 are satisfied. Further, if there exists $p \in M$ with $\dim rgf(p) \geq 3$ this will hold over some open neighbourhood U of p and then, since $rgf \subset \phi$, rgf must intersect $\overset{+}{S}$ in at least a 2$-$dimensional subspace at each such point and which must therefore contain a non-simple member. It follows that $\lambda = 0$ on U and so $\nabla' = \nabla$. Now suppose there exists a non-empty open subset $U \in M$ on which $\dim rgf = 2$ and at no point of which admits a non-simple member of rgf. Then for $p \in U$ and $K \in rgf(p)$, $K = S + \gamma(l \wedge n + L \wedge N)$ for $S \in \overset{+}{S}_p$, $\gamma \in \mathbb{R}$ and $\overset{*}{K} = S - \gamma(l \wedge n + L \wedge N)$ and so $K \cdot \overset{*}{K} = |S| + 4\gamma^2$. To make K simple one needs either $\gamma = 0$ and S totally null, or $\gamma \neq 0$ and $|S| < 0$. Thus, on U, rgf contains a totally null member of $\overset{+}{S}$ and, using lemma 7.5(i) in chapter 5, one of $l \wedge n$ or $L \wedge N$ (to avoid non-simple members since curvature class C is needed here). Thus rgf may be taken as being spanned by $l \wedge N$ and $l \wedge n$ on U (up to obvious isomorphisms). Then (7.57) shows that $l \wedge N$ and $l \wedge n$ are eigenspaces of a and hence, on U, $a_{ab} = \alpha g_{ab} + \beta N_a N_b$ for smooth α and β. A substitution into (7.20) and contractions with $l^a n^b$ and with $N^a L^b$ then give $\lambda = 0$ and so $\nabla' = \nabla$. The only other possibility is that $\dim rgf \leq 1$ on M and in this case it is possible that $\nabla' \neq \nabla$ [34, 67].

For type $4b = \mathrm{Sp}(\overset{+}{S}, l \wedge L + n \wedge N) = \mathrm{Sp}(\overset{+}{S}, x \wedge y - s \wedge t)$, $kerf$ contains, at each point of M $\mathrm{Sp}(G, H)$ where $G = x \wedge t + y \wedge s$ and $H = x \wedge s - y \wedge t$ and which, together with $K \equiv x \wedge y - s \wedge t$ constitute a basis for $\overset{-}{S}$. But then (7.53) holds for G and H and also for $[G, H] = -2K$ which means it holds for each member of $\overset{-}{S}$ since G and H cannot span a subalgebra of $\overset{-}{S}$ (see chapter 5). But this forces λ to be a multiple of g and the conditions of lemma 7.4 are satisfied. It can then be shown by similar arguments to the previous case that $\nabla' = \nabla$ unless $\dim rgf \leq 1$ on M (the 2$-$dimensional case, that is, curvature class C, is not possible here as it was for case $4a$ since the blades of a totally null and a spacelike bivector intersect only trivially). Again, examples of g and g' can be found for which $\nabla' \neq \nabla$ [67].

For type $4c = \mathrm{Sp}(l \wedge N, l \wedge n, l \wedge L, L \wedge N)$ and $ker f$ contains $\mathrm{Sp}(l \wedge N, l \wedge L)$ and the conditions of lemma 7.4 are not satisfied. However, a local parallel null vector field arises in some neighbourhood of each $p \in M$. This case has some similarities to the cases R_9 and R_{14} in the Lorentz signature discussion and examples can be found with $\nabla' \neq \nabla$ [67]. One thus has the following theorem.

Theorem 7.12 *Let $dim M = 4$ with g and g' projectively related metrics of neutral signature on M. Suppose g has holonomy algebra of type $3a$, $3b$, $3c$ (with $dim rg f(p) \geq 2$ for some $p \in M$), type $3d$ (with $\alpha \neq 0 \neq \beta$) or type $3d$ (with $0 = \alpha \neq \beta$ and with $dim rg f(p) \geq 2$ for some $p \in M$). Then in the usual notation $\nabla' = \nabla$. If the holonomy type of (M, g) is $4a$ or $4b$ one again gets $\nabla' = \nabla$ except possibly when $dim rg f \leq 1$ on M.*

It is noted that even in the cases $3c$, $3d$ (with $0 = \alpha \neq \beta$), $4a$ or $4b$ where, in each case, the restriction $dim rg f \leq 1$ is imposed on M one again arrives at $\nabla' = \nabla$ provided the blade of the spanning member of $rg f$ is one of certain special types listed above on some non-empty open subset of M. In fact, over all signatures, the types S_1, S_2, $\overset{+}{S}_3$, $\overset{-}{S}_3$ (positive definite signature), R_2, R_3, R_4, R_6, R_7, R_8 and R_{12} (Lorentz signature) and types $1a$–$1d$, $2a$–$2j$, $3a$, $3b$ and $3d(\alpha\beta \neq 0)$ (neutral signature) all lead immediately to $\nabla' = \nabla$ whilst the types S_3, $\overset{+}{S}_4$ and $\overset{-}{S}_4$ (positive definite), R_{10}, R_{11} and R_{13} (Lorentz) and $3c$, $3d(\alpha = 0 \neq \beta)$, $4a$ and $4b$ each lead to $\nabla' = \nabla$ provided there exists $p \in M$ such that $dim rg f(p) \geq 2$.

Bibliography

[1] Halmos P. R., Finite Dimensional Vector Spaces, Van Nostrand, 1958.

[2] Herstein I. N., Topics in Algebra, Blaisdell, 1964.

[3] Kelley J. L., General Topology, Van Nostrand, 1955.

[4] Dugundji J., Topology, Allyn and Bacon, 1974.

[5] Heisenberg M., Topology, Holt, Reinhart and Winston, 1974.

[6] Greenberg M. J., Euclidean and Non-Euclidean Geometries, Freeman, San Francisco, 1974.

[7] Martin G. E., The Foundations of Geometry and the Non-Euclidean Plane, Intext Educational Publishers, New York, 1972.

[8] Hilbert D., The Foundations of Geometry, Open Court Publishing, 1902.

[9] Brickell F. and Clark R.S., Differentiable Manifolds, Van Nostrand, (1970).

[10] Kobayashi S. and Nomizu K., Foundations of Differential Geometry, Vol 1, Interscience, New York, (1963).

[11] Helgason S., Differential Geometry, Lie Groups, and Symmetric Spaces. American Mathematical Society, Graduate Studies in Mathematics, Vol 34, (1978).

[12] Hicks N. J., Notes on Differential Geometry, Van Nostrand, (1971).

[13] Hall G. S., Symmetries and Curvature Structure in General Relativity. World Scientific, (2004).

[14] de Felice F. and Clarke C. J. S., Relativity on curved manifolds. C.U.P. Cambridge, (1990).

[15] Hawking S. W. and Ellis G. F. R., The large scale structure of space-time. Cambridge University Press, 1973.

[16] Stephani H. Kramer D. MacCallum M. A. H. Hoenselaers C. and Herlt E., Exact Solutions to Einstein's Field Equations. Second Edition, Cambridge University Press, 2003.

[17] Hall G. S., *J. Phys.*, **A9**, (1976), 541.

[18] Ihrig E. *Intern. Jn. Theor. Phys.*, **13**, (1976), 23.

[19] Edgar S. B. *J. Math. Phys.*, **32**, (1991), 1011.

[20] Ambrose W. and Singer I. M. *Trans. Am. Math. Soc.*, **75**, (1953), 428.

[21] Hall G. S. and Kırık B., *J. Geom. Phys.*, **98**, (2015), 262.

[22] Spivak M., A comprehensive introduction to differential geometry, Vol 1, Publish or Perish, Boston, Massachusetts, 1970.

[23] Hermann R, International Symposium on Nonlinear Differential Equations and Nonlinear Mechanics, Academic Press, New York, 1963, 325.

[24] Stefan P., *J. London. Maths. Soc.*, **21**, 1980, 544.

[25] Sussmann H. J., *Trans. Am. Maths. Soc.*, **180**, 1973, 171.

[26] Wolf J. A., Spaces of Constant Curvature. Publish or Perish, Boston, Massachusetts, 1974.

[27] Markus L., *Anns. Maths.*, **62**, (1955), 411.

[28] Dunajski M. and West S., Preprint DAMTP-2006-90.

[29] Hall G. S. and Rendall A. D., *Int. Jn. Theoret. Phys.*, **28**, (1989), 365.

[30] Weyl H., *Gottinger. Nachrichten.* (1921), 99-112.

[31] Hall G. S. and MacNay L. E. K., *Class. Quant. Grav.*, **22**, (2005), 1493.

[32] Hall G. S., *J. Geom. Phys.*, **60**, (2010), 1.

[33] Stephani H., General Relativity, Cambridge University.Press, (1982).

[34] Hall G. S. and Wang Z., *J. Geom. Phys.*, **62**, (2012), 449.

[35] Hall G. S., *Balkan.J. Geom.*, **23**, (2018), 44.

[36] Riemann G. F. B., On the hypotheses which lie at the foundations of geometry. Translated in William E., *From Kant to Hilbert*, **Vol 2**, Clarendon Press, Oxford, 1996.

[37] Eisenhart L. P., Riemannian Geometry, Princeton, (1966).

[38] Kulkarni R. S., *Ann. Maths.*, **91,** (1970), 311.

[39] Brinkmann H. W., *Math. Ann.*, **94**, (1925), 119.

[40] Sachs R. K., *Proc. Roy. Soc.*, **264**, (1961), 309.

[41] Penrose R., Gravitatsya. In A. Z. Petrov Festschrift volume, Naukdumka, Kiev, (1972), 203.

[42] Plebanski J., *Acta Phys Polon.*, **26**, (1964), 963.

[43] Churchill R. V.,*Trans. Am. Math. Soc.*, **34**, (1932), 784.

[44] Cormack W. J. and Hall G. S., *J. Phys. A.*, **A12**, (1979), 55.

[45] Crade R. F. and Hall G. S., *Acta. Phys. Polon.*, **B13**, (1982), 405.

[46] Hall G. S., *Int. Journ. Mod. Phys D.* **26**, (2017) 1741004.

[47] Shaw R., *Quart. J. Math. Oxford.*, **21**, (1970), 101.

[48] Schell J. F., *J. Math. Phys.,* **2**, (1961), 202.

[49] Petrov A. Z., Einstein Spaces. Pergamon Press, (1969).

[50] Petrov A. Z., *Gen. Rel. Grav.*, **32**, (2000).

[51] Penrose R. *Anns. Phys.,* **10**, (1960), 171.

[52] Bel L., *Cah. de. Phys.* **16**, (1962), 59. (in French). Translated in *Gen. Rel. Grav.* **30**, (2000), 2047.

[53] Lichnerowicz A., *Compt. Rend.,* **246**, (1958), 893.

[54] Debever R., *Compt. Rend.,* **249**, (1959), 1744.

[55] Hall G. S., *Gen. Rel. Grav.,* **15**, (1983), 581.

[56] McIntosh C.B.G. and Halford W.D., *J. Phys. A.,* **14**, (1981), 2331.

[57] Hall G. S., unpublished (1979).

[58] Hall G. S., *Gen. Rel. Grav.,* **16**, (1984), 79.

[59] Hall G. S., *Publications de l'Institut Mathematique,* **103**, (2018), 1.

[60] Cormack W. J. and Hall. G. S., *Int. J. Theor. Phys.J.,* **18**, (1979), 279.

[61] Hall G. S. and Rendall A. D., *Gen. Rel. Grav.,* **19**, (1987), 771.

[62] Thorpe J. *Math. Phys.,* **10**, 1969, 1.

[63] Ruh B., *Math. Z.,* **189**, (1985), 371.

[64] Ehlers J. and Kundt W., in Gravitation; an Introduction to current research, ed Witten L, Wiley, New York, (1962), 49.

[65] Hall G. S., *Arab Journ. Maths.,* doi.org/10.1007/s40065-018-0235-3, (2019).

[66] Wang Z. and Hall G. S., *J. Geom. Phys.*, **66**, (2013), 37-49.

[67] Wang Z., Projective Structure on 4-dimensional Manifolds. PhD thesis, University of Aberdeen, 2012.

[68] Hall G. S., *J. Geom. Phys.*, **111**, (2017), 111.

[69] Rohrlich F., Classical Charged Particles, Addison-Wesley, 1965.

[70] Dicke R. H., *Ann. Phys.*, **26**, 1964, 442.

[71] Wald R. M., General Relativity, University of Chicago Press, 1984.

[72] Einstein A., Relativity, Crown Publishers, Inc, New York, 1961.

[73] Hall G. S., *Coll. Math.*, **150**, (2017), 63.

[74] Ghanan R. and Thompson G., *J. Math. Phys.*, **42**, (2001), 2266.

[75] Law P., *J. Math. Phys.*, **32**, (1991), 3039.

[76] Law P., *J. Geom. Phys.*, **56**, (2006), 2093.

[77] Batista C., arXiv:1204.5133v4 (2013).

[78] Ortaggio M., *Class. Quan. Grav.*, **26**, (2009),195015.

[79] Coley A., Private Communication.

[80] Coley A. and Hervik S., *Class. Quant. Grav.*, **27**, (2010), 015002.

[81] Walker A. G., *Proc. Lond. Math. Soc.*, **52**, (1950), 36.

[82] Hall G. S. Preprint, University of Aberdeen, 2022.

[83] Schutz B., Geometrical Methods of Mathematical Physics. Cambridge University Press, 1980.

[84] Weyl H., *Math. Zeit.* **2**, (1918), 384. (in German). Translated in Gen. Rel. Grav., **54** (2022) https://doi.org/10.1007/s10714-022-02930-7.

[85] Hall G. S. and McIntosh C. B. G., *Inter. Journ. Theor. Phys.*, **22**, (1983), 469.

[86] Pirani F. A. E. *Phys. Rev.*, **105**, (1957), 1089.

[87] Jordan P. Ehlers J. and Kundt W., *Akad. Wiss. Mainz. Abh. Math.-Nat. Kl.*, Jahrg (1960), No 2 (in German). Translated in Gen. Rel. Grav. **41**, (2009), 2191.

[88] Hall G. S., *Publications de l'Institut Mathematique*, **94**, (2013), 55.

[89] Abraham R. Marsden J.E. and Ratiu T., Manifolds, Tensor Analysis and Applications. Springer, 2nd edition, 1988.

[90] Poor W., Differential Geometric Structures, McGraw Hill, 1981.

[91] Yano K., The Theory of Lie Derivatives and its Applications. North Holland, Amsterdam, 1957.

[92] Knebelman M. S., *Amer. J. Maths.*, **52**, (1930), 280.

[93] Paterson E. M., *Quart. J. Maths., Oxford* **22** (1951), 151.

[94] Datta D. K., *Tensor. N.S.* **15**, (1964), 61.

[95] Datta D. K., *Bull. Aust. Math. Soc.*, **10**, (1974), 71.

[96] Besse A. L., Einstein Manifolds. Springer, 1987.

[97] Mikes J., Vanzurova A. and Hinterleitner I., Geodesic Mappings. Palacky University, Olomouc, 2009.

[98] Hall G. S. and Lonie D. P., *Class. Quant. Grav.*, **24**, (2007), 3617.

[99] Barnes A., *Class. Quant. Grav.*, **10**, (1993), 1139.

[100] Thomas T. Y., The Differential Invariants of Generalized Spaces. Cambridge, 1934.

[101] Sinjukov N. S., Geodesic Mappings of Riemannian Spaces. Moscow: Nauka (in Russian).

[102] Hall G. S. and Lonie D. P., *Journ. Geom. Phys.*, **61**, (2011), 381.

[103] Hall G. S. and Lonie D. P., *Class. Quant. Grav.*, **26**, (2009), 125009.

[104] Kiosak V and Matveev V., *Comm. Math. Phys.*, **289** (2009), 383.

[105] Hall G. S. and Lonie D. P., *Class. Quant. Grav.*, **17**, (2000), 1369.

[106] Aichelburg P. C., *J. Math. Phys.*, **11**, (1970), 2485.

[107] Hall G. S. and Lonie D. P., *J. Math. Phys.*, **49**, (2008), 022502.

[108] Hall G. S. and Lonie D. P., *Sigma* **5**, (2009), 066.

[109] Hall G. S. and Lonie D. P., *Class. Quant. Grav.*, **28**, (2011), 083101 (17 pages).

[110] Hall G. S., *Class. Quant. Grav.*, **17**, (2000), 4637.

[111] Mikes J. Hinterleitner I. and Kiosak V., AIP conference proceedings **861**, (2006), 428.

[112] Sinjukov N. S., *Scientific Annual, Odessa University*, (1957), 133.

[113] Hall G. S. and Lonie D. P., *Class. Quant. Grav.*, **12**, (1995), 1007.

[114] Lonie D. P., Projective Symmetries, Holonomy and Curvature Structure in General Relativity. PhD thesis, University of Aberdeen, 1995.

[115] Einstein A., *Sitzungsberichte der Preussishen Akad.d. Wissenschaften*, (1917).

Index

Printed in the United States
by Baker & Taylor Publisher Services